RF Power Amplifiers

RF Power Amplifiers

Mihai Albulet

NOBLE
PUBLISHING

Noble Publishing Corporation
Atlanta, GA

Library of Congress Cataloging-in-Publication Data

Albulet, Mihai, 1962-
  RF power amplifiers / Mihai Albulet.
    p.cm.
  Includes bibliographical references and index.
  ISBN 1-884932-12-6
    1. Power amplifiers. 2. Amplifiers, Radio frequency. I. Title.

  TK7871.58.P6 A43 2001
  621.348'12--dc21

                                                    2001030069

N O B L E
P U B L I S H I N G

Printed in the Unites States of America

ISBN 1-884932-12-6

Contents

Preface

Many practicing engineers view RF, especially large-signal RF circuits, as a somewhat mysterious, "black magic" subject. This book attempts to show that there is nothing unusual or inexplicable about RF power amplifiers — understanding them is simply a matter of understanding several basic principles and their applications. Although accurate CAD modeling and/or optimization can become almost impossible, since mathematical modeling of RF power amplifiers is often too difficult or complex to provide useful practical results, yield design equations, or predict a circuit's performance, the main purpose of a theoretical approach is to provide a starting point for computer simulations or experimental tweaking, or simply a physical understanding of the circuit.

Given this relative obscurity of the subject, this book is certainly not as practical as some readers would undoubtedly prefer it to be. No "miraculous recipe" is given for the design of the perfectly suited RF power amplifier for a particular application. In some cases, readers may even decide that my book does not indicate a practical enough design method for a particular circuit nor suggest a way to approach the design. This book does not describe either because I did not intend to write a practical handbook on RF power amplifiers — I believe that this is not an appropriate area for cookbook solutions.

The primary purpose of this book is to present the basic concepts used in the analysis and design of RF power amplifiers. Detailed mathematical derivations reveal the assumptions and limitations of the presented results, allowing the reader to estimate their usefulness in practical applications. Theory is the best practice and a good theoretical understanding is the quickest way toward achieving practical results. A designer must know a priori the circuit topologies and the basic operation principles as well as limitations of the various amplification classes. Selecting the appropriate circuit topology and operating mode, knowing their pros and cons, and setting realistic goals for the expected performance are imperative for beginning a practical design. Then CAD simulators and/or experimental tweaking will be successful in optimizing the design.

This book covers the basics of the RF power amplifiers, such as amplification classes, basic circuit topologies, bias circuits and matching networks. An exhaustive coverage of the power amplifier area is beyond the scope of the book; therefore, applications, system architecture concepts, and linearization techniques are not discussed here.

Chapter 1 discusses several basic concepts, terminology and definitions. Chapter 2 is dedicated to classic RF power amplifiers. Included are the oldest and best-known classes of amplification: A, B, AB, and C. This classification is based on the conduction angle of the active device and also includes the so-called mixed-mode Class C. Separate sections treat bias circuits, narrowband and broadband matching networks, gain leveling and VSWR correction, amplitude modulation, stability, thermal calculations, and Class C frequency multipliers.

Chapter 3 focuses on switching-mode Class D amplifiers. Described are the idealized operation of these amplifiers as well as practical considerations (parasitics and non-ideal components, mistuning or frequency variation, drive considerations). Other sections in this chapter cover Class D circuits operating in intermediate classes (BD and DE) or as frequency multiplies. The last section focuses on computer simulation of Class D circuits.

Chapter 4 presents switching-mode Class E power amplifiers. The chapter begins with an outline of the idealized operation, followed by a discussion of the practical considerations. Additional sections describe amplitude modulation, Class E frequency multipliers, and computer simulation of this circuit.

Chapter 5 is dedicated to Class F amplifiers. This includes established techniques to improve efficiency using harmonic injection in Class B or C circuits, the so-called Class F1 amplifier (also known as "high-efficiency Class C" "Class C using harmonic injection," or "biharmonic or polyharmonic Class C"), but also more recent switching-mode circuits, such as Classes F2 and F3.

Chapter 6 comments on switching-mode Class S amplifiers and modulators. Although these circuits are audio- or low-frequency amplifiers, they are important subsystems in many high-efficiency transmitters. Finally, Chapter 7 presents several considerations regarding bipolar and MOS RF power transistors.

Acknowledgments

I would like to express my deep appreciation to a number of people who contributed to this book in many ways. My thanks to Mr. Nathan O. Sokal (president, Design Automation, Inc.), and to Dr. Frederick H. Raab (Green Mountain Radio Research Company) for taking the time to review several parts of this manuscript and making useful comments and suggestions. I owe much to Mr. Sokal for providing the HB-PLUS and HEPA-PLUS programs developed by Design Automation.

I am grateful to the Department of Telecommunications at the Technical University of Iasi, Romania, where I first became involved in the RF power amplifier field. A large part of this book comes from knowledge acquired and research conducted during my tenure at this university.

Last, and certainly not least, I am indebted to my wife, Lucretia, and my daughter, Ioana, for their support and understanding during the writing of this book.

1

Introduction

The understanding of RF power amplifiers is greatly helped by a background knowledge of the most important theoretical features of small-signal high-frequency amplifiers. A summary of the subject is well beyond the scope of this book; instead, a short overview is offered. For a full discussion of small-signal amplifiers, please refer to references [1–5].

"Small signal" implies that the signal amplitude is small enough such that a linear equivalent circuit (such as a hybrid-pi circuit or any linear two-port circuit with constant coefficients) can model the amplifier. RF power amplifiers function very differently from small-signal amplifiers. Power amplifiers operate with large signals, and the active devices display strong nonlinear behavior. The amplifier output may be modeled as an infinite power series consisting of nonlinear terms added to a linear term and a dc offset. The power series coefficients depend on the transistor operating point (dc bias point, or the average operating point) and are considered constant to changes in the input and output RF signal. A more realistic model could use the Volterra series, which allows for the inclusion of phase effects. However, all these models have a serious limitation in that they can only accurately model weak nonlinear circuits for which the power series coefficients are almost constant (a narrow operation zone around the dc-bias point). In a large-signal power amplifier, nonlinear effects are very strong because transistor parameters depend on many factors, including the input and output matching network configuration and the input and output signal amplitudes and waveforms. In addition, the active device may be driven into saturation or cut-off for a certain portion of the RF cycle. Modeling these strong nonlinear effects is a very difficult task, even if CAD models and tools are available.

Power amplifiers are identified by classes (named A, B, C...). The amplifier class of operation depends on circuit topology, operating principle, how the transistor is biased or driven, and the specific component values in the load network. Further, combinations of operating modes and intermediate classes are possible. In this book, the classification provided in Reference [6] is used.

1.1   Ideal Parallel-Tuned Circuit

An ideal parallel-tuned circuit is a paralleled LC circuit that provides zero conductance (that is, infinite impedance) at the tuning frequency, f_0, and infinite conductance (zero impedance) for any other frequency. When connected in parallel to a load resistor, R, the ideal parallel-tuned circuit only allows a sinusoidal current (with frequency f_0) to flow through the load. Therefore, the voltage across the RLC parallel group is sinusoidal, while the total current (that is, the sum of the current through load and the current through the LC circuit) may have any waveform.

A good approximation for the ideal parallel-tuned circuit is a circuit with a very high loaded Q (the higher the Q, the closer the approximation). Note that a high-Q parallel-tuned circuit uses small inductors and large capacitors, which may be a serious limitation in practical applications.

1.2   Ideal Series-Tuned Circuit

An ideal series-tuned circuit is a series LC circuit that provides zero impedance at the tuning frequency, f_0, and infinite impedance for any other frequency. When connected in series to a load resistor, R, the ideal series-tuned circuit only allows a sinusoidal current with frequency f_0 to flow through the load. Therefore, the current through the series RLC group is sinusoidal, while the voltage across the RLC group may have any waveform.

A good approximation for the ideal series-tuned circuit is a circuit with a very high loaded Q (the higher the Q, the closer the approximation). Note that a high-Q series-tuned circuit must use large inductors and small capacitors, which may be a serious limitation in practical applications.

1.3   Efficiency

Efficiency is a crucial parameter for RF power amplifiers. It is important when the available input power is limited, such as in battery-powered portable or mobile equipment. It is also important for high-power equipment where the cost of the electric power over the lifetime of the equipment and the cost of the cooling systems can be significant compared to the purchase price of the equipment.

Efficiency is output power versus input power. However, this definition is too broad, because "output power" and "input power" may have different meanings. Input power may include both the dc-input power (that is, the power supplied by the dc supply) and the RF input power (the drive power), or only the dc-input power. The most common definitions encountered are presented below [3–6].

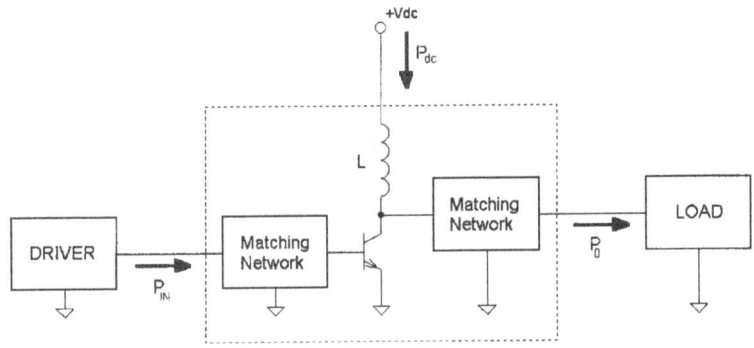

Figure 1-1 *Efficiency definitions in RF power amplifiers.*

1.4   Collector Efficiency

Collector efficiency is a term more appropriate for amplifiers using *bipolar transistors* (BJTs), although it is often used for any RF power amplifiers. Some authors prefer to use *plate efficiency* for amplifiers using vacuum tubes or *drain efficiency* for amplifiers using MOSFETs or, simply refer to it as *efficiency*. Collector efficiency is defined as (see Figure 1-1)

$$\eta = \frac{P_0}{P_{dc}} \qquad 1.1$$

where P_0 is the RF output power (dissipated into the load) and $P_{dc} = V_{dc} I_{dc}$ is the input power supplied by the dc supply to the collector (or drain / plate) circuit of the power amplifier. P_0 usually includes both the RF fundamental power and the harmonics power. In many applications, harmonic suppression filters are included in the output-matching network. Because the harmonic power is negligible, the RF fundamental power is a very good approximation for P_0. Unless stated otherwise, the definition above will be used in this book.

1.5   Overall Efficiency

Although it is a very convenient measure of a circuit's performance, collector efficiency does not account for the drive power required, which may be quite substantial in a power amplifier. Power gains (that is the ratio of output power to drive power) of 10 dB or less are common at high RF frequencies (and even at low frequencies in switching-mode amplifiers). In general, RF power amplifiers designed for high collector efficiency tend to achieve a low power gain, which is a disadvantage for the overall power budget.

From a practical standpoint, a designer's goal is to minimize the total dc power required to obtain a certain RF output power. The overall efficiency is defined as

$$\eta_{OVERALL} = \frac{P_0}{P_{dc} + P_{IN}} = \frac{P_0}{P_{dc} + \dfrac{P_0}{G_P}} \qquad 1.2$$

where

$$G_P = \frac{P_0}{P_{IN}} \qquad 1.3$$

is the power gain.

1.6   Power-Added Efficiency

Power-added efficiency is an alternative definition that includes the effect of the drive power used frequently at microwave frequencies and is defined as

$$\eta_{POWER-ADDED} = \frac{P_0 - P_{IN}}{P_{dc}} = \frac{P_0 - \dfrac{P_0}{G_P}}{P_{dc}} \qquad 1.4$$

The overall efficiency and the power-added efficiency, although related to each other, differ in their numerical values.

EXAMPLE 1.1
An RF power amplifier delivers P_0 = 100 W into the load resistance. The input power supplied by the dc power supply to the collector circuit is P_{dc} = 150 W and the power gain is G_p = 10 (that is 10 dB). Collector efficiency is 100/150 = 66.67%, overall efficien-

cy is 100/(150 + 100/10) = 62.50%, and power-added efficiency is
(100 − 100/10)/150 = 60%.

1.7 Power Output Capability

The power output capability, C_P, provides a means of comparing different types of power amplifiers or amplifier designs. The power output capability is defined as the output power produced when the device has a peak
collector voltage of 1 volt and a peak collector current of 1 ampere. If the
power amplifier uses two or more transistors (as in push-pull designs, or in
circuits with transistors connected in parallel, or using combiners), then
the number of devices is included in the denominator (thus allowing a fair
comparison of various types of amplifiers, both single-ended or using several transistors).

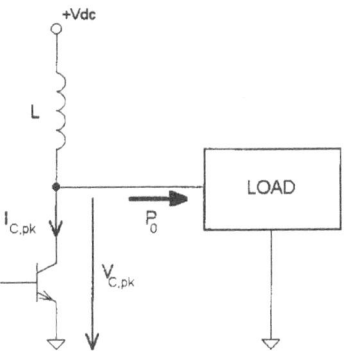

Figure 1-2 *Power output capability in RF power amplifiers.*

If P_0 is the RF output power, $I_{C,pk}$ is the peak collector current, $V_{C,pk}$ is
the peak collector voltage, and N is the number of transistors in circuit,
then the power output capability is given by

$$C_P = \frac{P_0}{N I_{C,pk} V_{C,pk}}$$

1.5

Power transistors are the most expensive components in power amplifiers. In cost-driven designs, designers are constrained to use the lowest cost
transistors. This means the devices have to be used as close as possible to
their maximum voltage and current ratings. Therefore, the larger the power
output capability of the circuit, the cheaper its practical implementation.

EXAMPLE 1.2

An RF power amplifier must deliver P_0 = 100 W into the load resistance. A Class D circuit using two active devices can achieve a maximum theoretical power output capability $C_{P,\,Class\,D}$ = 0.1592 (see Chapter 3). If the circuit is designed so that the peak collector voltage is $V_{C,pk}$ = 100 V, then the peak collector current is

$$I_{C,pk} = \frac{P_0}{N C_{P,Class\,D} V_{C,pk}} = \frac{100}{2 \cdot 0.1592 \cdot 100} = 3.14\,\text{amps}$$

and the required device ratings for the two transistors used in the Class D circuit are 100 volts and 3.14 amps. A single-ended Class E circuit can achieve a maximum theoretical power output capability $C_{P,Class\,E}$ = 0.0981 (see Chapter 4). If the circuit is designed so that the peak collector voltage is $V_{C,pk}$ = 100 V, then the peak collector current is

$$I_{C,pk} = \frac{P_0}{N C_{P,Class\,E} V_{C,pk}} = \frac{100}{0.0981 \cdot 100} = 10.19\,\text{amps}$$

and the required device ratings for the transistor used in the Class E circuit are 100 volts and 10.19 amps. Another possible alternative is to use a push-pull Class E circuit ($C_{P,\,Class\,E,\,push\text{-}pull}$ = 0.0981, see Chapter 4). In this case, the circuit could use two transistors with the required ratings of 100 volts and 5.1 amps. The Class D circuit is the best in terms of power output capability and could potentially be the lowest cost design. The Class E circuits have to use transistors with higher ratings, which are more expensive, in order to provide the same output power. However, in practical designs the tradeoffs are much more complicated because circuit complexity, gain, efficiency and overall cost are part of the equation.

1.8 References

1. Carson, R. S. *High-Frequency Amplifiers*. New York: John Wiley & Sons, 1982.
2. Vendelin, G., A. Pavio, and U. Rhode. *Microwave Circuit Design*. New York: John Wiley & Sons, 1982.
3. Terman, F. E. *Radio Engineering*. New York: McGraw-Hill, 1947.
4. Clarke, K. K. and D. T. Hess. *Communication Circuits: Analysis and Design*. Boston: Addison-Wesley, 1971.
5. Krauss, H. L., C. V. Bostian, and F. H. Raab. *Solid-State Radio Engineering*. New York: John Wiley & Sons, 1980.
6. Sokal, N. O., I. Novak, and J. Donohue. "Classes of RF Power Amplifiers A Through S, How They Operate, and When to Use Them." *Proceedings of RF Expo West 1995*, (San Diego, CA) (1995): 131–138.

2

Classic RF Power Amplifiers

Chapter 2 discusses several types of RF power amplifiers (PAs), most often called Class A, AB, B, and C. Class C amplifiers, in turn, are usually divided into three categories: a) current-source (or underdriven) Class C PAs, b) saturated (or overdriven) Class C PAs, and c) mixed-mode Class C PAs. With the exception of mixed-mode Class C PAs, which behave somewhat differently, the other circuits have the following common features:

- They have the same basic collector circuit schematic, as shown in Figure 2-1.

- The circuits are all driven with sinusoidal (or approximately sinusoidal) waveforms.

- The active device behaves, at least for a certain portion of the RF cycle, as a controlled-current source.

The portion of the RF cycle the device spends in its active region (i.e., behaves as a controlled-current source) is the *conduction angle* and is denoted by $2\theta_c$. Based on the conduction angle, the amplifiers are generally classified as [1-10]:

- Class A amplifiers, if $2\theta_c = 360°$. The active device is in its active region during the entire RF cycle.

- Class AB amplifiers, if $180° < 2\theta_c < 360°$.

- Class B amplifiers, if $2\theta_c = 180°$.

- Class C amplifiers, if $2\theta_c < 180°$. Note that, in saturated Class C amplifiers, the conduction angle includes the portion of the RF cycle when the active device is saturated.

All these amplifiers use the same basic collector circuit topology of Figure 2-1. This is a single-ended circuit, and the transistor operates in the common-emitter configuration; however, common-base configurations are possible. Variations among practical circuits operating in different classes may occur in the base-bias or drive circuits. The collector circuit includes an RF choke (RFC) that provides a constant (DC) input current, I_{dc}, a DC-blocking capacitor, C_d (short-circuit at the operating frequency and its harmonics), the load resistor, R, and a parallel resonant LC circuit tuned to the operating frequency.

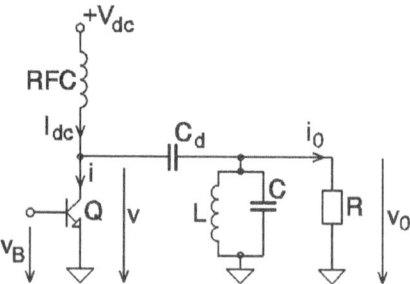

Figure 2-1 *Basic circuit of single-ended Class A, AB, B, or C amplifier.*

Figure 2-2 illustrates the push-pull circuit. As in Figure 2-1, a common-emitter configuration is depicted, although a common-base arrangement is also possible. Note that today's technologies do not allow the development of high-performance pnp BJTs or p-channel MOSFETs for RF power applications. As a result, complementary circuits like those used in Class AB audio-frequency PAs cannot be used in high-frequency RF PAs.

The push-pull circuit of Figure 2-2 includes an input transformer, T_1, that applies the signal to the two transistors so that they are driven 180 degrees out-of-phase. Output transformer T_2 combines the output powers of the two transistors. Although the push-pull circuit is most often used in Class B or AB wideband circuits, it can be also used to increase the output power in Class A or Class C circuits.

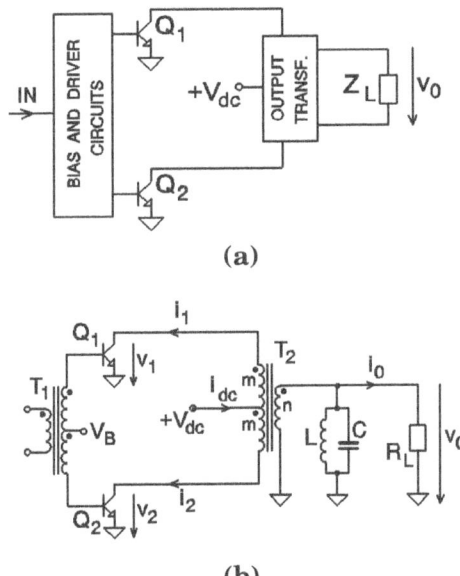

(a)

(b)

Figure 2-2 *Push-pull Class A, AB, B, or C amplifier; (a) basic circuit, (b) usual implementation.*

2.1 Class A Amplifiers

For Class A operation [1-8], the quiescent point (I_{dc}) must be selected to keep the transistor in its active region during the entire RF cycle, thus assuring a 360 degree conduction angle.

A simplified analysis of the single-ended Class A amplifier is based on the following assumptions:

a. The RF choke is ideal. It has no series resistance, and its reactance at the operating frequency is infinite. Consequently, the RF choke allows only a constant (DC) input current, I_{dc}, whose value is determined by the bias circuit (not shown in Figure 2-1).

b. C_d is a DC-blocking capacitor (short circuit at the operating frequency).

c. The active device behaves as an ideal controlled-current source. For this simplified analysis, disregard the saturation voltage and/or the saturation resistance of the transistor. The transfer characteristic of the active device is assumed to be perfectly linear, i.e., a sinusoidal drive signal determines a sinusoidal collector current.

Based on these assumptions, the collector current is given by

$$i(\theta) = I_{dc} - I_0 \sin\theta \qquad\qquad 2.1$$

where $\theta = \omega t = 2\pi f t$ is the angular time. The output current is sinusoidal.

$$i_0(\theta) = I_{dc} - i(\theta) = I_0 \sin\theta \qquad\qquad 2.2$$

The collector voltage is

$$v(\theta) = V_{dc} + V_0 \sin\theta = V_{dc} + RI_0 \sin\theta \qquad\qquad 2.3$$

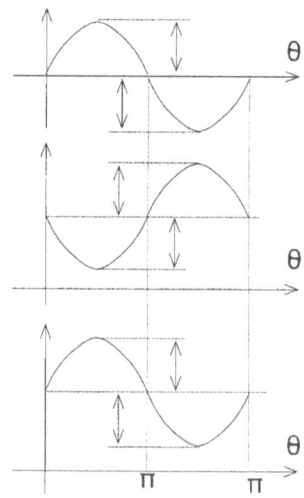

Figure 2-3 *Waveforms in a Class A amplifier.*

The corresponding waveforms of the Class A single-ended circuit are shown in Figure 2-3. Class A operation is assured by maintaining $i(\theta) \geq 0$ and $v(\theta) \geq 0$. Therefore, the transistor remains in the active region if the following conditions are satisfied

$$I_0 \leq I_{dc} \qquad V_0 \leq V_{dc} \qquad\qquad 2.4$$

As a result, the output power (dissipated in the load resistance R)

$$P_0 = \frac{I_0^2}{2} R = \frac{V_0^2}{2R} \qquad\qquad 2.5$$

has the maximum theoretical value

$$P_{0,max} = \frac{V_{dc}^2}{2R}$$ 2.6

This maximum value is obtained for $V_0 = V_{dc}$.

Equation 2.4 suggests that, in practice, the selection of a quiescent current equal to the peak output current is recommended.

$$I_{dc} = I_{0,max} = \frac{V_{dc}}{R}$$ 2.7

Thus, the DC input power and collector efficiency are given by

$$P_{dc} = V_{dc}I_{dc} = \frac{V_{dc}^2}{R}$$ 2.8

$$\eta = \frac{P_0}{P_{dc}} = \frac{1}{2}\frac{V_0^2}{V_{dc}^2} \leq \eta_{max} = \frac{1}{2} = 50\,\%$$ 2.9

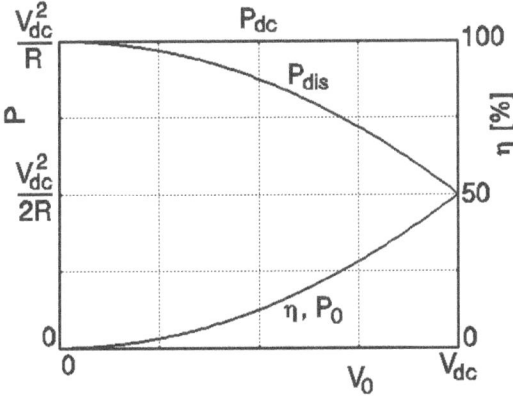

Figure 2-4 *Powers and collector efficiency versus the amplitude of the output voltage in a Class A amplifier.*

Figure 2-4 presents the variation of the collector efficiency, η, the output power, P_0, the DC input power, P_{dc}, and the power dissipated in the active device $P_{dis} = P_{dc} - P_0$, with the output voltage V_0. Without a drive signal applied to the circuit (i.e., with $V_0 = 0$), the DC input power is completely dissipated in the transistor. This is a major disadvantage of Class A circuits with respect to Class B or Class C circuits. Class B or C circuits do not dissipate any power in the transistor in the absence of the drive signal. Another drawback of the Class A amplifier is its low collector efficiency value — a maximum theoretical value of 50 percent. Thus, in practical circuits, the maximum obtainable collector efficiency is about 40 to 45 percent (taking into account power losses due to the nonideal character of the components).

If $V_0 = V_{dc}$, the peak instantaneous collector voltage is $2V_{dc}$, and the peak instantaneous collector current is $2I_{dc}$. The power output capability of the Class A amplifier is therefore

$$C_P = \frac{P_{0,max}}{(2V_{dc})(2I_{dc})} = \frac{1}{8} = 0.125. \qquad 2.10$$

Some observations and practical considerations

1. The resonant LC circuit is not necessarily required here. The Class A amplifier can use a resistive load and operate over a wide frequency range. However, the nonlinearity of the active device cannot be avoided at high signal levels and the output signal would be distorted. Thus, the load often includes tuned circuits, or band- or low-pass filters, to filter out the collector current harmonics.

2. The basic circuit and the operation of the Class A RF PAs are quite similar to those of the small-signal Class A amplifier. There is no dividing line between small-signal and Class A PAs.

3. The push-pull circuit in Figure 2-2 may be used to combine the output powers provided by two identical transistors (Class A operated). This cancels most of the even harmonic currents. The drawbacks are related to the circuit complications: transformers are usually bulky, expensive, and introduce additional losses. The power output capability of the push-pull Class A amplifier is 1/8, the same value as in the single-ended Class A amplifier.

4. The Class A amplifier presents a linear transfer characteristic and a high power gain (20 to 30 dB, even at high frequencies). However, because of their low efficiency level, Class A amplifiers are most often used as low-level drivers for more efficient PAs. In such applications, the Class A amplifier consumes only a small portion of the total DC power, and the overall efficiency of the amplifying chain is not significantly affected. Class A RF PAs are also used for laboratory equipment (for example, very low-distortion linear wideband amplifiers) or at microwave frequencies where it could be difficult to employ other classes of amplification.

5. The harmonic distortion of the load current can be calculated easily if the harmonic content of the collector current and the quality factor of the par-

allel resonant circuit (or the frequency characteristic of the output filter, if a more complicated band- or low-pass filter is used) are known [2].

6. RF BJTs have a high saturation voltage $V_{sat} = 1 \ldots 3$ V. This is an RF saturation voltage and its value is significantly higher than the DC or low-frequency value provided in the data sheets. The effect of the saturation voltage can be taken into account by replacing Equation 2.4 with

$$I_0 \leq I_{dc} \qquad V_0 \leq V_{dc} - V_{sat} \qquad\qquad 2.11$$

We obtain

$$P_{0,max} = \frac{(V_{dc} - V_{sat})^2}{2R}$$

$$I_{dc} = \frac{V_{dc} - V_{sat}}{R} \qquad P_{dc} = \frac{V_{dc}(V_{dc} - V_{sat})}{R} \qquad 2.12$$

$$\eta_{max} = \frac{1}{2}\left(1 - \frac{V_{sat}}{V_{dc}}\right) < \frac{1}{2} \qquad C_P = \frac{1}{4}\frac{V_{dc} - V_{sat}}{2V_{dc} - V_{sat}} < \frac{1}{8}$$

The output power, the collector efficiency, and the power output capability decrease with increasing V_{sat}. Note that although V_{sat} limits the maximum output power, collector efficiency, and power output capability, it does not affect the amplifier operation in the active region of the transistor.

EXAMPLE 2.1

Design a Class A amplifier that delivers $P_0 = 1$ W to a 50-ohm load. The DC-supply voltage is $V_{dc} = 15$ V. Assume that the transistor used in this circuit has $V_{sat} = 2$ V. According to Equation 2.5, the peak output voltage and current are given by $V_0 = 10$ V and $I_0 = 0.2$ A, respectively. Note that Equation 2.11 is satisfied and the transistor does not saturate. Therefore, it is convenient to choose $I_{dc} = I_0 = 0.2$ A. Now, $P_{dc} = 3$ W and $\eta = 33.3\%$. The peak collector voltage is $v_{max} = V_{dc} + V_0 = 25$ V, and the peak collector current is $i_{max} = 2I_0 = 0.4$ A resulting in $C_P = 0.1$. Under nominal power conditions, the transistor dissipates $P_{dis} = 2$ W. For safe operation, consider $P_{dis} = P_{dc} = 3$ W for the thermal calculations. The amplifier performance can be improved if a matching network (see Sections 2.6 and 2.7) is used, so the equivalent load of the amplifier is

$$R = \frac{(V_{dc} - V_{sat})^2}{2P_0} = 84.5 \ \Omega \qquad\qquad 2.13$$

With this value, $I_{dc} = 153.8$ mA, $P_{dc} = 2.31$ W, $\eta = 43.3\%$, and $C_P = 0.116$ from Equation 2.12.

---

7. During saturation, BJTs are characterized by an approximately constant saturation voltage, V_{sat}, and a usually negligible saturation resistance. MOSFETs are characterized only by an approximately constant saturation resistance, R_{ON}. Like V_{sat}, R_{ON} does not affect operation in the active region, but reduces the maximum output power, collector efficiency, and power output capability. Assuming $I_{dc} = I_0$, $\theta = 3\pi/2$ (see Figure 2-3).

$$V_{dc} - RI_0 = 2I_0 R_{ON} \qquad 2.14$$

The maximum output voltage is then given by

$$V_{0,max} = RI_0 = \frac{R}{R + 2R_{ON}} V_{dc} \qquad 2.15$$

Finally,

$$P_{0,max} = \frac{V_{dc}^2 R}{2(R + R_{ON})^2}$$

$$I_{dc} = \frac{V_{dc}}{R + R_{ON}} \qquad P_{dc} = \frac{V_{dc}^2}{R + 2R_{ON}} \qquad 2.16$$

$$\eta_{max} = \frac{1}{2} \frac{1}{1 + 2\frac{R_{ON}}{R}} < \frac{1}{2} \qquad C_P = \frac{1}{8} \frac{1}{1 + \frac{R_{ON}}{R}} < \frac{1}{8}$$

The output power, the collector efficiency, and the power output capability decrease with increasing R_{ON}.

EXAMPLE 2.2

Design a Class A amplifier that delivers $P_0 = 1$ W to a 50-ohm load. The DC-supply voltage is $V_{dc} = 15$ V; a MOSFET with $R_{ON} = 5\ \Omega$ is used in this circuit. According to Equation 2.5, the peak output voltage is given by $V_0 = 10$ V, and the peak output current is $I_0 = 0.2$ A. Because $V_{dc} - RI_0 = 5$ V $> 2I_0 R_{ON} = 2$ V, the MOSFET does not saturate. As in Example 2.1, $I_{dc} = 0.2$ A, $P_{dc} = 3$ W, $\eta = 33.3\%$, $v_{max} = 25$ V, $i_{max} = 0.4$ A, and $C_P = 0.1$. Amplifier performance can be improved if a matching network is used to transform the load resistance (50 ohms) into the optimum load resistance of the amplifier. Its value is found by solving the expression of $P_{0,max}$ for R in Equation 2.16. Thus, $R = 91.41\ \Omega$, or $R = 1.09\ \Omega$; however, the last value is unacceptable because it would give very small values for η and C_P. With $R = 91.41\ \Omega$, Equation 2.16 yields $I_{dc} = 147.9$ mA, $P_{dc} = 2.22$ W, $\eta = 45.1\%$, and $C_P = 0.119$.

8. A resistive load is assumed in the previous analyses. However, real amplifiers often operate into reactive loads. This is not desirable, but may result from variations of the load or the output filter impedance, mistuning, or parasitic reactances. In the analysis that follows, assume that a susceptance, B, (at the operating frequency) is connected in parallel with the load resistance, R, in Figure 2-1. Because the operation of the transistor as a controlled-current source is not affected, the output voltage is given by

$$V_0 = I_0 \frac{R}{\sqrt{1+(BR)^2}} = \rho I_0 R \qquad 2.17$$

where

$$\rho = \frac{1/R}{|Y|} = \frac{1}{\sqrt{1+(BR)^2}} \le 1 \qquad 2.18$$

Assuming that the active device is biased at $I_{dc} = V_{dc}/R$, the requirement $V_0 \le V_{dc}$ is satisfied. Therefore

$$P_0 = \frac{V_0^2}{2R} = \frac{I_0^2 R}{2}\rho^2 \qquad P_{0,max} = \frac{V_{dc}^2}{2R}\rho^2 \qquad P_{dc} = \frac{V_{dc}^2}{R}$$

$$\eta_{max} = \frac{1}{2}\rho^2 < \frac{1}{2} \qquad C_P = \frac{\rho^2}{4(1+\rho)} < \frac{1}{8} \qquad 2.19$$

Note that a reactive load significantly decreases the output power, collector efficiency, and power output capability.

EXAMPLE 2.3
A Class A amplifier delivers $P_0 = 1$ W to a 50-ohm load. The DC-supply voltage is $V_d = 10$ V and the saturation voltage and/or resistance are ignored. Under nominal conditions $V_0 = 10$ V, $I_0 = 0.2$ A, $I_{dc} = 0.2$ A, $P_{dc} = 2$ W, $\eta = 50\%$, $v_{max} = 20$ V, $i_{max} = 0.4$ A, and $C_P = 0.125$. If mistuning causes a susceptance of $B = 0.01$ S in parallel with the load resistance, then Equations 2.17 through 2.19 yield ρ

= 0.894, V_0 = 8.94 V, P_0 = 0.8 W, P_{dc} = 2 W, η = 40%, v_{max} = 18.94 V, i_{max} = 0.4 A, and C_P = 0.106.

9. A bypass local capacitor is recommended in the DC-power supply line to prevent RF currents from migrating along the power bus.

2.2 Class B and AB Amplifiers

This section concentrates on the push-pull circuit of Figure 2-2 [1-10]. because it is the circuit most used in Class B and AB amplifiers. The major results for a single-ended Class B or AB amplifier are listed in section *Current-Source Class C Amplifiers*.

A push-pull Class B RF PA operates much like a Class B audio frequency PA. The two active devices (Q_1 and Q_2) are driven 180 degrees out-of-phase so they are alternately active (i.e., behave as controlled-current sources) and cut off for each half of the RF cycle (see Figure 2-5).

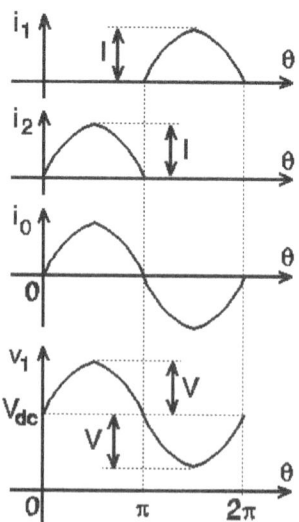

Figure 2-5 *Waveforms in push-pull Class B amplifiers.*

A simplified analysis of the Class B circuit is based on the usual assumptions about the ideal behavior of components given in the previous section. Moreover, the output transformer T_2 is ideal, having m turns in each half of the primary winding and n turns in the secondary winding. In each half-cycle, a half-sinusoidal current of peak value I is carried through one-half of the primary winding of T_2. As a result, the secondary current is sinusoidal

$$i_0(\theta) = I_0 \sin\theta = \frac{m}{n} I \sin\theta \qquad 2.20$$

and determines a sinusoidal output voltage

$$v_0(\theta) = V_0 \sin\theta = \frac{m}{n} I R_L \sin\theta \qquad 2.21$$

The voltages across the two transistors are given by

$$v_1(\theta) = V_{dc} + V \sin\theta = V_{dc} + \frac{m}{n} V_0 \sin\theta = V_{dc} + \frac{m^2}{n^2} I R_L \sin\theta$$

$$= V_{dc} + RI \sin\theta \qquad 2.22$$

$$v_2(\theta) = V_{dc} - V \sin\theta = V_{dc} - RI \sin\theta$$

where

$$R = \left(\frac{m}{n}\right)^2 R_L \qquad 2.23$$

is the resistance seen across one-half of the primary winding with the other half open (i.e., the equivalent load resistance of each transistor). The two transistors do not saturate if $V \le V_{dc}$. As a result, the output power (dissipated in the load resistance R_L)

$$P_0 = \frac{V_0^2}{2R_L} = \frac{V^2}{2R} \qquad 2.24$$

has a maximum value (for $V = V_{dc}$) given by

$$P_{0,max} = \frac{V_{dc}^2}{2R} \qquad 2.25$$

Because

$$i_{dc}(\theta) = i_1(\theta) + i_2(\theta) = I|\sin\theta| \qquad 2.26$$

the DC-input current is found as

$$I_{dc} = \frac{1}{2\pi} \int_0^{2\pi} i_{dc}(\theta) d\theta = \frac{2}{\pi} I = \frac{2}{\pi} \frac{V}{R} \qquad 2.27$$

and the DC input power and the collector efficiency are

$$P_{dc} = V_{dc}I_{dc} = \frac{2}{\pi}\frac{V_{dc}}{R}V \qquad\qquad 2.28$$

$$\eta = \frac{P_0}{P_{dc}} = \frac{\pi}{4}\frac{V}{V_{dc}} \le \eta_{max} = \frac{\pi}{4} \approx 78.5\% \qquad\qquad 2.29$$

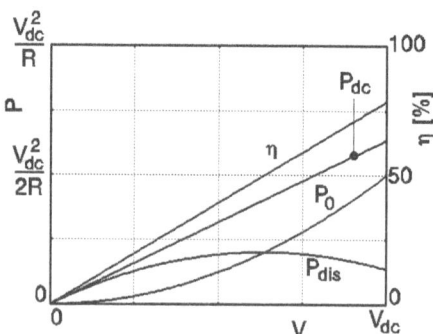

Figure 2-6 *Powers and collector efficiency versus the amplitude of the collector voltage in a push-pull Class B amplifier.*

Figure 2-6 shows collector efficiency η and powers P_0, P_{dc} and $P_{dis} = P_{dc} - P_0$ (the power dissipated in both transistors) versus the amplitude of collector voltage V. Without a drive signal, the transistors are both cut off, and there is no power dissipated in them. With an increase in the drive level, the output power and the efficiency increase. The maximum value of P_{dis} is obtained for $V = 2V_{dc}/\pi$.

$$P_{dis,max} = \frac{2}{\pi^2}\frac{V_{dc}^2}{R} = \frac{4}{\pi^2}P_{0,max} \approx 0.4053P_{0,max} \qquad\qquad 2.30$$

All thermal calculations should be based on this value.

The maximum theoretical value of collector efficiency is 78.5 percent. Therefore, a 60 to 70 percent efficiency is usually a reasonable value for real-life Class B amplifiers. Figure 2-5 shows that the peak collector voltage is $v_{max} = 2V_{dc}$ and the the peak collector current is $i_{max} = V_{dc}/R$. As a result, the power output capability is given by

$$C_P = \frac{P_{0,max}}{2(2V_{dc})I} = \frac{1}{8} = 0.125 \qquad\qquad 2.31$$

which is the same as in the Class A circuits.

Some observations and practical considerations

1. As with the Class A circuit, the push-pull Class B circuit does not necessarily require a tuned circuit or a band- or low-pass filter as part of the load circuit. However, they are often used in practical applications to filter out the unwanted output signal harmonics caused by the inherent nonlinearity of the active devices. This is a serious limitation because the obtainable bandwidth — the ratio of f_{max} to f_{min}— cannot exceed 1.5 to 1.7 without retuning.

2. The push-pull Class B circuit is a natural choice for linear, wideband RF PAs. Even if its linearity, output harmonic content, and intermodulation distortion are not as good as in a Class A circuit, the efficiency of a Class B amplifier is considerably higher.

3. Real transistors do not change abruptly from the cutoff region to the active region. For both BJTs and MOSFETs, the transition is gradual and nonlinear and involves an offset voltage (about 0.7 volt for BJTs and 2 to 4 volts for power MOSFETs). Therefore, if the active devices are not biased to produce a small quiescent collector current, the resulting crossover distortion will have a pronounced effect — especially on small signal levels [2, 4, 5]. Crossover distortion is reduced by biasing the transistors at a small quiescent current, typically 1 to 10 percent from the peak value of the collector current. Thus, the conduction angle is greater than 180 degrees and the active devices operate in Class AB. Note that it is almost impossible to obtain "true" Class B operation, i.e., $2\theta_c = 180°$. Without bias voltage, the active device would operate in Class C (with $2\theta_c < 180°$) due to the offset voltage. This regime is called Class B by many authors, implying that a quiescent current is used to reduce crossover distortion.

4. If the push-pull circuit is perfectly symmetrical, the output current does not contain even harmonics. In real amplifiers, even harmonics are caused by the circuit asymmetry (especially the two halves of the primary winding of T_2) and by the differences between the characteristics of the two transistors. The problem of selecting two matched transistors can be eliminated using single-chip push-pull RF power transistors. These are two transistors in a single package with separate base and collector leads and emitters common [11]. A theoretical estimation of the even harmonics level (due to the circuit asymmetry) is a difficult task and the results are often in poor agreement with the experimental measurements.

5. The effects of V_{sat} on the performance of push-pull Class B amplifiers are determined using a procedure similar to that used for Class A amplifiers.

$$P_0 = \frac{V^2}{2R} \qquad P_{0,max} = \frac{(V_{dc} - V_{sat})^2}{2R} \qquad P_{dc} = \frac{2}{\pi} \frac{V_{dc}}{R} V$$

$$\eta = \frac{\pi}{4} \frac{V}{V_{dc}} \le \eta_{max} = \frac{\pi}{4}\left(1 - \frac{V_{sat}}{V_{dc}}\right) < \frac{\pi}{4} \qquad C_P = \frac{1}{4} \frac{V_{dc} - V_{sat}}{2V_{dc} - V_{sat}} < \frac{1}{8} \qquad 2.32$$

EXAMPLE 2.4

Design a push-pull Class B amplifier that delivers P_0 = 8 W to a 50-ohm load. The DC-supply voltage is V_{dc} = 12 V. Assume that the transistors used in this circuit have V_{sat} = 2 V. To avoid saturation, the collector voltage swing must be kept below $V_{dc} - V_{sat}$ = 10 V. According to Equation 2.32

$$P_0 \le P_{0,max} = \frac{(V_{dc} - V_{sat})^2}{2R} \qquad R \le \frac{(V_{dc} - V_{sat})^2}{2P_0} = 6.25\,\Omega \qquad 2.33$$

Assuming that tuned matching networks are not to be used, the most convenient value of R is R = 5.56 W, obtained with m/n = 1/3.[1] With R = 5.56 Ω, Equations 2.24 and 2.27 through 2.29 yield V = 9.43 V, I_{dc} = 1.08 A, P_{dc} = 12.96 W, and η = 61.7%, respectively. The peak collector voltage is $v_{max} = V_{dc} + V$ = 21.43 V and the current is i_{max} = V/R = 1.70 A, resulting in C_P = 0.110. Under nominal power conditions, the dissipated power is $P_{dis} = P_{dc} - P_0$ = 4.96 W (in both transistors). However, according to Equation 2.30, the maximum dissipated power is $P_{dis,max}$ = 5.25 W, and this value must be considered for the thermal calculations. In many practical applications, a tuned matching network is used to filter out the harmonics and perform the impedance matching. For example, a matching network can be used to transform the 50-ohm load resistance into a 56.25-ohm resistance. Next, a transformer with m/n = 1/3 gives R = 6.25 W. With this value, V = 10 V, I_{dc} = 1.02 A, P_{dc} = 12.22 W, η = 65.4%, v_{max} = 22 V, i_{max} = 1.60A, C_P = 0.114, $P_{dis,max}$ = 4.67 W.

---

6. The effects of R_{ON} on the performance of push-pull Class B amplifiers are determined using a procedure similar to that for Class A amplifiers.

$$P_{0,max} = \frac{V_{dc}^2 R}{2(R + R_{ON})^2} \qquad P_{dc,max} = \frac{2}{\pi} \frac{V_{dc}^2}{R + R_{ON}}$$

$$\eta_{max} = \frac{\pi}{4} \frac{1}{1 + \frac{R_{ON}}{R}} < \frac{\pi}{4} \qquad C_P = \frac{1}{8} \frac{1}{1 + \frac{R_{ON}}{2R}} < \frac{1}{8} \qquad 2.34$$

EXAMPLE 2.5

The circuit in Example 2.4 uses MOSFETs with $R_{ON} = 1\ \Omega$ instead of BJTs with $V_{sat} = 2$ V. By solving the expression of $P_{0,max}$ in Equation 2.34 for R, we obtain $R = 6.85\ \Omega$ or $R = 0.15\ \Omega$. (The last value is unacceptable.) $R = 6.85\ \Omega$ can easily be found with a matching network that transforms the 50-ohm load resistance into a 61.65-ohm resistance, and a transformer with $m/n = 1/3$. The results are as follows: $V = 10.47$ V, $P_{dc} = 11.68$ W, $\eta = 68.5\%$, $v_{max} = 22.47$ V, $i_{max} = 1.53$A, $C_P = 0.116$, $P_{dis,max} = 4.26$ W.

---

7. The effects of a reactive load on the performance of push-pull Class B amplifiers are determined using a procedure similar to that for Class A amplifiers. Assuming that a susceptance B is connected in parallel with the load R_L:

$$V = \frac{m}{n}V_0 = \rho \frac{m^2}{n^2}IR_L = \rho RI \qquad \rho = \frac{1/R_L}{|Y|} = \frac{1}{\sqrt{1+(BR_L)^2}} \leq 1$$

$$P_0 = \frac{V^2}{2R} \qquad P_{0,max} = \frac{V_{dc}^2}{2R} \qquad P_{dc} = V_{dc}I_{dc} = V_{dc}\frac{2}{\pi}I = V_{dc}\frac{2}{\pi}\frac{V}{\rho R} \qquad 2.35$$

$$\eta = \frac{\pi}{4}\rho\frac{V}{V_{dc}} \leq \eta_{max} = \rho\frac{\pi}{4} < \frac{\pi}{4} \qquad C_P = \frac{1}{8}\rho < \frac{1}{8}$$

2.3 Class C Amplifiers

Many applications that do not require linear amplification of the input signal as obtained in Class A, push-pull Class B, or AB amplifiers. Examples of such applications are the amplification of CW or FM signals or amplifiers for AM signals using collector amplitude modulation. Class C amplifiers have an important advantage because their collector efficiency is higher than that obtained in Class A, B, or AB amplifiers. The major disadvantages, with respect to the previously discussed amplification classes, are a higher harmonic content of the output that may require additional filtering and a lower power gain.

The literature on Class C amplifiers contains a large number of models and analyses of this circuit [1-10, 12-24]. Some of these analyses use complicated models for the active device (usually modified small-signal hybrid-pi or Ebers-Moll models). For the most part, analytic solutions for the resulting equations cannot be found and numerical computation is often required. Despite their complexity, these analyses have limited utility because some of their initial assumptions cannot be adapted for real circuits. Also, none of these analyses provide circuit design equations.

One of the most popular analysis techniques approximates the collector current waveform with a portion of a sine wave when the transistor is active [1, 2, 4, 5, 8, 9, 12, 13, 14]. The collector current is zero when the transistor is cut off. If the transistor does not saturate during the RF cycle, the regime is called *current-source Class C* or *underdriven Class C*. This is described in section **Current-Source Class C Amplifiers**.

If the active device saturates for a portion of the RF cycle, the collector current waveform may be described as zero when the transistor is cut off, a portion of a sine wave when it is active, and another portion of a sine wave when it is saturated [1, 4, 5, 25]. This is called *saturated Class C* or *overdriven Class C* (see section **Saturated Class-C Amplifiers**).

Although both models are based on some assumptions that are unacceptable in practical circuits (for example, the collector current waveform is a portion of a sine wave only at low frequencies [4, 5, 14]), they allow a simple understanding of the Class C circuit and provide simple design equations. These models are derived directly from the corresponding vacuum-tube Class C circuits. However, there are considerable differences between the operation of a vacuum-tube Class C amplifier and the operation of a solid-state Class C amplifier. The latter is significantly more complex and extremely difficult to analyze with reasonable accuracy. It is usually expected that solid-state Class C amplifiers operate in the *Class C mixed-mode regime* [4, 10, 18], as discussed in section **Class C Mixed Mode Amplifiers**.

Current-Source Class C Amplifiers

In a current-source (or underdriven) Class C amplifier, the active device operates alternately in the cutoff and active regions, and does not saturate during the RF cycle. In the active region, the transistor operates as a controlled-current source.

The following analysis refers to the single-ended Class C amplifier. The basic circuit is presented in Figure 2-1. A push-pull circuit, shown in Figure 2-2, may be also used because it has the same advantages and disadvantages, as discussed in Section 2.1, for the Class A amplifier. Class A and push-pull Class B and AB amplifiers do not need a parallel-tuned circuit for operation. However, in a Class C circuit, the parallel-tuned circuit (or an equivalent low- or band-pass filter) is required to assure correct circuit operation. For simplicity's sake, assume that the parallel-tuned circuit is ideal. Other assumptions provided in Section 2.1 are also considered in this analysis.

The DC component of the collector current $i(\theta)$ flows through the RF choke and then through the DC-power supply. The variable component of

$i(\theta)$ flows through DC-blocking capacitor C_d and through the parallel RLC tuned circuit. The tuned circuit provides a zero impedance path to ground for the harmonic currents contained in $i(\theta)$ and only the fundamental component of $i(\theta)$ flows through the load resistance. As a result, the output voltage is a sinusoidal waveform. This requires the use of a parallel resonant circuit (or an equivalent band- or low-pass filter), rather than a series-tuned (or equivalent) circuit.

The current and voltage waveforms in a current-source Class C amplifier are shown in Figures 2-7 and 2-8. Assuming that $2\theta_c$ is the conduction angle, the transistor is in its active region for $-\theta_c \le \theta \le \theta_c$ and the collector current is a portion of a sine wave. The peak value of the collector current pulse is I_M (see Figures 2-7 and 2-8). If the waveform of the collector current pulse were extended to a full 360 degree cycle, the amplitude of that cycle would be I. I_Q of Figure 2-7 may be considered as analogous to the quiescent current in a Class A or AB PA. (This so-called quiescent collector current in the Class C amplifier is actually a negative current, $-I_Q$.) The analogy is simply a formal one and is not supported by any physical reality.

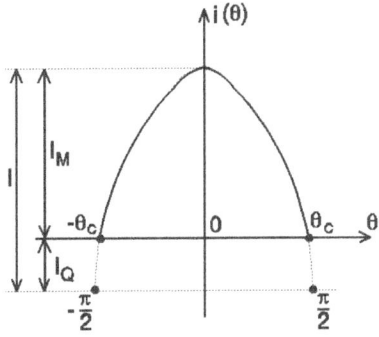

Figure 2-7 *Waveform of collector current pulse in current-source class C amplifier.*

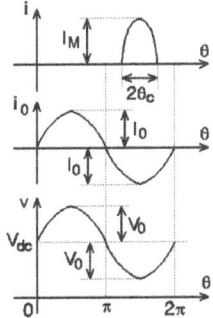

Figure 2-8 *Waveforms in a current-source Class C amplifier.*

The collector current is a periodical waveform described by

$$i(\theta) = \begin{cases} \dfrac{I_M(\cos\theta - \cos\theta_c)}{1 - \cos\theta_c} & -\theta_c + 2k\pi \le \theta \le \theta_c + 2k\pi \quad k \in Z \\ 0 \quad \text{otherwise} \end{cases} \qquad 2.36$$

Its Fourier analysis results in [1, 2, 4, 5, 8]

$$i(\theta) = I_M \sum_{n=0}^{\infty} \alpha_n(\theta_c)\cos n\theta \qquad 2.37$$

where

$$\alpha_0(\theta_c) = \frac{\sin\theta_c - \theta_c\cos\theta_c}{\pi(1 - \cos\theta_c)} \qquad \alpha_1(\theta_c) = \frac{\theta_c - \sin\theta_c\cos\theta_c}{\pi(1 - \cos\theta_c)}$$

$$\alpha_n(\theta_c) = \frac{\dfrac{\sin(n-1)\theta_c}{n-1} - \dfrac{\sin(n+1)\theta_c}{n+1}}{n\pi(1 - \cos\theta_c)} \qquad n = 2, 3, \dots \qquad 2.38$$

An equivalent form for this Fourier decomposition is

$$i(\theta) = I \sum_{n=0}^{\infty} \gamma_n(\theta_c)\cos n\theta \qquad 2.39$$

where

$$\gamma_0(\theta_c) = \frac{\sin\theta_c - \theta_c\cos\theta_c}{\pi} \qquad \gamma_1(\theta_c) = \frac{\theta_c - \sin\theta_c\cos\theta_c}{\pi}$$

$$\gamma_n(\theta_c) = \frac{1}{n\pi}\left[\frac{\sin(n-1)\theta_c}{n-1} - \frac{\sin(n+1)\theta_c}{n+1}\right] \qquad n = 2, 3, \dots \qquad 2.40$$

Equations 2.37 through 2.40 provide the collector current harmonics as a function of the conduction angle. These equations are valid for any $0 \le 2\theta_c \le 360°$; that is, for all Class A, AB, B, and C PAs. The variation of the Fourier coefficients α_0, α_1, α_2, and α_3 (giving the DC component, the fundamental, and the second and third harmonic, respectively), with the conduction angle $2\theta_c$, is depicted in Figure 2-9. Figure 2-10 presents the variation of the Fourier coefficients γ_0, γ_1, γ_2, and γ_3 with the conduction angle $2\theta_c$. For $180° \le 2\theta_c \le 360°$, α_3 and γ_3 are negative, and the third harmonic is 180 degrees out-of-phase to the fundamental.

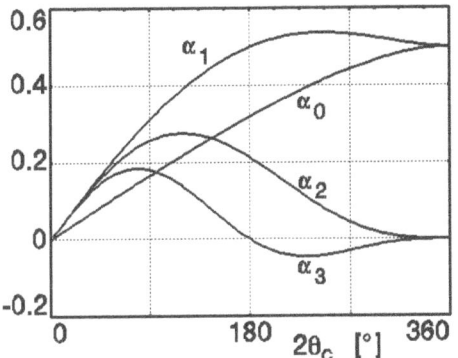

Figure 2-9 *Fourier series coefficients $\alpha_0 \ldots \alpha_3$ versus the conduction angle $2\theta_c$.*

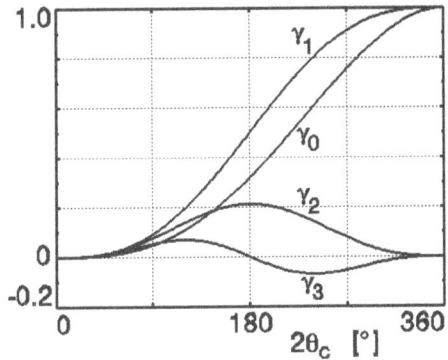

Figure 2-10 *Fourier series coefficients $\gamma_0 \ldots \gamma_3$ versus the conduction angle $2\theta_c$.*

Due to the ideal tuned circuit, the output current (flowing through the load resistance R) is sinusoidal and its amplitude is given by

$$I_0 = I_M \alpha_1(\theta_c) \qquad 2.41$$

As a result, the output voltage is also sinusoidal, with the amplitude $V_0 = RI_0$. The collector voltage is

$$v(\theta) = V_{dc} + V_0 \cos\theta = V_{dc} + RI_M \alpha_1(\theta_c)\cos\theta \qquad 2.42$$

The DC input power P_{dc} and the collector efficiency η are

$$P_{dc} = V_{dc}I_{dc} = V_{dc}I_M \alpha_0(\theta_c) \qquad 2.43$$

$$\eta = \frac{P_0}{P_{dc}} = \frac{V_0^2}{2RV_{dc}I_M\alpha_0(\theta_c)} = \frac{V_0}{V_{dc}}\frac{\alpha_1(\theta_c)}{2\alpha_0(\theta_c)} =$$
$$= \frac{V_0}{V_{dc}}\frac{\theta_c - \sin\theta_c \cos\theta_c}{2(\sin\theta_c - \theta_c \cos\theta_c)} \tag{2.44}$$

$$\eta_{max} = \frac{\theta_c - \sin\theta_c \cos\theta_c}{2(\sin\theta_c - \theta_c \cos\theta_c)} \tag{2.45}$$

The last equation shows that η increases with the output voltage V_0. The maximum possible value of V_0 is $V_0 = V_{dc}$ (neglecting V_{sat} and/or R_{ON}). This is the most convenient regime for a current-source Class C amplifier. It is sometimes called the *critical regime* [5]. The maximum theoretical collector efficiency (obtained for $V_0 = V_{dc}$) varies with the conduction angle, as shown in Figure 2-11.

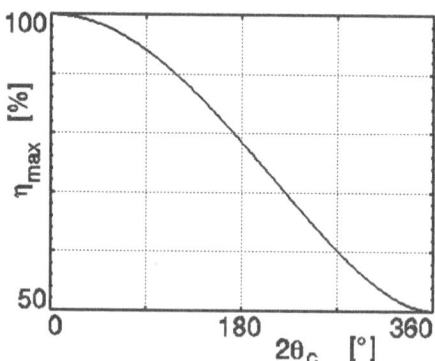

Figure 2-11 *Maximum theoretical collector efficiency η versus conduction angle $2\theta_c$.*

If $V_0 = V_{dc}$, the peak collector voltage is $v_{max} = 2 V_{dc}$ and the peak collector current is given by

$$i_{max} = i_M = \frac{V_{dc}}{R\alpha_1(\theta_c)} \tag{2.46}$$

The output power P_0 and the power output capability C_P are

$$P_0 = \frac{V_0^2}{2R} = \frac{V_{dc}^2}{2R} \tag{2.47}$$

$$C_P = \frac{P_0}{v_{max}i_{max}} = \frac{\alpha_1(\theta_c)}{4} \qquad\qquad 2.48$$

The power output capability varies with the conduction angle, as shown in Figure 2-12.

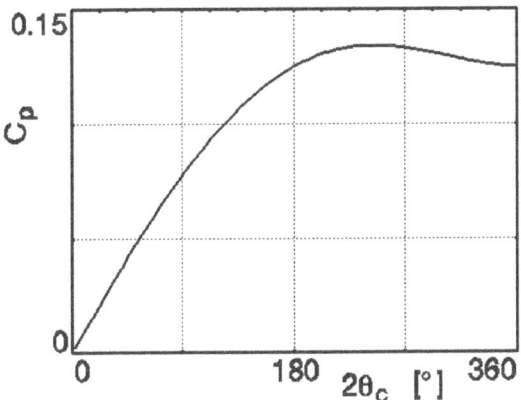

Figure 2-12 *Power output capability C_P versus conduction angle $2\theta_c$.*

The analysis of these results shows that

a. The collector efficiency is higher in Class C amplifiers than in Class A, AB, or B amplifiers, and it increases as the conduction angle decreases. If $\theta_c \to 0$, then $\eta \to 100\%$.

b. The power output capability of Class C amplifiers is lower than 0.125 (as obtained in Class A or B circuits) and decreases as the conduction angle decreases (see Figure 2-12).

c. The maximum power output capability is obtained in a Class AB circuit: if $2\theta_c \approx 245.2°$, then $C_P = 0.1341$.

d. In most practical applications, it is important to obtain a given output power. In this case, a decrease of the conduction angle determines an increase of the peak collector current. The variation of I and I_M with $2\theta_c$ (for constant output power and load resistance) is shown in Figure 2-13. Because I varies proportionally with the drive level (as a first approximation), the power gain of the amplifier decreases as the conduction angle decreases. As a result, the choice of conduction angle would be a tradeoff among collector efficiency, peak value of the collector current, and power gain.

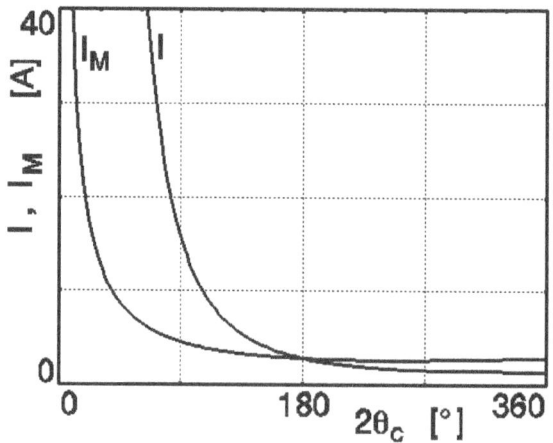

Figure 2-13 *Currents I and I_M (required to obtain an output power of 1 watt into a load resistance of 1 ohm) versus conduction angle $2\theta_c$.*

e. A comparison of the performance of Class A, B, and C amplifiers is provided in Table 2-1.

Table 2-1 *Theoretical performance of single-ended classic PAs.*

Class	η_{max}	$\dfrac{P_0}{V_{dc}^2 / R}$	$\dfrac{v_{max}}{V_{dc}}$	$\dfrac{i_{max}}{I_{dc}}$	C_P
A	50%	0.5	2	2	0.125
B	78.5%	0.5	2	π	0.125
C	$\dfrac{\alpha_1(\theta_c)}{2\alpha_0(\theta_c)}$	0.5	2	$\dfrac{1}{\alpha_0(\theta_c)}$	$\dfrac{\alpha_1(\theta_c)}{4}$

The conduction angle in a Class C amplifier is controlled by a DC-bias voltage V_B applied to the base, and an amplitude V_b of the signal across the base-emitter junction. A simple base circuit of a Class C PA using BJT devices is shown in Figure 2-14; its corresponding waveforms are shown in 2-15. C_b is a DC-blocking capacitor and RFC is an RF choke. The waveforms depicted in Figure 2-15 are valid only at low frequencies, where sinusoidal waveforms can be obtained in practical circuits and the base and collector currents are in phase.

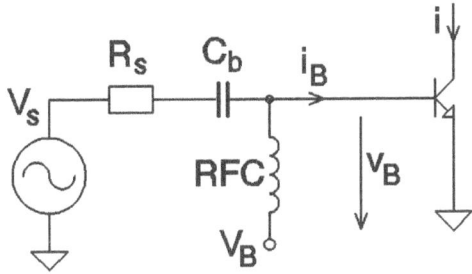

Figure 2-14 *Base circuit of a Class C PA using BJTs.*

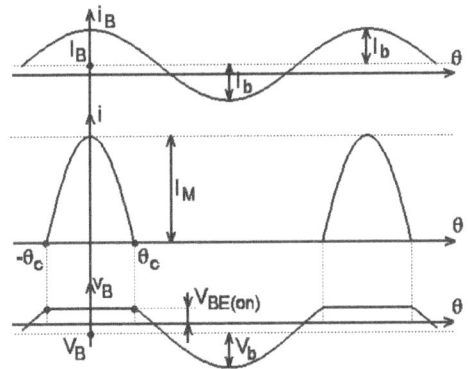

Figure 2-15 *Waveforms in the base circuit of the Class C amplifier.*

The amplitude of base current I_b and DC base current I_{B0} are given by

$$I_{B0} = \frac{I_{dc}}{\beta_0} = \frac{I}{\beta_0}\gamma_0(\theta_c) \qquad I_b = \frac{I}{\beta} = \frac{I_0}{\beta\gamma_1(\theta_c)} \qquad\qquad 2.49$$

For $-\theta_c \le \theta \le \theta_c$, the transistor is in its active region. Consequently, the voltage across its base-emitter junction is $V_{BE(on)} \approx 0.7$ and

$$V_{BE(on)} = v_B(\theta_c) = V_B + V_b\cos\theta_c \qquad\qquad 2.50$$

Equation 2.50 allows calculation of the required bias voltage, V_B, in the base circuit. (V_b can be calculated knowing I_b and the input impedance.) Because $\gamma_1(\theta_c)$ decreases as θ_c decreases, the amplitude of the base current I_b increases as θ_c decreases (see Equation 2.49). As a result, the power gain decreases as the conduction angle decreases. On the other hand, both V_b and $\cos\theta_c$ increase as the conduction angle decreases. According to

Equation 2.50, V_B, usually a negative voltage, as suggested in Figure 2-15, decreases, which creates an additional problem for small conduction angles. The peak reverse base-to-emitter voltage may exceed the emitter-base breakdown voltage (usually, 4 volts for RF power BJTs) and destroy the transistor or degrade its RF characteristics [26].

EXAMPLE 2.6

Design a Class C amplifier that delivers $P_0 = 8$ W to a 50-ohm load with 85-percent efficiency. The DC-power supply is $V_{dc} = 12$ V; the saturation voltage and/or resistance are ignored. For maximum collector efficiency, $V_0 = V_{dc}$. The equivalent load resistance of the amplifier must be $R = 9 \, \Omega$ (a matching circuit is required to transform the load resistance of 50 ohms into $R = 9 \, \Omega$). According to Figure 2-11, a collector efficiency $\eta_{max} = 85\%$ requires $2\theta_c \approx 147°$. The collector current fundamental is $I_0 = V_{dc}/R = 1.33$ A and, according to Equation 2.41, the peak collector current is $I_M = 2.97$ A. The DC input power is $P_{dc} = P_0/\eta = 9.41$ W. Thus, $I_{dc} = P_{dc}/V_{dc} = 0.78$ A. The peak collector voltage is $v_{max} = V_{dc} = 24$ V. The power output capability, $C_P = 0.112$, is given by Equation 2.48 or can be obtained from the values above.

Some practical considerations

1. The effects of V_{sat} on the performance of current-source Class C amplifiers are determined using a procedure similar to that for Class A amplifiers.

$$P_{0,max} = \frac{(V_{dc} - V_{sat})^2}{2R} \qquad \eta_{max} = \left(1 - \frac{V_{sat}}{V_{dc}}\right) \frac{\theta_c - \sin\theta_c \cos\theta_c}{2(\sin\theta_c - \theta_c \cos\theta_c)}$$

$$C_P = \frac{\alpha_1(\theta_c)}{2} \frac{V_{dc} - V_{sat}}{2V_{dc} - V_{sat}} \qquad\qquad\qquad\qquad 2.51$$

2. The effects of R_{ON} on the performance of current-source Class C amplifiers can be estimated numerically using Equation 2.65.
3. A reactive circuit load determines important decreases in output power, collector efficiency, and power output capability (as in Class A or B amplifiers).

Saturated Class C Amplifiers

In many practical Class C amplifier applications, it is convenient to drive the transistor into saturation for a portion of each RF cycle. Saturated operation has two important advantages. First, some increases in the output power, collector efficiency, and power output capability are possible. Second, the amplitude of the output signal depends primarily on the DC-supply voltage and is almost insensitive to the drive level.

A simplified analysis of the saturated Class C amplifier (sometimes called *overdriven Class C amplifier*) is based on the assumptions provided in the previous section. Further, the active device is modeled during saturation as a constant resistance, R_{ON}. The effect of V_{sat} may be taken into account by replacing V_{dc} with $V_{dc} - V_{sat}$ in Equations 2.52 through 2.61. Equations 2.62 through 2.64 remain unchanged.

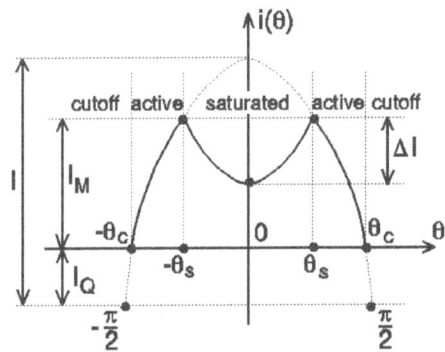

Figure 2-16 *Waveform of collector current pulse in the saturated Class C amplifier.*

The waveform of the collector current is approximated, as shown in Figure 2-16, and is described by

$$i(\theta) = \begin{cases} \dfrac{I_M(\cos\theta - \cos\theta_c)}{\cos\theta_s - \cos\theta_c} = I(\cos\theta - \cos\theta_c) \text{ transistor active} \\[3mm] \dfrac{V_{dc} - V_0\cos\theta}{R_{ON}} \text{ transistor saturated} \\[3mm] 0 \text{ transistor cut off} \end{cases} \qquad 2.52$$

The conduction angle is $2\theta_c$, although the transistor is either active or saturated during this period. For an angular time interval, $2\theta_s$, the transistor is saturated; $2\theta_s$ is called the *saturation angle*. As in Figure 2-7, consider (for a formal analogy with the other amplification classes) that the quiescent current in the Class C amplifier has a negative value, $-I_Q$.

The Fourier analysis of the collector current waveform results in

$$i(\theta) = \sum_{n=0}^{\infty}\left[I_M\alpha'_n(\theta_c,\theta_s) - \Delta I\alpha_n(\theta_s)\right]\cos n\theta \qquad 2.53$$

where

$$\alpha'_n(\theta_c,\theta_s) = \frac{\gamma_n(\theta_c) - \gamma_n(\theta_s)}{\cos\theta_s - \cos\theta_c} \tag{2.54}$$

and α_n is given in Equation 2.38; γ_n is given in Equation 2.40. Equation 2.53 gives the fundamental and the harmonics of the collector current as functions of θ_c and θ_s. The tuned parallel circuit provides a zero impedance path to ground for the harmonic currents contained in $i(\theta)$. The fundamental component of $i(\theta)$ flows through the load resistance, giving a sinusoidal output voltage with the amplitude

$$V_0 = R[I_M \alpha'_1(\theta_c,\theta_s) - \Delta I \alpha_1(\theta_s)] \tag{2.55}$$

Using Equation 2.52, for $\theta = 0$ (during saturation)

$$\Delta I = I_M - \frac{V_{dc} - V_0}{R_{ON}} = I(\cos\theta_s - \cos\theta_c) - \frac{V_{dc} - V_0}{R_{ON}} \tag{2.56}$$

and for $\theta = \theta_s$, Equation 2.52 yields

$$V_{dc} = V_0 \cos\theta_s + I_M R_{ON} = V_0 \cos\theta_s + I R_{ON}(\cos\theta_s - \cos\theta_c) \tag{2.57}$$

In the active region, the transistor behaves as a controlled-current source; thus, I and $\cos\theta_c = I_Q/I$ are determined by the driving signal parameters. If the values of V_{dc}, R, R_{ON}, and I_Q are known, Equations 2.55 through 2.57 allow analysis of the circuit to find V_0 and θ_s. From Equation 2.57, we obtain

$$V_0 = \frac{V_{dc} - R_{ON} I(\cos\theta_s - \cos\theta_c)}{\cos\theta_s} \tag{2.58}$$

Substituting V_0 in Equation 2.56 results in

$$\Delta I = \frac{V_{dc} - R_{ON} I(\cos\theta_s - \cos\theta_c)}{R_{ON}} \frac{1 - \cos\theta_s}{\cos\theta_s} \tag{2.59}$$

Finally, by substituting V_0 given by Equation 2.58 and ΔI given by Equation 2.59 in Equation 2.55, we obtain

$$\pi[V_{dc} - I R_{ON}(\cos\theta_s - \cos\theta_c)] =$$
$$= I R \cos\theta_s \left(\theta_c - \frac{1}{2}\sin 2\theta_c\right) - R\left(\frac{V_{dc}}{R_{ON}} + I\cos\theta_c\right)\left(\theta_s - \frac{1}{2}\sin 2\theta_s\right) \tag{2.60}$$

This equation will yield θ_s. It is required that $0 \le \theta_s \le \theta_c$. Next, from Equations 2.58 and 2.59, find V_0 and ΔI. With these values, calculate the output power

$$P_0 = \frac{V_0^2}{2R} \qquad\qquad 2.61$$

the DC input power

$$P_{dc} = V_{dc}I_{dc} = V_{dc}\left[I_M\alpha_0'(\theta_c,\theta_s) - \Delta I\alpha_0(\theta_s)\right] \qquad\qquad 2.62$$

the collector efficiency

$$\eta = \frac{P_0}{P_{dc}} \qquad\qquad 2.63$$

and the power output capability

$$C_P = \frac{P_0}{v_{max}i_{max}} = \frac{V_0^2}{2R(V_{dc} + V_0)I_M} \qquad\qquad 2.64$$

This model is valid only for $V_0 \le V_{dc}$. If $V_0 > V_{dc}$, Equation 2.56 would give $\Delta I > I_M$. In this case, the collector current would be zero during saturation and Equation 2.52 and those that follow could not be applied. In some references this operating regime is called *heavily overdriven Class C*, and the operating regime discussed before is called *slightly overdriven Class C* [5, 9]. The heavily overdriven Class C operation has no practical use and is not discussed here.

The behavior of the saturated Class C amplifier is illustrated by two examples [4].

1. A single-ended Class B amplifier biased at $I_Q = 0$ (i.e., $2\theta_c = 180°$ for any I) and with $R_{ON} = 0.1$ R.[2]

2. A single-ended Class C amplifier biased at $-I_Q = -0.9$ R/V_{dc} and with $R_{ON} = 0.1$ R.

Voltages are normalized to V_{dc}, resistances to R, currents to V_{dc}/R, and powers to V_{dc}^2/R. In both cases, the effects of varying I (due to the variation of the drive level) between 0 and 8 V_{dc}/R are analyzed. The results are shown in Figure 2-17 (the conduction and the saturation angle, the output voltage, the peak collector current, the output power, the power output capability, and the collector efficiency).

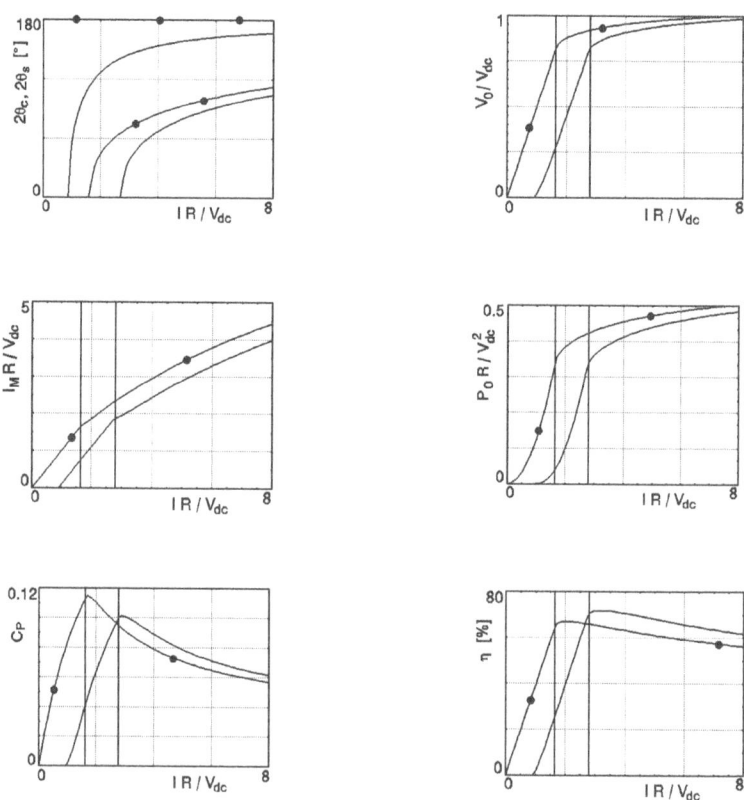

Figure 2-17 *Circuit performance of Class B amplifier (— • —) and Class C amplifier (—) versus the normalized amplitude of the collector current* ($R_{ON}/R = 0.1$).

For low values of I, the active device does not saturate and the analysis procedure detailed in section **Current-Source Class C Amplifiers** is used. Once the collector current I is higher than a *critical value*, or I_{cr}, the transistor saturates for a portion of the RF cycle; therefore, the previous equations must be used. The critical value is found from

$$V_{dc} = V_0 + I_M R_{ON} = R I_M \alpha_1(\theta_c) + I_M R_{ON} =$$

$$= (I_{cr} - I_Q)[R\alpha_1(\theta_c) + R_{ON}]$$

2.65

where

$$\theta_c = \arccos\left(\frac{I_Q}{I_{cr}}\right)$$ 2.66

Substituting θ_c from Equation 2.66 into Equation 2.65 yields an equation that can be solved numerically for I_{cr}. For the two cases considered here, $I_{cr}(R/V_{dc}) \approx 1.67$ (the Class B circuit) and $I_{cr}(R/V_{dc}) \approx 2.75$ (the Class C circuit). Two vertical lines in Figure 2-17 mark these values in order to separate the current-source and the saturated operation.

Analysis of the results given in Figure 2-17 shows that

a. In current-source operation, the output voltage and the output power increase very quickly as the drive level is increased (i.e., with I). In saturated operation, they continue to increase, but significantly more slowly. The peak collector current also increases more slowly in saturated operation.

b. Because I is, as a first approximation, proportional with the drive level, the power gain in saturated operation decreases very quickly as the drive level increases.

c. The collector efficiency and the power output capability reach maximum values just after the saturation occurs. This is the optimum operating point of the Class C amplifier.

d. V_0/V_{dc} is almost constant with the drive level in saturated operation. If V_{dc} varies, V_0 varies almost proportionally with V_{dc}. This property is used to obtain AM signals by varying the DC-supply voltage (collector amplitude modulation), as discussed in Section 2.9.

Class C Mixed-Mode Amplifiers

Many Class C solid-state PA references describe the operation of this circuit in a manner similar to that given in above. This procedure is derived directly from the classic analysis of vacuum-tube Class C amplifiers [1, 5, 9] and allows a simple explanation of the basic principles of the Class C circuit. Other, more complex models are based on the approximation that the collector current is a portion of a sine wave (possibly, slightly distorted) [5, 9, 13 – 16, 20 – 22]. These models take into account some of the most important parameters of real circuits (the resistances and reactances associated with the active device or the nonlinearity of its characteristics) and assume that the transistor does not saturate during the RF cycle.

Results based on these models (including those given in the previous two sections) must be applied cautiously in solid-state Class C circuits. These theories are especially useful in a qualitative approach (providing more insight into the operation of the circuit than other complicated models), but almost useless for practical design because

1. Many parameters required for analysis and design (for example, the base bulk resistance or the equivalent emitter balancing resistor) are not provided in the data sheets and are very difficult to measure or estimate. Other parameters (for example, the active device capacitances — also not available in the data sheets) are strongly dependent on the signal characteristics such as level, waveform, or frequency and cannot be estimated or measured with accuracy for a particular application.

2. It is extremely difficult to control a practical circuit (component values, parasitics, or drive level) so the transistor does not saturate during the RF cycle. On the other hand, experience shows that better results are possible if the transistor is driven into saturation.

3. It is almost impossible to obtain a sinusoidal collector voltage, as required by these models. This is caused by the voltage-variable collector-emitter capacitance that can be a significant part of the total capacitance in the resonant circuit and by the very low saturation resistance that dominates the resonant circuit (if the transistor is driven into saturation) and flattens the collector voltage waveform.

4. A very important assumption in these analyses is that the circuit uses a high-Q parallel-tuned circuit or an equivalent band- or low-pass filter. It provides a zero impedance path for the harmonics of the collector current and forces a sinusoidal collector voltage. However, a parallel-tuned circuit [or an equivalent, like a pi matching network (see Section 2.6)] is somewhat difficult to implement because it requires low inductance and high capacitance values.[3] A series-tuned circuit (or an equivalent, such as a T network) would be much more convenient in applications that require reasonable component values.[4] However, such a circuit cannot be used because it would not provide a zero impedance path for the harmonics of the collector current. A series-tuned circuit would force a sinusoidal collector current and allow the presence of harmonic components in the collector voltage.

5. It is very difficult to bias and drive a BJT for Class C operation. Such an operation would require very small inductances, making it extremely difficult to control the conduction angle and the collector current waveform [4].

Class C operation, described in the previous sections, is often called *classic Class C* or *true Class C* [4] and is mostly encountered in vacuum-tube amplifiers. Practical solid-state Class C amplifiers operate in a different regime called *Class C mixed-mode* [4, 10, 18]. Some references use the terms *Class D* or *Class C-D* to describe this concept.

The operation of a Class C mixed-mode amplifier is very complex, considerably different from one amplifier to another, and extremely difficult to describe accurately in mathematical terms. A simplified description of the circuit operation is based on the one given in Figure 2-18. Figure 2-19 presents several typical waveforms [4].

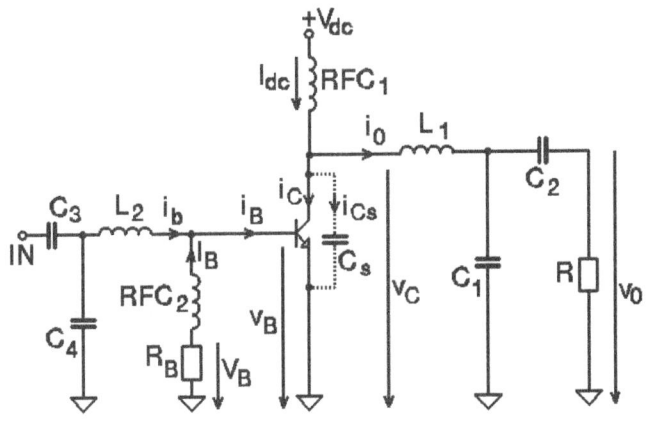

Figure 2-18 *Mixed-mode Class C amplifier.*

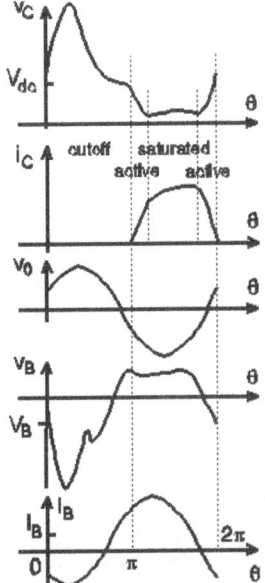

Figure 2-19 *Typical waveforms in mixed-mode Class C amplifier [4].*

The circuit in Figure 2-18 uses a T matching network to transform load resistance R into the equivalent load required to obtain the desired output power. The T network is functionally equivalent to a series resonant circuit; it will allow the presence of harmonic components in the collector voltage, and the current flowing through L_1 will be forced to be sinusoidal. C_s is not a discrete circuit element. It is a voltage-variable capacitance consisting of the output capacitance of the active device and the parasitic capacitances, and plays an important role in circuit operation.

When the transistor is in its active region, the collector voltage is determined by the current i_{Cs} flowing through C_s. i_{Cs} is the difference between the DC current, I_{dc}, the output sinusoidal current, i_0, and the collector current, i_C. (i_C is determined by the controlled-current source operation of the transistor in the active region.) When the transistor is saturated, the collector voltage is almost constant (V_{CEsat}), $i_{Cs} \approx 0$, and the collector current is the difference between I_{dc} and i_0. Finally, when the transistor is cut off, $i_C(\theta) = 0$ and $i_{Cs}(\theta) = I_{dc} - i_0(\theta)$ charges C_s, giving the collector voltage.

This simple description suggests that the mathematical model of the collector circuit is very complex and an analytical solution cannot be found. However, it is possible to define numerical solutions for the differential equations that describe the circuit [23]. These solutions provide the steady-state waveforms of the currents and voltages in the circuit. Their Fourier analysis further provides circuit performance values such as powers, efficiency, and others. Note that the obtained results vary considerably from one amplifier to another, depending on the circuit topology, component values, and transistor parameters.

Figure 2-18 also presents a typical base circuit of a Class C mixed-mode PA. L_2, C_3, and C_4 form a T matching network. Assuming the drive is a sinusoidal current source, a positive base current, i_B, determines $v_B(\theta) \approx 0.7$ V (the forward-biased base junction) and the transistor will be active or saturated, depending on the collector voltage. The AC base current flows entirely through L_2, because the RF choke forces only a DC current flow. The transistor remains active (or saturated) until $i_B(\theta)$ becomes negative and removes the charge stored in the base. The base junction becomes reverse biased. The base voltage is determined by the currents flowing through C_4 and the base-to-emitter capacitance. Typical waveforms of $i_B(\theta)$ and $v_B(\theta)$ are depicted in Figure 2-19. Note that the base current contains a DC component, I_B, and a DC path must be provided in the base circuit (RFC$_2$ and possibly R_B). In many practical applications, RFC$_2$ is connected to ground; thus, the DC voltage across the base-emitter junction is zero. R_B (if present) causes a negative DC voltage across the base-emitter junction. In theory, this may produce an increase in the collector efficiency because the conduction angle decreases. However, the practical utility of R_B is not very clear because it is difficult to control the conduction angle in real-life circuits.

Also, note that the input matching network should have a capacitor connected to ground; it provides an AC current path when the transistor is cut off.

2.4 Bias Circuits

RF PA bias circuits [4, 5, 7, 27, 28] are especially useful in Class A and AB amplifiers intended for linear operation. Requirements imposed on the bias circuits include reproducibility and/or possibilities for individual adjustment and temperature stability. Because a theoretical estimation is complicated, the amount of quiescent current required for minimum IMD in Class AB amplifiers is usually determined experimentally. Typically, minimum IMD is obtained with a quiescent collector current of 1 to 10 percent of the peak collector current. Note that MOSFETs are much more sensitive than BJTs to the quiescent current value.

A common feature of the bias circuits for BJTs and MOSFETs is the absence of emitter resistors. They are almost always used in small-signal amplifiers and assure a stable quiescent point. However, their use is usually avoided in RF PAs because it is extremely difficult to provide a satisfactory RF decoupling of the emitter (or source) to ground. The impedance of the common terminal (emitter or source) must be kept as low as possible to maximize power gain.

Bias Circuits for BJTs

A simple bias source for BJTs is shown in Figure 2-20(a). It uses a clamping diode to provide a low impedance voltage source. A large value capacitor is often connected across the diode to further reduce the AC impedance. The quiescent current can be adjusted by varying R. Usually, the diode is mounted on the heatsink of the transistor, Q, to perform a temperature compensation function. Obviously, it is desirable that the diode and the RF transistor have similar DC characteristics. The main disadvantage of this circuit is its low efficiency. The diode must be able to carry a high forward current (depending on the required base current), and R must dissipate a large amount of power (especially if V_{dc} is high). These drawbacks can be overcome by using an emitter follower to amplify the clamping diode current, as shown in Figure 2-20(b). The forward diode current now can be $h_{FE(Q)}$ times lower than in the previous case. One of the two diodes may be used for temperature compensation of the RF transistor.

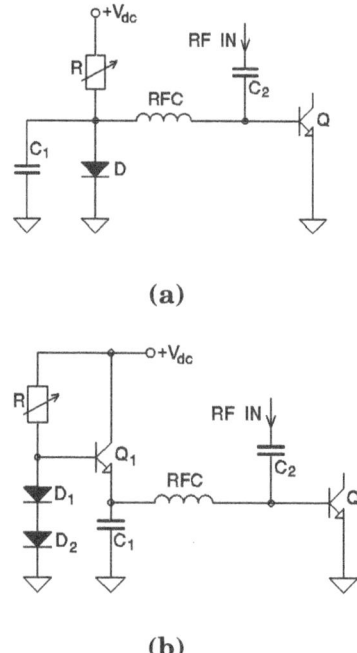

(a)

(b)

Figure 2-20 *Simple biasing sources for BJTs.*

For perfect temperature tracking, the diode and the RF power transistor must have similar DC parameters. Although this need for similar parameters presents a problem, it can be solved easily with a component called the *byistor* [7]. The byistor is a RF power transistor. It includes a small value resistor on the same substrate (see Figure 2-21). An additional variable resistor is required to allow for adjustment of the quiescent collector current of the RF transistor.

Another simple bias source for BJTs is presented in Figure 2-22. L is the collector inductance (possibly an RF choke) of the RF power transistor, Q. C_2 assures DC-supply decoupling. This circuit is identical to a closed loop system. If the collector current of the RF transistor tends to increase, the voltage drop across R also increases and the base-to-emitter voltage of Q_1 decreases. This causes the collector current of Q_1 to decrease. As a result, the base current of Q decreases, holding the collector current of Q at its original value. A similar self-adjustment occurs in the opposite direction. A practical implementation of this principle is shown in Figure 2-22(b). Choosing a higher value for V_z will improve the stability of the quiescent point, but this will increase the power dissipated in R. R_2 can be used to reduce the power dissipated in Q_1. The overall effect of this circuit is similar to that of a resistor connected in the emitter of the RF power transistor.

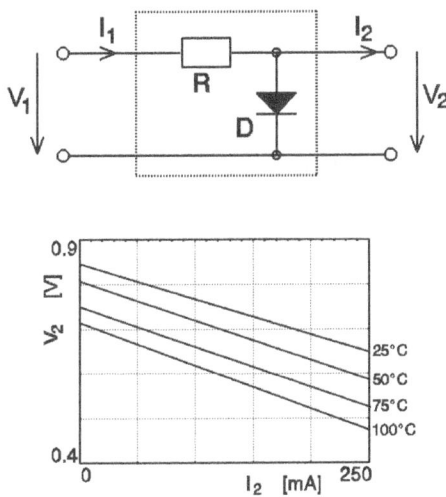

Figure 2-21 *Byistor device used for thermal compensation of the BJT's bias and its typical characteristics with temperature as a parameter.*

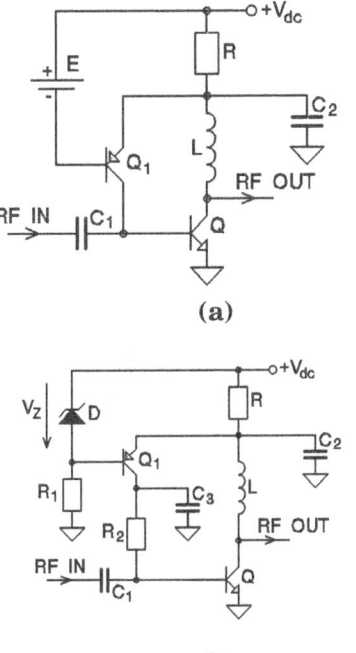

Figure 2-22 *Biasing BJTs in Class A (AB) amplifiers: (a) basic circuit; (b) practical implementation.*

Circuits similar to those in Figure 2-20 can be used to bias BJTs in push-pull amplifiers. The basic circuit and a simple practical implementation are shown in Figure 2-23.

Class C PAs using BJTs are usually operated with zero bias. An RF choke (or an inductance that is a part of the input matching circuit) connects the base to ground, setting $V_{BE} = 0$. In most cases, manufacturers' data sheets only present the parameters of the RF power transistors for this particular case (see section 2.5). A negative base-bias voltage can be introduced in a Class C circuit to reduce the conduction angle and improve the collector efficiency. Most often, this is obtained using a resistor in series with the RF choke in the base circuit (see Figure 2-18).

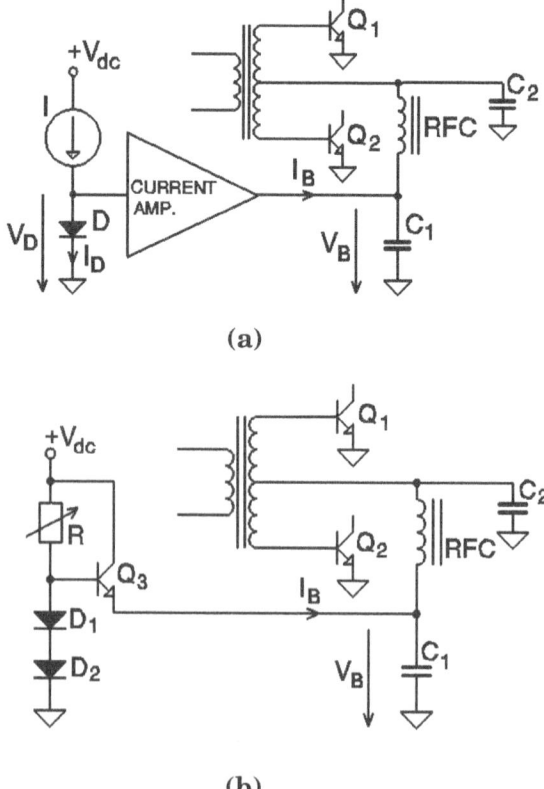

Figure 2-23 *Biasing BJTs in push-pull Class A (AB) amplifiers; (a) basic circuit, (b) practical implementation.*

Bias Circuits for MOSFETs

A simple bias circuit for MOSFETs is shown in Figure 2-24. The gate bias voltage value is provided by Zener diode D. R_G is used for signal source impedance matching and to improve amplifier stability.

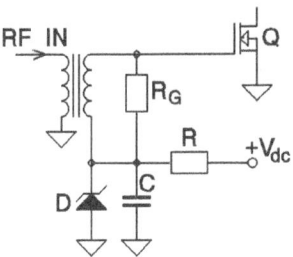

Figure 2-24 *Simple MOSFET bias circuit.*

Temperature compensation of the quiescent drain current (I_{DQ}) can be achieved with networks of thermistors and resistors. The threshold voltage of MOSFETs decreases about 1mV/°C with an increase in temperature, affecting I_{DQ}. On the other hand, the transconductance of the MOSFETs decreases with an increase in temperature and also affects I_{DQ}. The overall effect is that, for lower values of I_{DQ}, the temperature coefficient of the gate bias voltage must be negative. A convenient solution is to use a *negative temperature coefficient* (NTC) thermistor mounted in close thermal contact with the RF power MOSFET. The bias circuit is shown in Figure 2-25. The temperature coefficient of the gate bias voltage can be adjusted by the ratio of R_1 and R_3 (a NTC thermistor). However, most MOSFET data sheets provide only typical characteristics, and significant variations of some parameters can be expected from unit to unit (for example, the transconductance can sometimes vary as much as 100 percent). This may require individual checking of the temperature tracking of each amplifier.

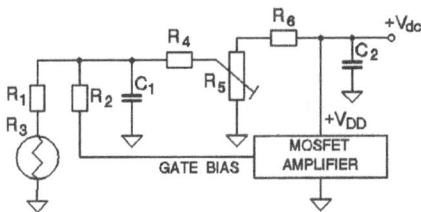

Figure 2-25 *MOSFET bias circuit with temperature compensation.*

An IC voltage regulator can be used for biasing MOSFETs. Such a circuit has two advantages: a low source impedance and a good bias voltage regulation against changes in the power supply voltage. A practical circuit is suggested in Figure 2-26. The ratio of R_5 and R_6 (a NTC thermistor) adjusts the temperature coefficient; R_4 adjusts the bias voltage.

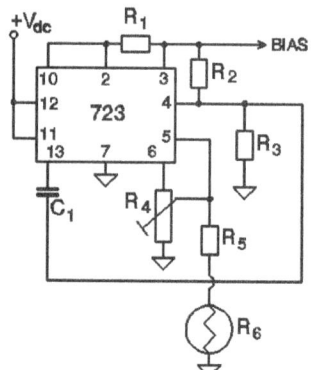

Figure 2-26 *MOSFET bias circuit with an integrated voltage regulator.*

The circuit in Figure 2-27 is a closed loop system for MOSFET biasing. The main advantage is the automatic and precise temperature compensation to any MOSFET regardless of its characteristics. Moreover, this circuit does not need thermistors (or other temperature sensors), and a reasonable spreading of the threshold voltage around a typical value is allowed. The quiescent drain current I_{DQ} is set by R_8. Assuming that V_{dc} is a stable voltage source, a stable voltage reference is applied to the negative input of operational amplifier IC_1. The voltage across R_1 is fed to the positive input of IC_1, assuring the feedback required for the stabilization of I_{DQ}. The values of R_5, R_6, and D_1 are chosen to ensure that the bias voltage of the gate varies in a reasonable range (usually ±0.5 volts), enabling full voltage swing at the output of IC_1. This large voltage swing is required to provide a negative bias to the gate of T_1, which functions as a voltage-controlled resistor. Its resistance decreases when the current flowing through R_1 increases.

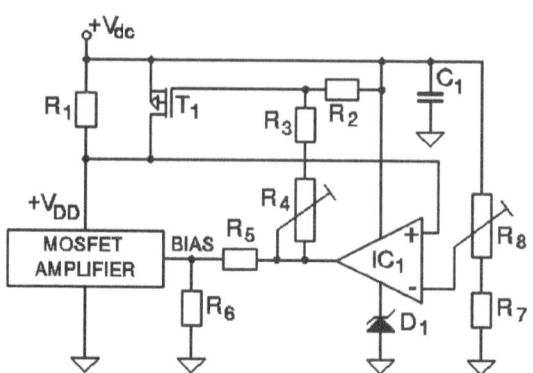

Figure 2-27 *A high-performance MOSFET bias circuit.*

Class C amplifiers using MOSFETs can be operated with zero gate bias. However, in this case, the input voltage amplitude must be high enough to overcome the threshold voltage. The collector efficiency is high, but the power gain is low. Consequently, some positive gate bias voltage can be applied in Class C circuits to increase the power gain.

2.5   Large Signal Parameters

Small-signal low-frequency amplifiers are often designed using some type of equivalent circuit for the transistor. (The most popular, by far, is the small-signal hybrid-pi model.) This technique also may be used at high frequencies. However, it is very difficult to accurately estimate the model parameters of the equivalent circuit, and this model is therefore rarely used. The most common RF small-signal amplifier design procedure is based on small-signal two-port parameters (y- or s-parameters). This is a systematic, mathematical procedure with an exact solution and can be applied to any active linear device. Unfortunately, these parameters are useless for the design of very nonlinear RF PAs. Table 2-2 illustrates this with a comparison of the parameters of a BJT measured for small-signal linear operation and large-signal Class C operation [29].

Table 2-2 *Small- and large-signal parameters at f = 300 MHz, for the 2N3948 BJT.*

	Small Signal Class A V_{CE} = 15 V, I_C = 80 mA	Large Signal Class C V_{CE} = 13.6 V, P_0 = 1 W
input impedance	9 Ω || 12 nH	38 Ω || 21 pF
output impedance	199 Ω || 4.6 pF	92 Ω || 5 pF
power gain	12.4 dB	8.2 dB

A considerable change occurs for the input impedance. The output resistance and power gain are also considerably different for the two modes of operation. This example clearly shows that small-signal parameters of the active device are not useful in high-power applications.[5] The small-signal s-parameter design method has led to the large-signal s-parameter technique [30, 31]. However, practical use is limited by two factors.

1. Large-signal s-parameters are much more difficult to measure accurately than small-signal s-parameters, especially for high-power devices.

2. There is no evidence of successful use of large-signal s-parameters in designing RF PAs above a few watts of output power.

RF power transistors can be also characterized by the *load-pull tech-nique* [31]. Performance parameters of the active device (gain, efficiency, input return loss, and others) are graphically represented versus the load impedance as surface or contour plots. Load-pull data can be used only at the operating conditions (output impedance, output power, frequency, or bias voltages) at which they are measured. This technique provides useful information about the behavior of the active device, but requires laborious and extensive measurements.

The most popular RF PA design technique is based on the large-signal input and output impedances of the active device [3, 4, 29, 31]. Almost every RF power transistor data sheet includes a section identifying the large-signal impedances and power gain. This system is universally accepted by transistor manufacturers and circuit designers and provides a systematic RF PA design procedure.

Large-signal input and output impedances refer to the transistor terminal impedances operating in a matched amplifier at specified values of frequency, DC supply voltage, and output power. It is important to be very careful when using these data, because they are only valid for the specified circuit parameter values. In addition to frequency, DC supply voltage, and output power values, other conditions such as temperature, input power, bias voltages or currents, and harmonic current levels must be considered. Further, the following two factors must be considered:

1. The term *impedance* has significance only for sinusoidal signals. However, in an RF PA, both input and output signals are not sinusoidal. Consider that the fundamental components of the currents and voltages are taken into account when large-signal impedances are defined. The large-signal input and output impedances are actually measured with sinusoidal signals. Figure 2-46 and the explanations related provide a better understanding of this concept.

2. The term *output impedance* is somewhat misleading because its significance is different from that used in small-signal amplifiers. In the RF PA context, the output impedance is almost always referred to as the complex conjugate of the optimum load impedance into which the device output operates at a given output power, voltage, and frequency. This parameter is not related to the standard small-signal output impedance of the active device and, as detailed in Table 2-2, their numerical values differ considerably. Figure 2-46 and its related explanations provide a better understanding of the large-signal impedance concept.

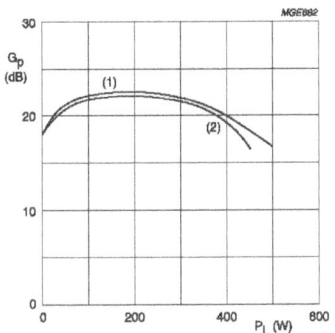

Class-B operation; $V_{DS} = 50$ V; $I_{DQ} = 2 \times 0.1$ A; f = 108 MHz;
$Z_L = 3.2 + j4.3$ Ω (per section); $R_{GS} = 4$ Ω (per section).
(1) $T_h = 25$ °C.
(2) $T_h = 70$ °C.

Figure 2-28 *Power gain as a function of load power, typical values, BLF278, Class B operation. (From BLF278 VHF push-pull power MOS transistor B Product Specification, October 21, 1996, courtesy of Philips Semiconductors.)*

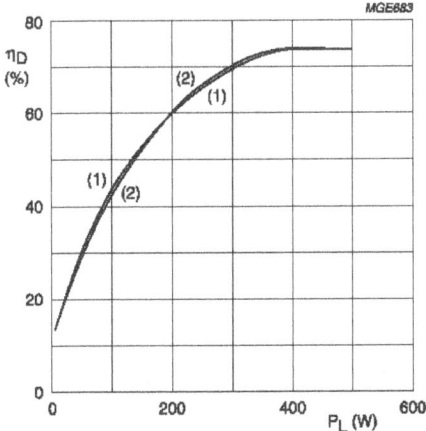

Class-B operation; $V_{DS} = 50$ V; $I_{DQ} = 2 \times 0.1$ A; f = 108 MHz;
$Z_L = 3.2 + j4.3$ Ω (per section); $R_{GS} = 4$ Ω (per section).
(1) $T_h = 25$ °C.
(2) $T_h = 70$ °C.

Figure 2-29 *Efficiency as a function of load power, typical values, BLF278, Class B operation. (From BLF278 VHF push-pull power MOS transistor B Product Specification, October 21, 1996, courtesy of Philips Semiconductors.)*

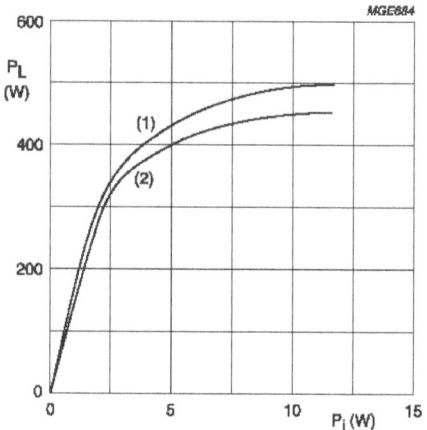

Class-B operation; V_{DS} = 50 V; I_{DQ} = 2 × 0.1 A; f = 108 MHz;
Z_L = 3.2 + j4.3 Ω (per section); R_{GS} = 4 Ω (per section).
(1) T_h = 25 °C.
(2) T_h = 70 °C.

Figure 2-30 Load power as a function of input power, typical values BLF278, Class B operation. (From BLF278 VHF push-pull power MOS transistor B Product Specification, October 21, 1996, courtesy of Philips Semiconductors.)

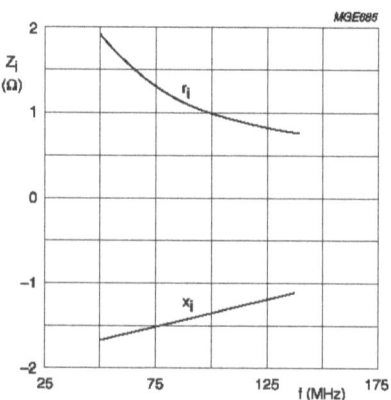

Class-B operation; V_{DS} = 50 V; I_{DQ} = 2 × 0.1 A;
R_{GS} = 4 Ω (per section); P_L = 300 W.

Figure 2-31 Input impedance as a function of frequency (series components), typical values per section, BLF278, Class B operation. (From BLF278 VHF push-pull power MOS transistor B Product Specification, October 21, 1996, courtesy of Philips Semiconductors.)

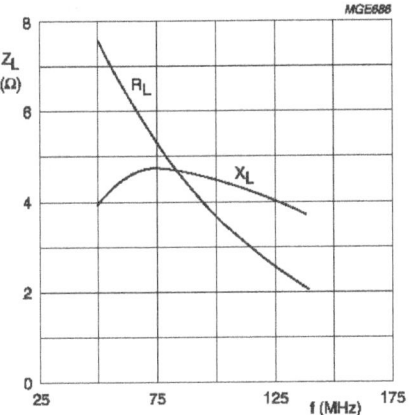

Class-B operation; V_{DS} = 50 V; I_{DQ} = 2 x 0.1 A;
R_{GS} = 4 Ω (per section); P_L = 300 W.

Figure 2-32 *Load impedance as a function of frequency (series components) typical values per section BLF278, Class B operation. (From BLF278 VHF push-pull power MOS transistor B Product Specification, October 21, 1996, courtesy of Philips Semiconductors.)*

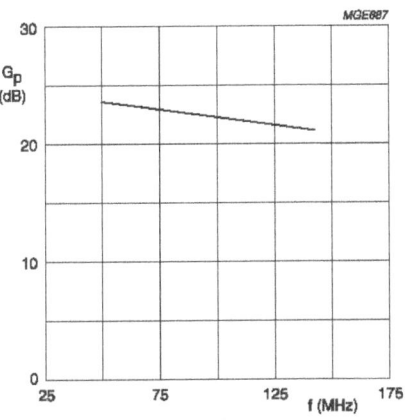

Class-B operation; V_{DS} = 50 V; I_{DQ} = 2 x 0.1 A;
R_{GS} = 4 Ω (per section); P_L = 300 W.

Figure 2-33 *Power gain as a function of frequency, typical values per section, BLF278, Class B operation. (From BLF278 VLF push-pull power MOS transistor B Product Specification, October 21, 1996, courtesy of Philips Semiconductors.)*

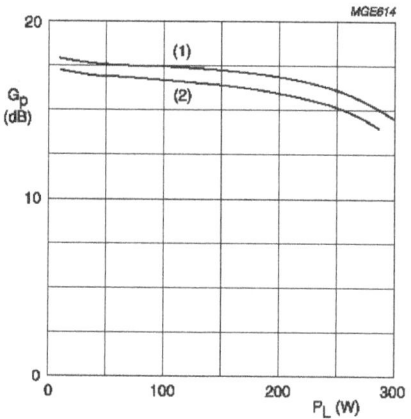

Class-AB operation; V_{DS} = 50 V; I_{DQ} = 2 × 0.5 A; f = 225 MHz;
Z_L = 0.74 + j2 Ω (per section); R_{GS} = 2.8 Ω (per section).
(1) T_h = 25 °C.
(2) T_h = 70 °C.

Figure 2-34 *Power gain as a function of load power, typical values, BLF278, Class AB operation. (From BLF278 VHF push-pull power MOS transistor B Product specification, October 21, 1996, courtesy of Philips Semiconductors.)*

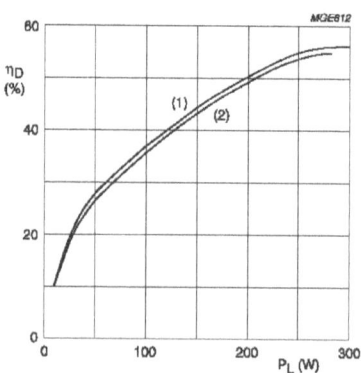

Class-AB operation; V_{DS} = 50 V; I_{DQ} = 2 × 0.5 A; f = 225 MHz;
Z_L = 0.74 + j2 Ω (per section); R_{GS} = 2.8 Ω (per section).
(1) T_h = 25 °C.
(2) T_h = 70 °C.

Figure 2-35 *Efficiency as a function of load power, typical values, BLF278, Class AB operation. (From BLF278 VHF push-pull power MOS transistor B Product Specification, October 21, 1996, courtesy of Philips Semiconductors.)*

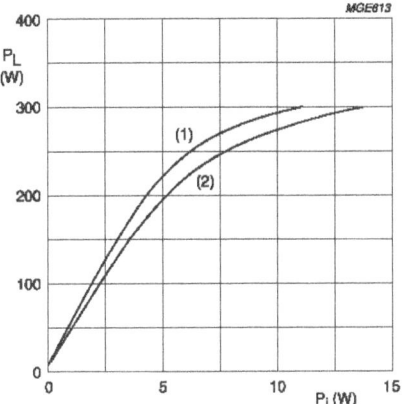

Class-AB operation; V_{DS} = 50 V; I_{DQ} = 2 x 0.5 A; f = 225 MHz;
Z_L = 0.74 + j2 Ω (per section); R_{GS} = 2.8 Ω (per section).
(1) T_h = 25 °C.
(2) T_h = 70 °C.

Figure 2-36 Load power as a function of input power, typical values, BLF278, Class AB operation. (From BLF278 VHF push-pull power MOS transistor B Product Specification, October 21, 1996, courtesy of Philips Semiconductors.)

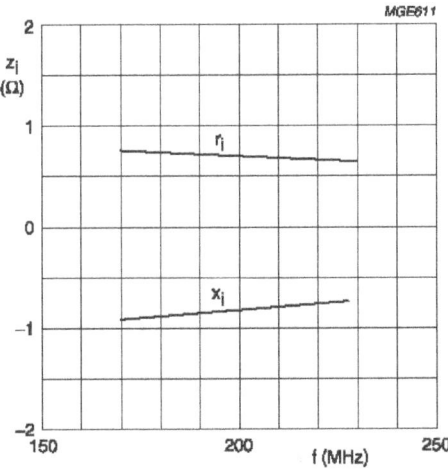

Class-AB operation; V_{DS} = 50 V; I_{DQ} = 2 x 0.5 A;
R_{GS} = 2.8 Ω (per section); P_L = 250 W.

Figure 2-37 Input impedance as a function of frequency (series components), typical values per section, BLF278, Class AB operation. (From BLF278 VHF push-pull power MOS transistor B Product Specification, October 21, 1996, courtesy Philips Semiconductors.)

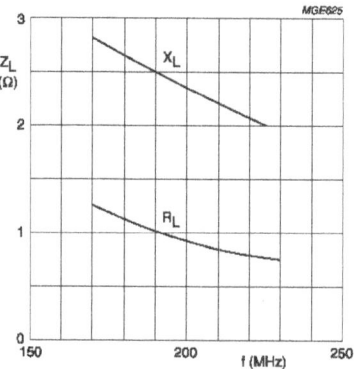

Class-AB operation; V_{DS} = 50 V; I_{DQ} = 2 x 0.5 A;
R_{GS} = 2.8 Ω (per section); P_L = 250 W.

Figure 2-38 *Load impedance as a function of frequency (series components), typical values per section, BLF278 VHF push-pull power MOS transistor B Product Specification, October 21, 1996, courtesy of Philips Semiconductors.)*

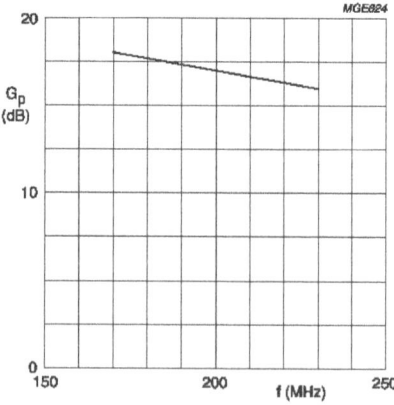

Class-AB operation; V_{DS} = 50 V; I_{DQ} = 2 x 0.5 A;
R_{GS} = 2.8 Ω (per section); P_L = 250 W.

Figure 2-39 *Power gain as a function of frequency, typical values per section, BLF278, Class AB operation. (From BLF278 VHF push-pull power MOS transistor B Product Specification, October 21, 1996, courtesy of Philips Semiconductors.)*

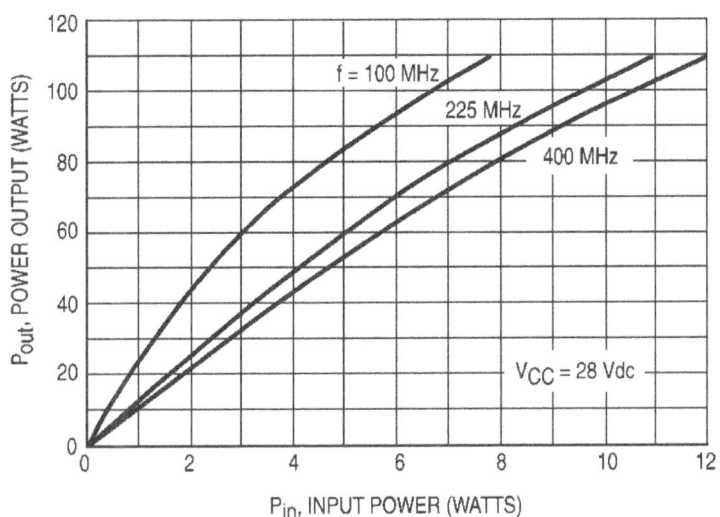

Figure 2-40 *Output power as a function of input power, typical values, MRF329, Class C operation, $V_{dc} = 28$ V. (From Motorola RF Device Data DL110/D Rev 8, 1997, courtesy of Motorola Semiconductors.)*

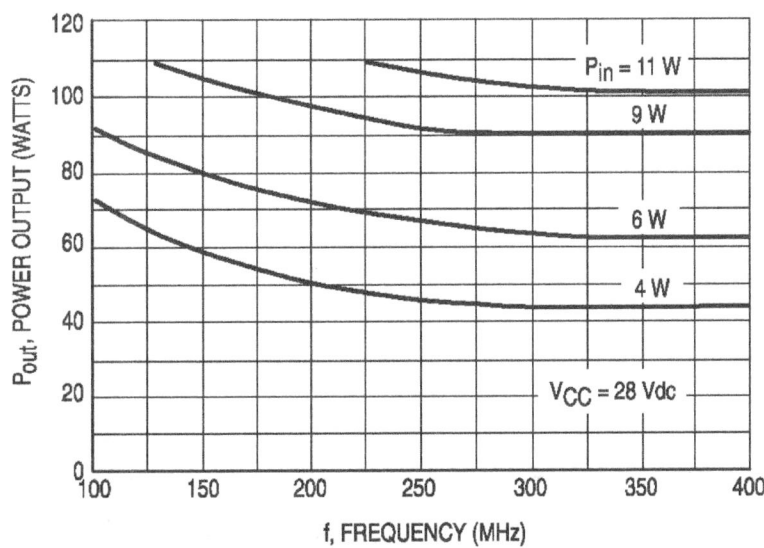

Figure 2-41 *Output power as a function of frequency, typical values, MRF329, Class C operation, $V_{dc} = 28$ V. (From Motorola RF Device Data DL110/D Rev 8, 1997, courtesy of Motorola Semiconductors.)*

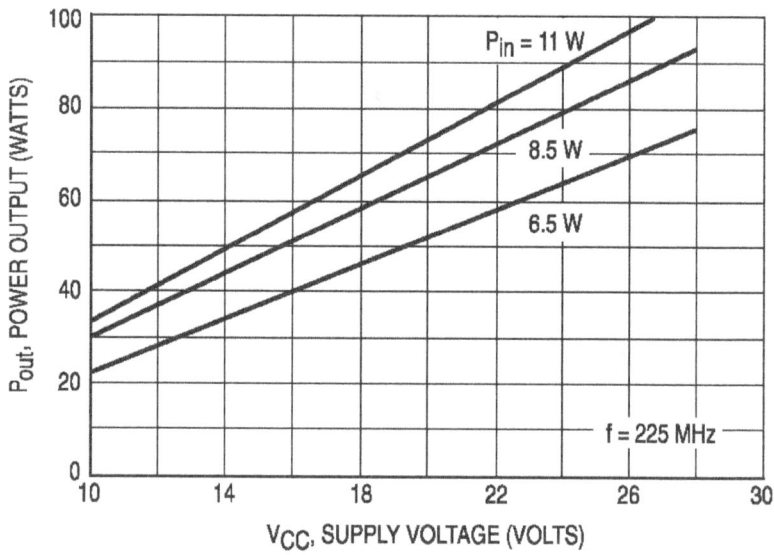

Figure 2-42 *Output power as a function of supply voltage, typical values, MRF329, Class C operation, f = 225 MHz. (From Motorola RF Device Data DL110/D Rev 8, 1997, courtesy of Motorola Semiconductors.)*

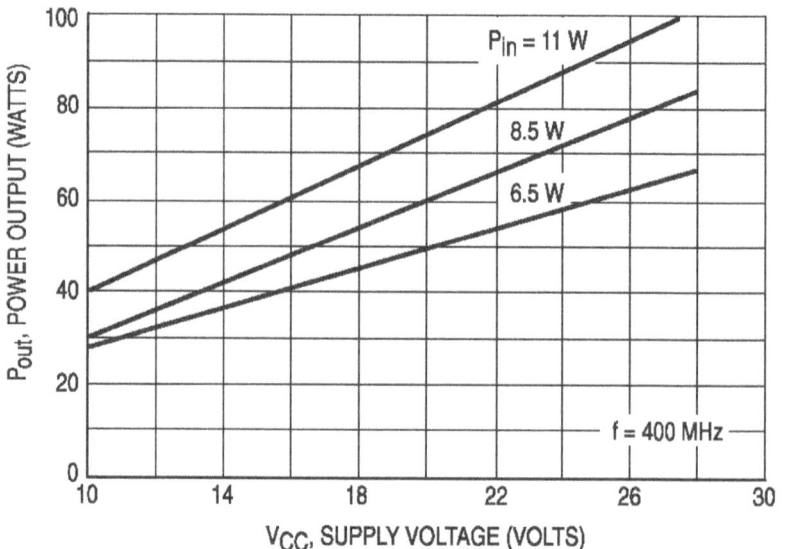

Figure 2-43 *Output power as a function of supply voltage, typical values, MRF329, Class C operation, f = 400 MHz. (From Motorola RF Device Data DL110/D Rev 8, 1997, courtesy of Motorola Semiconductors.)*

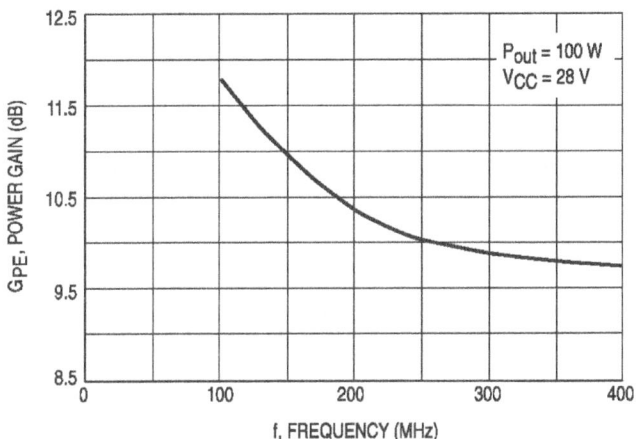

Figure 2-44 *Power gain as a function of frequency, typical values, MRF329, Class C operation, V_{dc} = 28 V, P_{out} = 100 W. (From Motorola RF Device Data DL110/D Rev 8, 1997, courtesy of Motorola Semiconductors.)*

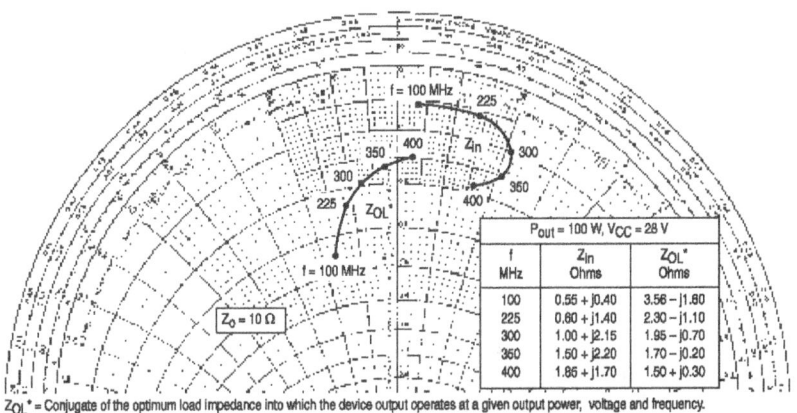

Figure 2-45 *Series equivalent input/output impedances as a function of frequency, typical values, MRF239, Class C operation, V_{dc} = 28 V, P_{out} = 100 W. (From Motorola RF Device Data DL110/D Rev 8, 1997, courtesy of Motorola Semiconductors.)*

Figures 2-28 through 2-45 show typical data sheet characteristics of an RF power transistor. Figures 2-28 through 2-33 provide the parameters of the BLF278 MOSFET for Class B operation with a quiescent drain current of 0.1 amp. Figures 2-34 through 2-39 provide the parameters of the same BLF278 MOSFET for Class AB operation with a quiescent drain current of 0.5 amp.

Finally, Figures 2-40 through 2-45 present the parameters of the MRF329 BJT for Class C operation with zero base-to-emitter bias voltage. The other measuring conditions are specified for each figure. Some comments:

a. Data sheets include information about the power gain and the efficiency of the active device (see Figures 2-28 through 2-30, 2-33 through 2-36, and 2-39 through 2-44). They are usually presented in graphical form as functions of input power, output power, or frequency. Tables with numerical values may also be provided.

b. Large-signal impedances may be presented in these forms:

- *Series resistance-reactance components as a function of frequency* (see Figures 2-31, 2-32, 2-37, and 2-38). At low frequencies, series resistance-capacitance components may be provided instead of the series resistance-reactance components.

- *Series resistance-reactance components as a function of frequency plotted on a Smith chart* (see Figure 2-45). This form is especially popular for use with UHF devices due to the extensive application of the Smith chart in microstrip matching network design.

- *Tables with numerical values* (usually series resistance-reactance components, see Figure 2-45).

- *Parallel resistance-reactance components as a function of frequency* are sometimes provided in the data sheets.

Caution: Figures 2-32 and 2-38 present the optimum load impedance in the drain circuit $Z_L = R_L + jX_L$. Figure 2-45 gives the complex conjugate of the optimum load impedance $Z_{0L}{}^* = R_L - jX_L$.

c. Significant differences can be observed between Class B and Class AB data of the same transistor. As mentioned earlier, large-signal characteristics are strongly dependent on the measuring conditions.

d. Large-signal parameters of RF power transistors operating in Class C amplifiers are usually given for common emitter configuration, with both the emitter and base at DC ground potential. The emitter is connected directly to the ground and the base is DC grounded through an RF choke or an inductive element of the input matching network (as shown in Figure 2-46). Note that circuits operating with zero base-to-emitter bias voltage are sometimes called Class B circuits, although the conduction angle is smaller than 180 degrees. Be aware that applying bias to the base not only changes the conduction angle, the power gain, and the collector efficiency, but also changes the large signal impedances of the transistor. For transistors to be used in common base configuration, the data sheets present the large-signal parameters measured in a common base circuit.

e. Large-signal impedance data of HF and VHF transistors are sometimes published without collector load resistance information. Based on the following assumptions, this resistance can be easily calculated:

- V_{CEsat} is ignored.

- The collector voltage is a sinusoidal waveform (i.e., the output tuned circuit is a high Q parallel resonant circuit or an equivalent network).

- There is no DC voltage drop in the collector RF choke.

Denoting the RF power with P_0, the load resistance is

$$R_L = \frac{V_{dc}^2}{2P_0} \qquad 2.67$$

Some of the above assumptions seem not to be valid. Consider the second assumption, which requires the use of a high Q parallel-tuned circuit or an equivalent network. If V_{CEsat} is known, a more accurate expression for R_L would be

$$R_L = \frac{\left(V_{dc} - V_{CEsat}\right)^2}{2P_0} \qquad 2.68$$

Equations 2.67 and 2.68 can be used successfully to design HF and VHF PAs. At these low frequencies, lumped component matching networks are used and generally have sufficient tuning possibilities to compensate for errors associated with Equations 2.67 or 2.68. At higher frequencies (UHF), the load resistance (or the output resistance) is always provided in the data sheets because the equations above cannot be used to calculate R_L. On the other hand, the microstrip matching networks most often used in UHF PAs do not have the same tuning range as lumped element matching networks. Finally, note the tendency of today's RF power transistor manufacturers to include collector load resistance data among the other data sheet parameters, even at low frequencies (HF, VHF).

f. The harmonic current levels in the input and output matching networks significantly affect the large-signal impedances. Be aware that the device impedances in the data sheets are measured in a specified test fixture and any change in circuit configuration, component values, or circuit layout affects the large-signal parameters to some extent. Data sheets values must be used as a first approximation for a particular circuit design. Still, some experimental work is always required to optimize the circuit performance, especially in broadband amplifiers.

The remainder of this section provides a brief discussion of the measurement procedure[6] of large-signal parameters [4, 29, 31]. The transistor under test is placed in a circuit that includes matching networks designed to provide wide tuning capabilities. These wide tuning capabilities are required because the large-signal impedances of the active device are not known a priori. Of course, reasonable precise estimates based on the manufacturers' experience are possible. On the other hand, it is desirable to use the same circuit for measurements at different power output levels.

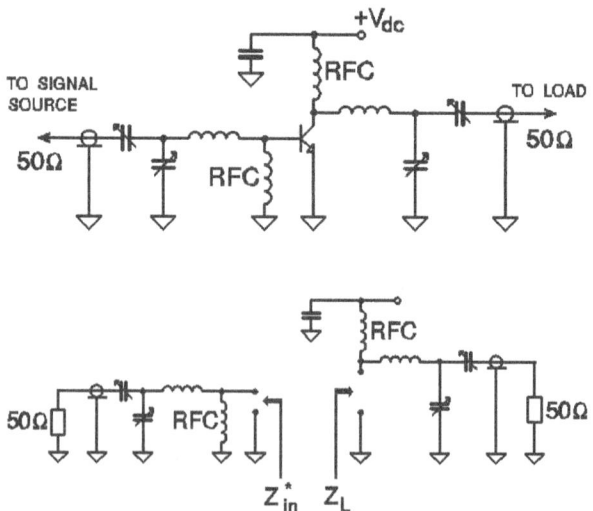

Figure 2-46 *Circuit used to measure large-signal parameters.*

The circuit in Figure 2-46 is typical for HF and VHF measurements. Similar circuits, using microstrip matching networks, are employed at higher frequencies. The input and output matching circuits are tuned for a careful impedance match. Impedance matching is achieved when the amplifier provides the desired output power with maximum gain,[7] and the reflected power at the input is equal to zero (a dual-directional coupler or directional power meter is used to indicate the reflected power). After the circuits have been carefully tuned, the transistor is removed. The DC power supply is disconnected and the signal source is replaced by a 50-ohm passive termination. Complex impedances are then measured in the base and collector circuit, as shown in Figure 2-46. The input impedance is the complex conjugate of the impedance measured toward the signal source at the base circuit connection of the test transistor. Impedance Z_L measured toward the load at the collector circuit connection of the test transistor is sometimes directly indicated in the data sheet (as shown in Figures 2-32 and 2-38). Its complex conjugate is shown in Figure 2-45.

2.6   Narrowband Matching Networks

Impedance matching consists of transforming a load impedance, Z_L, in the optimal working impedance of the signal source Z (Figure 2-47). Depending on the specific purpose of the circuit, the optimal working impedance (Z) may assure maximum power delivered to the load, maximum efficiency or power gain, minimum distortion of the signal across the load and more. In a specific case, this optimal impedance may be the complex conjugate of the source impedance,[8] assuring a maximum power transfer, as is usual in small-signal amplifiers. In RF power amplifiers, the optimal working impedance of the signal source is different in most cases from the complex conjugate of the source impedance. However, as an almost general rule, the reactive component of the source impedance must be compensated by a convenient reactance seen at the input of the matching network, so the signal source operates into a purely resistive load. As was discussed, mismatching in RF power amplifiers may cause reduced efficiency and/or output power, increased stresses of the active devices, distortion of the output signal and so on.

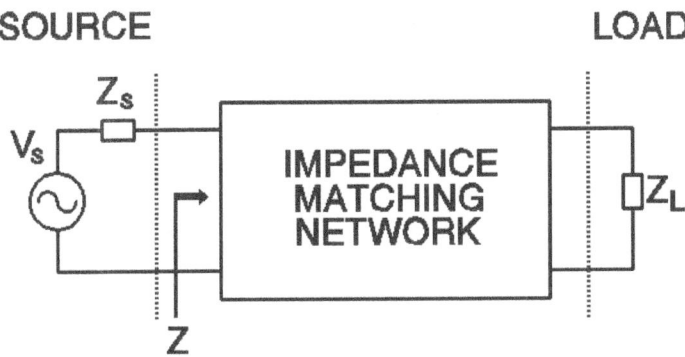

Figure 2-47 *The general problem of impedance matching.*

Considering the specific problems of RF PAs, the matching circuits must meet, as much as possible, the following requirements.
a.   Transform the load impedance in the optimal working impedance of the signal source.

b. Maintain the specified amplitude- and phase-frequency response over a certain frequency range in accordance with what is being transmitted, and in compliance with relevant standards.

c. Attenuate (filter out) the higher harmonics so any of the harmonic's power delivered to the load (the input of the subsequent stage or the antenna of the final stage) will not exceed the maximum safe value. Thus, only low- or band-pass circuits are usually employed in RF PAs.

d. Have insignificant power loss, i.e., maintain high efficiency. This can be done by using only lossless circuit elements in the matching networks.

e. Other requirements related to cost, size, weight, reliability and practicality.

If the RF PA operates at a fixed frequency or over a narrow frequency band in comparison with the carrier frequency, the above requirements must be met at only one frequency, and narrowband matching networks should be used. If the amplifier operates over a wide frequency band, the matching requirements (or at least some of them) must be met over the entire frequency range. This requires the use of wideband matching circuits.

At low frequencies (HF and VHF), the narrowband impedance matching is usually achieved with lumped element circuits. At higher frequencies, distributed element networks are most often required.

Lumped Element Narrowband Matching Networks

There are several different methods for the analysis and design of lumped element narrowband matching networks. The most popular technique is based on successive conversions of series combinations of resistance and reactance to equivalent parallel combinations of the same elements, and vice versa. Any series resistance-reactance group is equivalent, at a given frequency, to a parallel resistance-reactance group.

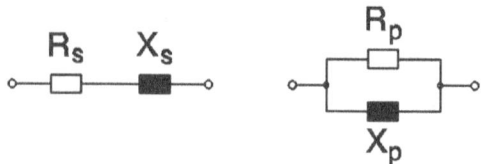

Figure 2-48 *Parallel and series resistance-reactance groups.*

Assuming that X_s and X_p in Figure 2-48 are similar elements (i.e., both are either capacitances or inductances), the impedances of the two circuits are given by

$$Z_p = \frac{R_p\left(\pm jX_p\right)}{R_p \pm jX_p} \quad Z_s = R_s \pm jX_s \qquad 2.69$$

The two circuits are equivalent if

$$Z_p = Z_s \qquad 2.70$$

This conditions yields

$$R_p = R_s\left[1+\left(\frac{X_s}{R_s}\right)^2\right] \quad X_p = X_s\left[1+\left(\frac{R_s}{X_s}\right)^2\right] \qquad 2.71$$

or, in an equivalent form,

$$R_s = \frac{R_p}{1+\left(\dfrac{R_p}{X_p}\right)^2} \quad X_s = \frac{X_p}{1+\left(\dfrac{X_p}{R_p}\right)^2} \qquad 2.72$$

Note that taking into account that the quality factor

$$Q = \frac{X_s}{R_s} = \frac{R_p}{X_p} \qquad 2.73$$

has the same value if the circuits are equivalent, the following simple relations can be obtained

$$R_p = R_s\left(1+Q^2\right) \quad X_p = X_s\left(1+\frac{1}{Q^2}\right) \qquad 2.74$$

The method of successive conversions may be used to analyze and find the design equations for any narrowband matching circuit and will be employed in the following sections.

Two-Reactance Matching Networks

The simplest lumped matching networks are the L circuits shown in Figures 2-49 and 2-50 [1-9, 12, 32].

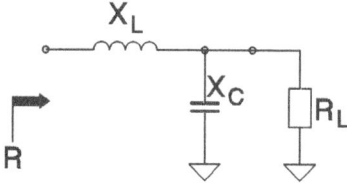

Figure 2-49 *L matching network* ($R < R_L$).

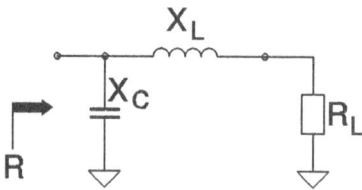

Figure 2-50 *L matching network* ($R > R_L$).

Assume the impedances to be matched are purely resistive ones (R and R_L in Figures 2-49 and 2-50). X_L and X_C are the reactances of the reactive elements at the operating frequency f. The design equations for the network in Figure 2-49 are obtained by transforming the parallel group (R_L, X_C) in its series equivalent group (R'_L, X'_C), as shown in Figure 2-51.

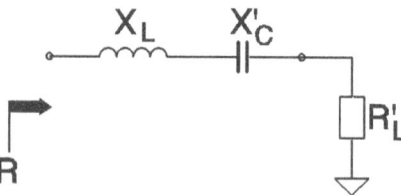

Figure 2-51 *Series-parallel conversion in the circuit of Figure 2-49.*

The conditions for impedance matching are

$$R = R'_L \qquad X_L = X'_C \qquad\qquad 2.75$$

Using Equations 2.72 and 2.75, the following design equations are determined:

$$X_L = R\sqrt{\frac{R_L}{R} - 1} \qquad X_C = \frac{R_L}{\sqrt{\frac{R_L}{R} - 1}} \qquad\qquad 2.76$$

The circuit in Figure 2-49 can be used only if $R < R_L$. If $R > R_L$, the input and the output port of the L matching network must be interchanged, as shown in Figure 2-50. For this circuit, the design equations are given by

$$X_L = R_L\sqrt{\frac{R}{R_L} - 1} \qquad X_C = \frac{R}{\sqrt{\frac{R}{R_L} - 1}} \qquad\qquad 2.77$$

EXAMPLE 2.7
The optimal load impedance of an RF PA operating at $f = 10$ MHz is $R = 10\ \Omega$. An L matching network is used to deliver the output power to an $R_L = 50\ \Omega$ resistor. Because $R < R_L$, the circuit in Figure 2-49 should be used. According to Equation 2.76, $X_L = 20\ \Omega$ and $X_C = 25\ \Omega$; thus, $L = X_L/(2\pi f) \approx 318$ nH and $C = 1/(2\pi f X_C) \approx 637$ pF.

The circuits in Figures 2-49 and 2-50 assume purely resistive matched impedances. However, the impedances to be matched in practical circuits often have reactive components. A general procedure to deal with the reactive components of the matched impedances is described below and can be applied to any lumped element matching network.

a. A suitable matching circuit configuration is chosen. The impedances to be matched are converted to an appropriate form — parallel or series resistance-reactance groups — according to the network topology.

b. The reactive components are ignored and the matching circuit is designed considering just the resistive terminations.

c. Finally, the reactive components are considered. They are included in (subtracted from) the corresponding elements of the matching network.

The examples below illustrate this procedure. In all cases, the operating frequency of the circuits is $f = 10$ MHz.

EXAMPLE 2.8
$Z_L = 50\ \Omega\,||\,20$ pF and $Z = (10 + j2)\Omega$. Choose the L matching network depicted in Figure 2-49. This topology includes a parallel capacitor with the load impedance and a series inductor at the input. It is therefore convenient to represent Z_L as a parallel resistance-reactance group and Z as a series resistance-reactance group. Ignoring the reactive components, design the circuit for $R_L = 50\ \Omega$ and $R = 10\ \Omega$. The result (see Example 2.7) is $X_L = 20\ \Omega$, $X_C = 25\ \Omega$, $L \approx 318$ nH, and $C \approx 637$ pF. Finally, consider the reactive components of the matched impedances. The load includes a parallel 20 pF capacitance.

This value must be subtracted from C and, as a result, a 617 pF capacitor is required in the circuit. A series-inductive reactance of 2 ohms, which is an inductance of about 32 nH at 10 MHz, must be added at the input; thus, a 350 nH inductor is required in the circuit. The design procedure is illustrated in Figure 2-52.

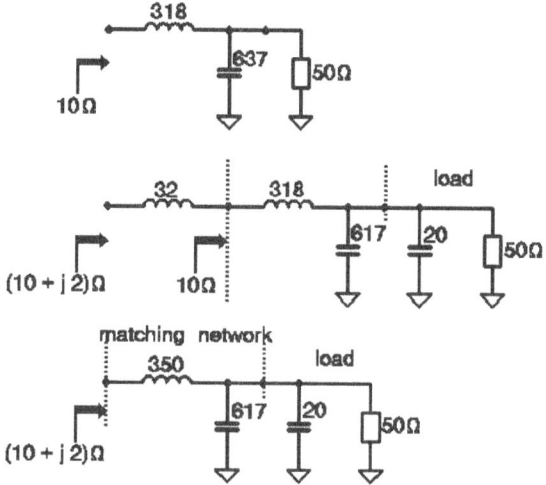

Figure 2-52 *The matching circuit in Example 2.8 ($f = 10$ MHz, inductances in nH, and capacitances in pF).*

EXAMPLE 2.9

$Z_L = 50\ \Omega\,||\,20$ pF and $Z = (10 - j2)\ \Omega$. As in the previous example, a 617 pF capacitor is used in the matching circuit. A series inductive reactance of 2 ohms (32 nH at 10 MHz) must be subtracted at the input and a 286 nH inductor used in the L matching network.

EXAMPLE 2.10

$Z_L = (50 + j25)\Omega$ and $Z = (30\,||\,j30)\Omega$. The matched impedances are in the appropriate form (Z_L as a series resistance-reactance group, Z as a parallel resistance-reactance group) to use the circuit in Figure 2-50. However, this circuit cannot be used because $R = 30\ \Omega < R_L = 50\ \Omega$.

Next, try the L matching network depicted in Figure 2-49. It includes a parallel capacitor with the load, which requires that the load impedance be converted to a parallel resistance-reactance group. Equation 2.71 gives $Z_L = (62.5\,||\,j125)\Omega$ or $Z_L = 62.5\ \Omega\,||\,1.99$ mH (at 10 MHz). The matching network includes a series inductor, so Z must be converted to a series form. Equation 2.72 gives $Z = (15 + j15)\ \Omega$. Next, Equation 2.76 can be applied for $R_L = 62.5\ \Omega$ and $R = 15\ \Omega$. The result is $X_L = 26.7\ \Omega$, $X_C = 35.1\ \Omega$, $L \approx 425$ nH, and $C \approx 453$ pF.

Finally, consider the reactive components of the matched impedances. The load includes a parallel 1.99 mH inductance that must be compensated by a 127 pF capacitance (127 pF resonates with 1.99 mH at 10 MHz). A series inductance of 239 nH (that is, an inductive reactance of 15 ohms at 10 MHz) must be added at the input. The design procedure and the final circuit are illustrated in Figure 2-53.

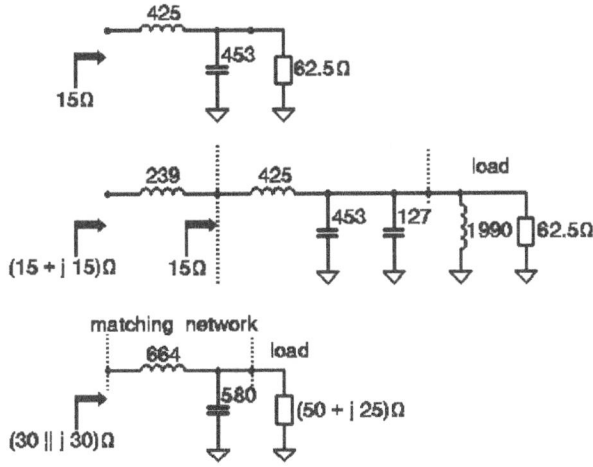

Figure 2-53 *The matching circuit in Example 2.10 (f = 10 MHz, inductances in nH, and capacitances in pF).*

A graphical approach to matching network design is possible and provides results identical to those of the mathematical approach. The graphical approach is based on the use of the Smith chart; it combines series and parallel reactances. It is possible to obtain a plot on the chart that shows the role played by each component in the impedance transformation.

This procedure can be illustrated using the circuit presented in Example 2.10. The normalized circuit and corresponding impedance-admittance chart are shown in Figures 2-54 and 2-55.

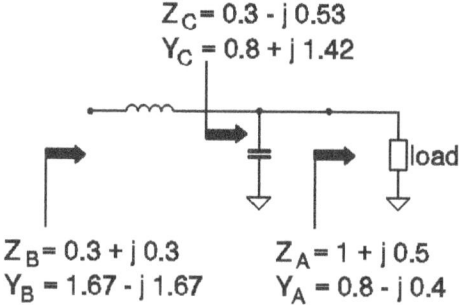

$Z_C = 0.3 - j0.53$
$Y_C = 0.8 + j1.42$

load

$Z_B = 0.3 + j0.3$ $Z_A = 1 + j0.5$
$Y_B = 1.67 - j1.67$ $Y_A = 0.8 - j0.4$

Figure 2-54 *Normalized impedances of the circuit in Example 2.10.*

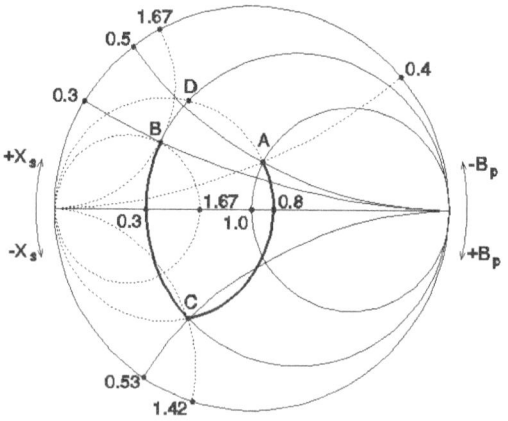

Figure 2-55 *Impedance-admittance chart for the matching network in Example 2-10 (constant conductance and constant susceptance circles are indicated with dotted lines).*

The load impedance $Z_L = (50 + j25)\Omega$ is plotted on the Smith chart at point A — normalized impedance $Z_A = 1 + j0.5$ ($Z_0 = 50\ \Omega$). The corresponding normalized admittance is $Y_A = 0.8 - j0.4$. The input impedance $Z = (15 + j15)\Omega$ is plotted at point B, normalized impedance $Z_B = 0.3 + j0.3$. (Because the input impedance is given in a parallel form, $Z = (30|\ |j30)\Omega$, it is easier to convert it to an admittance and represent it on the Smith chart using the normalized admittance coordinates. The normalized input admittance is $Y_B = 1.67 - j1.67$.) A capacitive susceptance is added by the matching circuit in parallel to the load. The resulting admittance is placed on the constant conductance circle $g = 0.8$, on the arc starting clockwise from A. This arc intersects the constant resistance circle $r = 0.3$ at point C, $Z_C = 0.3 - j0.53$ and $Y_C = 0.8 + j1.42$. An inductive reactance is added in series to the parallel group. The resulting impedance is placed on the constant resistance circle $r = 0.3$, on the arc starting counterclockwise from C. The desired input impedance is obtained in point B.

The series inductance of the matching circuit is represented by the total positive reactance change from C to B (from –0.53 to 0.3); that is, 0.83. Denormalizing this value, $X_L = 0.83 \times 50 = 41.5\ \Omega$, $L = X_L/(2\pi f) \approx 660$ nH. The parallel capacitance of the matching circuit is given by the total positive susceptance change from A to C (from –0.4 to 1.42), or 0.82. Denormalizing this value, $Y_C = 1.82 \times 0.02 = 0.0364$ S, $C = Y_C/(2\pi f) \approx 579$ pF.

The diagram in Figure 2-55 shows that

1. Another valid circuit solution is provided by point D (the other intersection point between the constant resistance circle $r = 0.3$ and the constant conductance circle $g = 0.8$). This circuit uses a series capacitor and a parallel inductor which, being a high-pass filter, is less attractive for use in RF PAs.

2. The circuit from Figure 2-50 cannot be used for the matched impedance values provided in this example. The constant resistance circle $r = 1$ and the constant admittance circle $g = 1.67$ do not intersect (see Figure 2-55).

Three-Reactance Matching Networks

The L matching networks in the previous section have several drawbacks:
a. The design problem has no solution for some combinations of matched impedances.
b. The values obtained may be impractical; the values of the capacitors and inductors may be too large or too small.
c. There is no design flexibility. Designers may wish to optimize their designs for other parameters of practical interest, such as harmonic attenuation, power losses, or bandwidth.

The three-reactance matching networks are most widely used because they are simple and provide flexibility [1-9, 12, 32-36]. Although each network has limitations, one of the circuits usually meets the design requirements with practical component values.

Several different approaches can be used to design a three-reactance matching circuit. The most popular technique, shown here for the pi network (see Figure 2-56), is based on choosing the circuit quality factor. The quality factor is related to the circuit bandwidth, harmonic attenuation, and circuit power losses. For other networks, only the design equations and the basic limitations and characteristics of the circuit will be presented here.

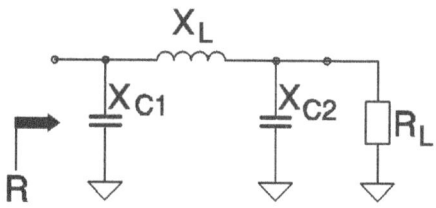

Figure 2-56 *Pi matching network.*

Successive conversions of resistance-reactance series combinations to equivalent parallel combinations of the same elements, and vice versa, are used to transform the pi network (see Figure 2-56) in an equivalent parallel RLC circuit (see Figure 2-57). The parallel (R_L, X_{C2}) group is converted to its series equivalent (R'_L, X'_{C2}). According to Equation 2.72,

$$R'_L = \frac{R_L}{1+\left(\dfrac{R_L}{X_{C2}}\right)^2} \qquad X'_{C2} = \frac{X_{C2}}{1+\left(\dfrac{X_{C2}}{R_L}\right)^2} \qquad 2.78$$

X_L and X'_{C2} are then combined in X'_L

$$X'_L = X_L - X'_{C2} \qquad\qquad 2.79$$

and the series (R'_L, X'_L) group is converted to its parallel equivalent (R'_L, X''_L). According to Equation 2.71, the values of R''_L and X''_L are given by

$$R''_L = R'_L\left[1+\left(\frac{X'_L}{R'_L}\right)^2\right] \qquad X''_L = X'_L\left[1+\left(\frac{R'_L}{X'_L}\right)^2\right] \qquad 2.80$$

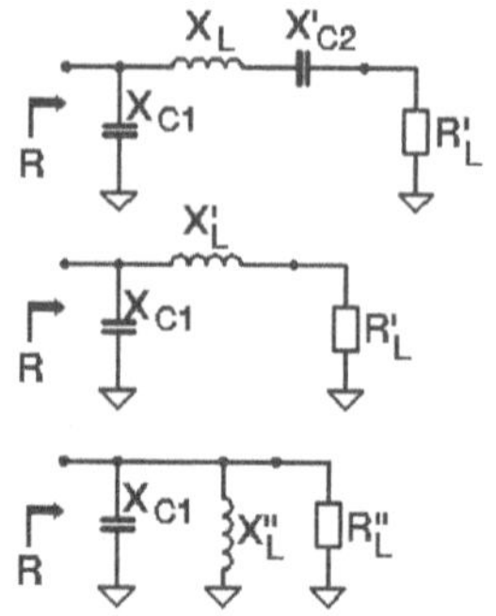

Figure 2-57 *Series-parallel transformations in the pi network of Figure 2-56.*

These transformations yield the equivalent parallel RLC circuit depicted in Figure 2-57. The matching conditions are

$$R = R_L'' \qquad X_{C1} = X_L'' \qquad\qquad 2.81$$

Because there are only two design equations and three unknown elements (X_{C1}, X_{C2}, and X_L), an additional design condition is required. The quality factor of the matching network is defined as

$$Q = \frac{R}{X_{C1}} \qquad\qquad 2.82$$

Equations 2.81 and 2.82 form an equation system with three unknowns. Substituting Equations 2.78 through 2.80 in Equation 2.81 results in

$$X_{C1} = \frac{R}{Q} \qquad X_{C2} = \frac{R_L}{\sqrt{\dfrac{R_L}{R}\left(1+Q^2\right)-1}}$$

$$X_L = \frac{QR}{1+Q^2}\left[1 + \frac{1}{Q}\sqrt{\frac{R_L}{R}\left(1+Q^2\right)-1}\right] \qquad\qquad 2.83$$

This circuit can be used only if

$$R_L\left(1+Q^2\right) > R \qquad\qquad 2.84$$

Some observations and practical considerations
1. The pi matching network is widely used in vacuum-tube transmitters to match large resistance values. For small resistance values, the inductance of L becomes impractically small, while the capacitance of both C_1 and C_2 becomes very large. This circuit is generally not useful in solid-state RF PAs where the matched resistances are often small.

EXAMPLE 2.11
Pi matching network, $R_L = 10\ \Omega$, $R = 5\ \Omega$, $f = 100$ MHz. We choose $Q = 3$ and from Equation 2.83 $X_{C1} = 1.67\ \Omega$, $X_{C2} = 2.29\ \Omega$, $X_L = 3.68\ \Omega$. The required component values are $C_1 \approx 953$ pF, $C_2 \approx 695$ pF (both impractically large), and $L \approx 5.9$ nH (impractically small).

---

2. Quality factor Q is related to the circuit bandwidth, harmonic attenuation, and efficiency. A general qualitative result taken from the theory

of resonant circuits remains valid in this case. Increasing Q will increase the filtering effect of the matching network (i.e., increase the harmonic attenuation), but will reduce the bandwidth and efficiency.

The Q factor of a three-reactance matching network is defined for an equivalent RLC circuit (for a pi network, the parallel RLC circuit in Figure 2-57). Some of the elements in this circuit have frequency-dependent values and, thus, the well-known equations relating Q to bandwidth and efficiency are not valid for three-reactance matching networks. The quality factor does not offer a priori accurate quantitative information about circuit performance. Choosing the quality factor to meet some specific requirements related to harmonic attenuation, bandwidth, or efficiency is a matter of experience, and the recommended values of Q usually range from 1 to 10. After a first-cut design, it is important to estimate the circuit performance and proceed to a new calculation (if required), adjusting the value of Q as needed.

3. It is relatively easy to find mathematical relations to calculate circuit performance (for example, harmonic suppression, bandwidth, efficiency). The use of appropriate simulation tools is probably the simplest solution. Note that the characteristics of the signal source (for example, its output impedance) are decisive for circuit performance. References [33] through [36] provide alternative methods for designing three-reactance matching networks and evaluating their performance.

4. Another pi matching network design method (that may be applied to many three-reactance or more complex matching networks) appears in Figure 2-58. The pi network is decomposed in a series combination of two L-section networks that first transform R_L into some resistance R_0, and then R_0 into R. The two L-section networks are designed using the procedure provided in section ***Two-Reactance Matching Networks***. Next, X_{L1} and X_{L2} are combined in X_L. This design technique is used less in practice, because R_0 (a parameter chosen by the designer) has no clear physical meaning. Note that the filtering effect of the pi network will improve in proportion to the difference between R_0, on the one hand, and R and R_L, on the other. At the same time, the efficiency and bandwidth of the network will decrease. These effects are similar to those generated by increasing the network Q. The choice R_0 as a factor of 1.5 . . .10 smaller than R and R_L is recommended.

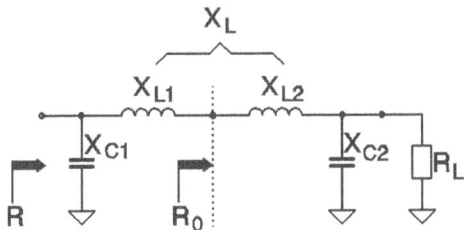

Figure 2-58 *Pi matching network as a series combination of two L-section networks.*

5. Pi matching networks have excellent filtering properties. In many applications, the desired degree of harmonic suppression can be achieved using a double-pi network (a series combination of two pi networks). A similar solution, in which the Q of the first section is very low, may be used to yield practical component values when low values of resistance are matched.

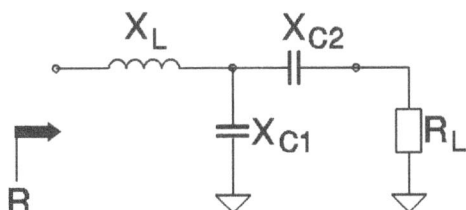

Figure 2-59 *T matching network.*

The T matching network in Figure 2-59 is applicable to most solid-state RF PAs. Its design equations are

$$X_L = QR$$

$$X_{C1} = \frac{R\left(1+Q^2\right)}{Q - \sqrt{\dfrac{R}{R_L}\left(1+Q^2\right) - 1}} \qquad X_{C2} = R_L\sqrt{\frac{R}{R_L}\left(1+Q^2\right) - 1} \tag{2.85}$$

and the circuit can be used only if

$$R < R_L \quad R\left(1+Q^2\right) > R_L \tag{2.86}$$

EXAMPLE 2.12

T matching network, $Z = 10\ \Omega\,||\,100$ pF, $Z_L = 20\ \Omega\,||\,50$ nH, $f = 100$ MHz. The two impedances must first be converted to their series form by using Equation 2.72 $Z = (7.17 - j4.5)\ \Omega$ and $Z_L = (14.23 + j9.06)\ \Omega$. With $R = 7.17\ \Omega$, $R_L = 14.23\ \Omega$, and by choosing $Q = 3$, $X_L = 21.51\ \Omega$, $X_{C1} = 72.4\ \Omega$, and $X_{C2} = 28.6\ \Omega$ from Equation 2.85. Now the reactive components of Z and Z_L, X_L must be reduced by 4.5 ohms, while a capacitive reactance of 9.06 Ω must be added to XC2. The final values of the circuit elements are $L \approx 27.1$ nH, $C_1 \approx 22$ pF, and $C_2 \approx 42.3$ pF.

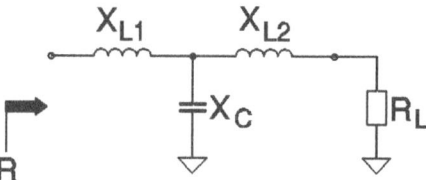

Figure 2-60 *Two-inductance T matching network.*

Another T matching network is depicted in Figure 2-60 and is also applicable to many solid-state RF PAs. The design equations are

$$X_{L1} = QR$$

$$X_{L2} = R_L\sqrt{\frac{R}{R_L}\left(1+Q^2\right)-1} \qquad X_C = \frac{R\left(1+Q^2\right)}{Q+\sqrt{\dfrac{R}{R_L}\left(1+Q^2\right)-1}} \qquad \text{2.87}$$

and the circuit can be used only if

$$R\left(1+Q^2\right) > R_L \qquad\qquad \text{2.88}$$

EXAMPLE 2.13

Two-inductance T matching network, $Z = 10\ \Omega\,||\,100$pF, $Z_L = 20\ \Omega\,||\,50$ nH, $f = 100$ MHz. Example 2.12 has shown that $Z = (7.17 - j4.5)\Omega$ and $Z_L = (14.23 + j9.06)$W. With $R = 7.17\ \Omega$, $R_L = 14.23\ \Omega$, and by choosing $Q = 3$, $X_{L1} = 21.51\ \Omega$, $X_{L2} = 28.6\ \Omega$, and $X_C = 14.31\ \Omega$ from Equation 2.87. As a result, the reactive components of Z and Z_L, X_{L1} must be reduced by 4.5 ohms and X_{L2} by 9.06 ohms. The final values of the circuit elements are $L_1 \approx 27.1$ nH, $L_2 \approx 31.1$ nH, and $C \approx 111$ pF.

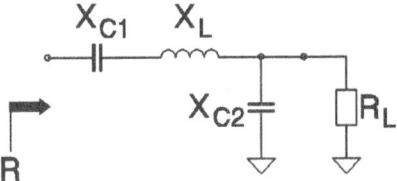

Figure 2-61 *Three-reactance L matching network.*

A three-reactance L matching network is presented in Figure 2-61. This network is also very useful in solid-state RF PAs because it yields practical components for low values of matched resistances. The design equations are

$$X_{C1} = QR \qquad X_{C2} = R_L \sqrt{\frac{R}{R_L - R}} \qquad X_L = QR + \sqrt{R(R_L - R)} \qquad \text{2.89}$$

and the circuit can be used only if

$$R < R_L \qquad\qquad \text{2.90}$$

EXAMPLE 2.14
Three-reactance L matching network, $Z = 10\ \Omega\,||\,100$ pF, $Z_L = 20\ \Omega\,||\,50$ nH, $f = 100$ MHz. Example 2.12 has shown that $Z = (7.17 - j4.5)\Omega$. With $R = 7.17\ \Omega$, $R_L = 20\ \Omega$, and by choosing $Q = 3$, $X_{C1} = 21.51\ \Omega$, $X_{C2} = 14.95\ \Omega$, and $X_L = 31.1\ \Omega$ from Equation 2.89. Now the reactive components of Z and Z_L, X_L must be reduced by 4.5 ohms and a 50.66 pF capacitance must be added in parallel with C_2 to compensate the reactive part of Z_L (50.66 pF resonates with 50 nH at 100 MHz). The final values of the circuit elements are $C_1 \approx 74$ pF, $C_2 \approx 157$ nH, and $L \approx 42.3$ nH.

Four-Reactance Matching Networks
    Four-reactance matching networks are rarely used for narrowband impedance matching. A four-reactance circuit allows exact impedance matching at two distinct frequencies and is appropriate for broadband impedance matching.
    Often in narrowband matching circuits, the fourth reactance is used to compensate the reactance of one of the matched impedances. This may provide more practical components in the final design. An example is given in Figure 2-62, where X_{L2} is used to compensate the output capacitance of the transistor, C_{out}. The design equation set is 2.89 and X_{L2} is given by

$$X_{L2} = X_{Cout} \qquad\qquad 2.91$$

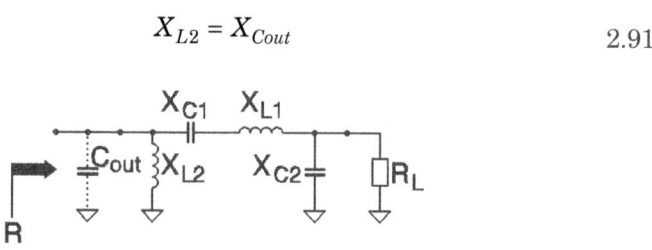

Figure 2-62 *Four-reactance narrowband matching network.*

Four-reactance narrowband matching networks are also obtained by combining two L-section networks or a three-reactance network with an L network. These combinations may be useful in practice if the matched resistances are very different, or a high harmonic suppression is desired.

Distributed Element Narrowband Matching Networks

Resonant circuits built around lumped LC elements may be used at frequencies below 1 GHz. However, even at lower frequencies, the designer may face serious difficulties when using lumped LC elements. On the other hand, when the operating frequency is sufficiently high, the use of transmission-line sections as network elements becomes practical due to the short lengths required. Although any type of transmission line can be used, the most popular is probably the microstrip. Today's RF power transistors generally use the *stripline opposed emitter* (SOE) package, which provides very low inductance strip line leads that interface very well with the microstrip lines.

The general theory and the practical design of circuits using microstrip lines is a vast subject and well covered by the literature. Some basic principles and several examples that illustrate the possibilities for using transmission lines in narrowband impedance matching applications follow.

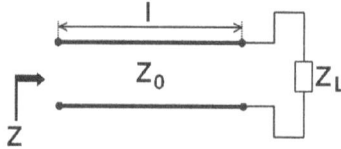

Figure 2-63 *Input impedance of a transmission line.*

The input impedance of a lossless transmission line operating into a load impedance of Z_L (see Figure 2-63) is given by references [4], [5], and [9].

$$Z = Z_0 \frac{Z_L + jZ_0 \tan \beta l}{Z_0 + jZ_L \tan \beta l} \tag{2.92}$$

where Z_0 is the characteristic impedance of the line, l is the physical length of the line, and

$$\beta = \frac{2\pi}{\lambda} = \frac{2\pi}{\lambda_0} \sqrt{\varepsilon_{eff}} \tag{2.93}$$

In Equation 2.93, λ_0 is the free-space wavelength, and λ is the wavelength in the dielectric medium of the transmission line with the dielectric constant ε_{eff}.

Rearranging Equation 2.92 in admittance terms results in an equivalent form.

$$Y = Y_0 \frac{Y_L + jY_0 \tan \beta l}{Y_0 + jY_L \tan \beta l} \tag{2.94}$$

Several interesting properties and possible applications of transmission lines in practical matching circuits may be obtained from Equations 2.92 and 2.94.

1. If Z and Z_L have a certain relationship, a single transmission line with appropriate characteristic impedance and electrical length can be used for narrowband impedance matching. Considering that

$$Z = R + jX \qquad Z_L = R_L + jX_L \tag{2.95}$$

and equating the real and imaginary parts of both sides of Equation 2.92

$$Z_0 = \sqrt{RR_L + \frac{R_L X^2 - R X_L^2}{R - R_L}} \qquad \tan \beta l = \frac{Z_0 (R - R_L)}{R X_L + R_L X} \tag{2.96}$$

Since Z_0 must be a real number, a theoretical solution of the matching problem can be found only if

$$RR_L + \frac{R_L X^2 - R X_L^2}{R - R_L} > 0 \tag{2.97}$$

Other practical constraints occur due to physical limitations in transmission-line fabrication: both the minimum and maximum achievable values of Z_0 are limited.[9]

EXAMPLE 2.15
Design a single transmission-line matching network to transform a load $Z_L = (50 + j10)\Omega$ into $Z = (20 - j10)\Omega$. With $R_L = 50\ \Omega$, $X_L = 10\ \Omega$, $R = 20\ \Omega$, and $X = -10\ \Omega$, Equation 2.96 yields the required characteristic impedance $Z_0 = 30\ \Omega$ and electrical length $\beta l \approx 71.6°$.

---

2. If Z and Z_L are purely resistive impedances, Equation 2.96 becomes

$$Z_0 = \sqrt{RR_L} \qquad l = (2n + 1)\frac{\lambda}{4} \quad n = 0, 1, \ldots \qquad 2.98$$

This is the well-known quarter-wave transformer. Its properties are illustrated in Figure 2-64.

For example, a short-circuited $\lambda/4$ stub is a useful element to provide a bias voltage to an active device similar to an RF choke (see Figure 2-65).

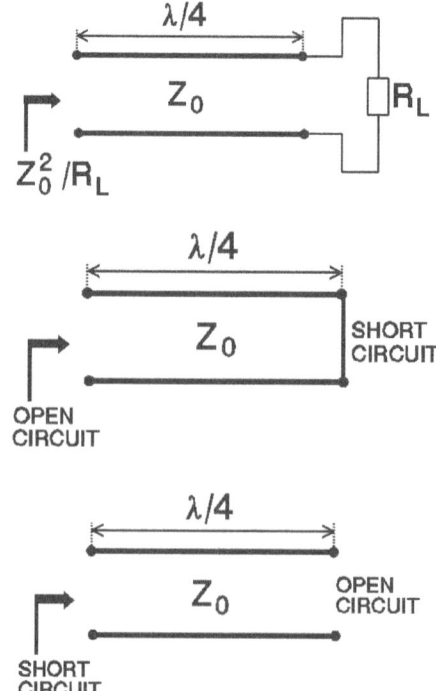

Figure 2-64 *Basic properties of a quarter-wave transformer.*

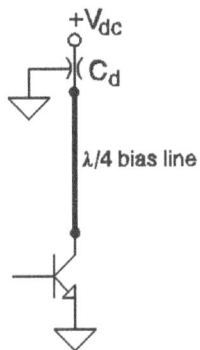

Figure 2-65 *Biasing an active device using a λ/4 short-circuited line.*

3. If $Z_L = 0$, Equation 2.92 becomes

$$Z = jZ_0 \tan \beta l \qquad 2.99$$

If $l < \lambda/4$ or $\lambda/2 < l < 3\lambda/4$, the input impedance appears to be inductive. If $\lambda/4 < l < \lambda/2$ or $3\lambda/4 < l < \lambda$, the input impedance appears to be capacitive. In practice, it is convenient to use transmission lines as short as possible and replace a shunt-mounted inductor by a short-circuited transmission line with $l < \lambda/4$.

EXAMPLE 2.16
At 870 MHz, a 5 nH inductance can be replaced by a section of short-circuited transmission line, with characteristic impedance $Z_0 = 50 \, \Omega$ and electrical length $\beta l \approx 28.7°$.

---

4. If $Y_L = 0$, Equation 2.94 becomes

$$Y = jY_0 \tan \beta l \qquad 2.100$$

A shunt-mounted capacitor can be replaced by an open-circuited transmission line with $l < \lambda/4$.

5. If both the real and imaginary part of $jZ_L \tan \beta l$ are negligible compared with Z_0, Equation 2.92 yields

$$Z \approx Z_L + jZ_0 \tan \beta l \qquad 2.101$$

As a result, a short length of the high Z_0 transmission line terminated by a low load impedance behaves as a series inductor with the load (see Figure 2-66). Another explanation is provided by the Smith chart. Note that on the extreme left side of the Smith chart, the constant r circles are almost congruent with the constant reflection coefficient circles.

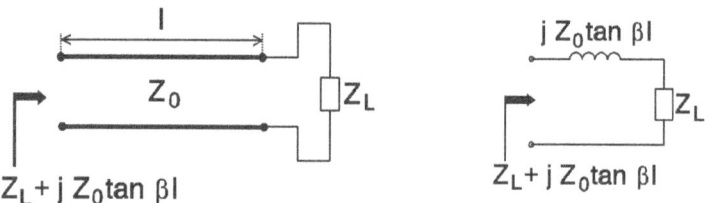

Figure 2-66 *A short transmission line terminated by a low impedance (left) and its lumped element equivalent (right).*

EXAMPLE 2.17
A 100-ohm transmission line with an electrical length of 10 degrees is terminated into a 20-ohm resistor. According to Equation 2.101, the input impedance is $Z \approx (20 + j17.6)$ Ω, and the transmission line is equivalent to a 3.2 nH series inductor at 870 MHz. Note that the exact value of the input impedance given by Equation 2.92 is $Z = (20.6 + j16.9)$Ω.

---

6. If both the real and imaginary part of $jY_L \tan\beta l$ are negligible in comparison with Y_0, Equation 2.94 yields

$$Y \approx Y_L + jY_0 \tan \beta l \qquad 2.102$$

A short length of the low Z_0 transmission line terminated by a high load impedance, behaves as a parallel capacitor with the load, as shown in Figure 2-67.

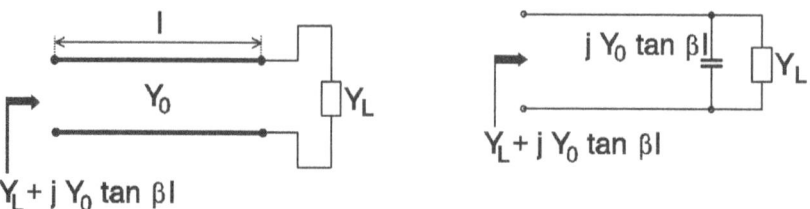

Figure 2-67 *A short transmission line terminated by a high impedance (left) and its lumped element equivalent (right).*

EXAMPLE 2.18

A 20-ohm transmission line with an electrical length of 10 degrees is terminated into a 100-ohm resistor. According to Equation 2.102, the input admittance is $Y \approx (10 + j8.82)$ mS and the transmission line is equivalent to a 1.6 pF parallel capacitor at 870 MHz. The exact value of the input admittance given by Equation 2.94 is $Y = (10.3 + j8.44)$ mS.

A direct application, extremely useful for practical implementation of distributed networks, is presented in Figure 2-68. A short length of high Z_0 microstrip line, terminated at both ends by low impedances, behaves as a series inductor in the circuit. (Remember that a narrow microstrip has a high Z_0; a wide microstrip has a low Z_0.) A short length of low Z_0 microstrip, terminated at both ends by high impedances behaves as a shunt capacitor in the circuit.

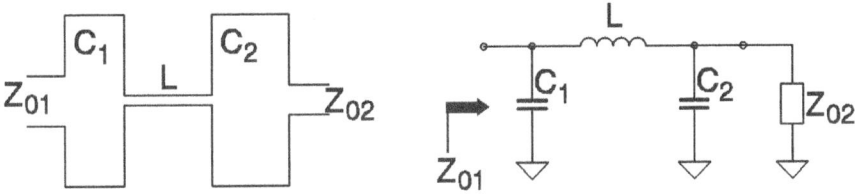

Figure 2-68 *Microstrip structure (left) and its lumped element circuit equivalent (right).*

7. If Z and Z_L are related, narrowband impedance matching can be achieved using a single transmission line whose characteristics are provided by Equation 2.96. In many practical applications, a more convenient solution is to use a transmission line aided by a short- or open-circuited stub.

A mathematical design procedure is presented below.

a. $Y = G + jB$ and $Y_L = G_L + jB_L$ are the initial data of the design. A convenient value is chosen for the characteristic admittance Y_0 of the transmission lines to be used in the design.

b. A transmission-line section is employed to obtain the desired value of the input conductance, G. By equating the real parts of both sides of Equation 2.94, the necessary electrical length of the transmission line is obtained from

$$\left[G\left(G_L^2 + B_L^2\right) - G_L Y_0^2\right]\tan^2 \beta l - 2GB_L Y_0 \tan \beta l + (G - G_L)Y_0^2 = 0 \qquad 2.103$$

Note that this equation may have two, one, or no solution, depending on the values of G, G_L, B_L, and Y_0.

c. The input susceptance of the transmission line is then calculated from Equation 2.94.

$$B_{in} = G\frac{\left(Y_0^2 - G_L^2 - B_L^2\right)\tan \beta l + Y_0 B_L\left(1 - \tan^2 \beta l\right)}{G_L Y_0\left(1 + \tan^2 \beta l\right)} \qquad 2.104$$

d. The required value of susceptance B is obtained using a short- or open-circuited stub. The susceptance seen at the input of this stub must be $B - B_{in}$.

EXAMPLE 2.19

$Y_L = (50 + j20)$mS and $Y = (10 - j6)$mS. Choosing $Y_0 = 20$ mS, Equation 2.103 yields two solutions, $\tan \beta l \approx 0.961$ and $\tan \beta l \approx 1.850$.

a. For $\tan \beta l \approx 0.96$ 1, the electrical length of the transmission line is 136.1° (≈ 0.378 l). The input conductance is found from Equation 2.104, $B_{in} \approx 12.65$ mS. Consequently, a short-circuited stub with an input susceptance of –18.65 mS is required. According to Equation 2.99, its electrical length must be 47° (≈ 0.131 l). The resulting circuit is depicted in Figure 2-69(a).

b. For $\tan \beta l \approx 1.850$, the electrical length of the transmission line is 61.6° (≈ 0.171 l). The input conductance is found from Equation 2.104, $B_{in} \approx -12.65$ mS. Consequently, an open-circuited stub with an input susceptance of 6.65 mS is required. According to Equation 2.100, its electrical length must be 18.4° (≈ 0.051 l). The resulting circuit is depicted in Figure 2-69 (b). This second solution is much more convenient because it requires shorter transmission lines.

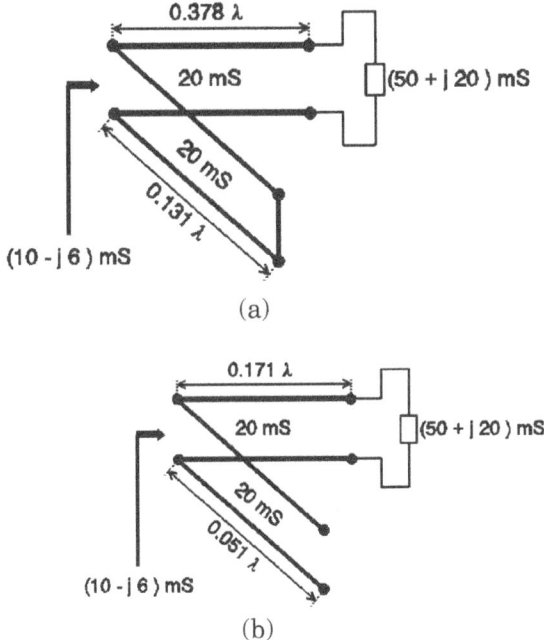

(a)

(b)

Figure 2-69 *The circuits in Example 2.19.*

A Smith chart may provide a more intuitive approach to this design. The procedure is illustrated in Figure 2-70.

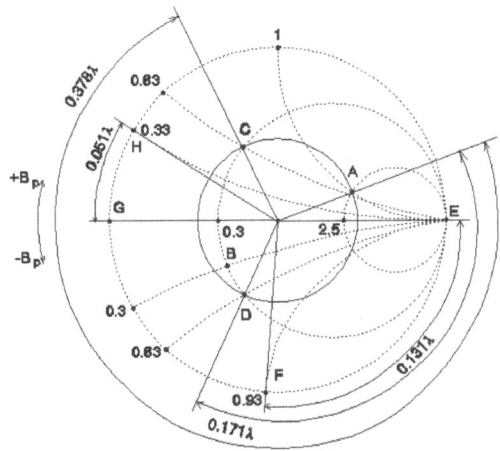

Figure 2-70 *Admittance chart for the circuits in Example 2.19.*

The load admittance is plotted on the Smith chart at point A — normalized admittance $Y_A = 2.5 + j1$ ($Y_0 = 20$ mS). The desired normalized admittance at the input of the matching circuit $Y_B = 0.5 - j0.3$ is plotted at point B. The input admittance of the transmission line loaded with Y_A is placed on a circle including point A and having its center in the middle of the chart. This circle intersects the constant conductance circle $g = 0.5$ at two points: C with $Y_C = 0.5 + j0.63$ and D with $Y_D = 0.5 - j0.63$. These points also indicate the required length of the transmission-line section, as depicted in Figure 2-70. Stub lengths can be determined from the Smith chart, as suggested in Figure 2-70. Point E correspond to $Y_E = \infty$ (short-circuit), F to $Y_F = -j0.93$, G to $Y_G = 0$ (open-circuit), and H to $Y_H = j0.33$.

2.7 Broadband Matching Circuits

Most designers are familiar with the basic principles of the conventional (wire-wound) transformer. Although it seems the best option for broadband impedance matching, the conventional transformer is almost useless in RF power amplifiers. This section briefly discusses the basic limitations of the conventional transformer, then looks at the most popular circuit for broadband impedance matching in RF PAs — the transmission-line transformer.

Basic Limitations of the Conventional Transformer

The conventional transformer consists of two coils (see Figure 2-71) placed so a magnetic field in one coil links the other coil, mostly via a magnetic core [4, 5, 37, 38].

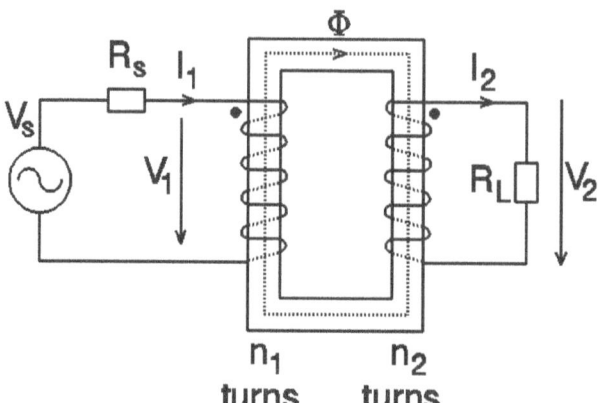

Figure 2-71 *Physical schematic of conventional transformer.*

The voltages induced in the primary and secondary coils are proportional to the number of turns in that coil; the currents through the coils are inversely proportional to the number of turns. For an ideal transformer (that is, a transformer with perfect magnetic coupling and no loss of energy), the resistance seen across the primary winding (see Figure 2-72) is given by

$$R = \left(\frac{n_1}{n_2}\right)^2 R_L = n^2 R_L \qquad 2.105$$

where

$$n = \frac{n_1}{n_2} \qquad 2.106$$

is the turns ratio.

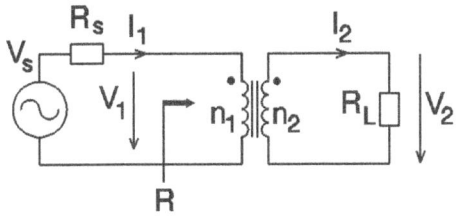

Figure 2-72 *Conventional transformer.*

Because the number of turns in both primary and secondary coils may be arbitrary, the conventional transformer may be used to assure a broadband matching of any impedance value. However, the use of a conventional transformer for broadband impedance matching raises difficulties due to its frequency limitations.

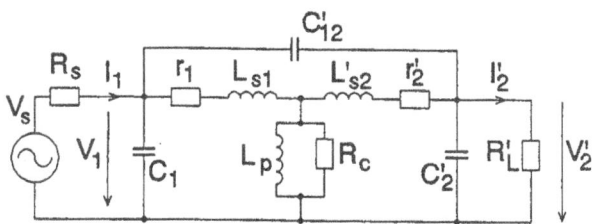

Figure 2-73 *Complete T equivalent circuit (referred to the primary winding) of a conventional transformer.*

The following presentation of these frequency limitations is based on the lumped element equivalent circuit of the wideband transformer shown in Figure 2-73 [8]. In the T equivalent circuit of Figure 2-73, all elements are referred to the primary winding as follows:

- $R'_L = n^2 R_L$ is the load resistance referred to the primary winding.

- r_1 is the series resistance of the primary winding.

- $r'_2 = n^2 r_2$ is the series resistance of the secondary winding referred to the primary winding (r_2 is the series resistance of the secondary winding).

- R_c models the power loss in the magnetic core.

- $L_p \approx k L_1$, where L_1 is the inductance of the primary winding and $0 \le k \le 1$ is the magnetic coupling coefficient between the primary and secondary winding.

- L_{s1} is the leakage inductance of the primary winding.

- $L'_{s2} = n^2 L_{s2}$ is the leakage inductance of the secondary winding, referred to the primary winding (L_{s2} is the leakage inductance of the secondary winding).

- C_1 is the parasitic capacitance of the primary winding and may include other parasitic capacitances at the input).
- $C'_2 = C_2/n^2$ is the parasitic capacitance of the secondary winding, referred to the primary winding (C_2 is the capacitance of the secondary winding and may include other parasitic capacitances at the output).

- $C'_{12} = C_{12}/n$ is the parasitic capacitance between the input terminals and the output terminals referred to the primary winding. (C_{12} is the capacitance between the input terminals and the output terminals.)

The complete circuit is rather complicated and its rigorous analysis is often too complex for practical purposes. The equivalent circuit is usually simplified by dividing the operating frequency range into low, middle, and high frequency domains. Thus, three equivalent circuits are obtained (see Figures 2-74 through 2-76).

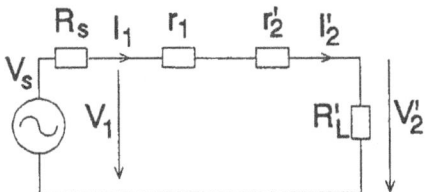

Figure 2-74 *Mid-band equivalent circuit (referred to the primary winding) of a conventional transformer.*

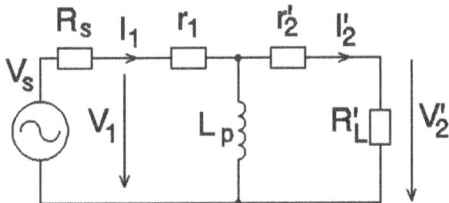

Figure 2-75 *Low-frequency equivalent circuit (referred to the primary winding) of a conventional transformer.*

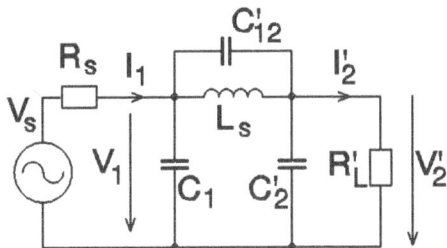

Figure 2-76 *High-frequency equivalent circuit (referred to the primary winding) of a conventional transformer.*

In the mid-band, the reactances are negligible. R_c may be ignored, although it should be included in the equivalent circuit if the power level is high and/or the power loss in the magnetic core is significant.

At lower frequencies, the capacitances and the leakage inductances are negligible. Shunt inductance L_p becomes important because it tends to shunt the output. L_p is the main cause of bandwidth limiting in the conventional transformer at low frequencies.

At high frequencies, the capacitances and the leakage inductances become important. L_p and R_c are negligible, and r_1 and r'_2 are usually ignored. However, the circuit in Figure 2-76 is still too complicated for practical purposes. It is usually further simplified by taking into account that C'_{12} is small in most circuits and may be ignored. Depending on the values of R_s and R'_L, C_1 or C'_2 may also be ignored. Reference [37] provides detailed analyses and expressions of the transfer functions in all practical cases. From a qualitative point of view, it is important to observe that the leakage inductances and the parasitic capacitances limit the bandwidth of the conventional transformer at high frequencies.

The considerations above show that to extend the passband, the inductance of the windings must be increased, while the leakage inductances and the parasitic capacitances must be decreased. These requirements are contradictory, and it is very difficult to obtain a satisfactory tradeoff in many practical applications. Moreover, because solid-state RF PAs operate with low input and load impedances, they require transformers with extremely low leakage inductances. For these reasons, conventional transformers cannot be used at frequencies above 30 MHz. Even at lower frequencies, transmission-line transformers are usually more convenient, especially for low impedances and high power levels.

Transmission-Line Transformer

In a transmission-line transformer [4-8, 11, 37, 39-52] the coils are arranged so that the interwinding capacitance combines with the inductance to form a transmission line. As a result, the high-frequency response is limited by the parasitics which have not been absorbed into the characteristic impedance of the transmission line or by the deviation of the characteristic impedance from its optimum value. In some particular configurations, for example, in the Ruthroff transformer, the length of the transmission line limits the high-frequency response. In the transmission-line transformer, the magnetic flux is canceled out in the ferrite core and a very high efficiency is possible over a large portion of the passband.

Guanella's Transmission-Line Transformer
The basic building block of a transmission-line transformer is the bifilar winding shown in Figure 2-77. The bifilar coil concept was introduced by Guanella in his earliest presentation on the devices [40].

The bifilar coil is usually constructed by coiling a transmission line (a wire pair, a twisted-wire pair, or a coaxial line) around a ferrite core (a rod or a toroid), or by threading the line through ferrite beads. A non-magnetic core can also be used, as Guanella did in his first design. (At that time, ferrites were not available.)

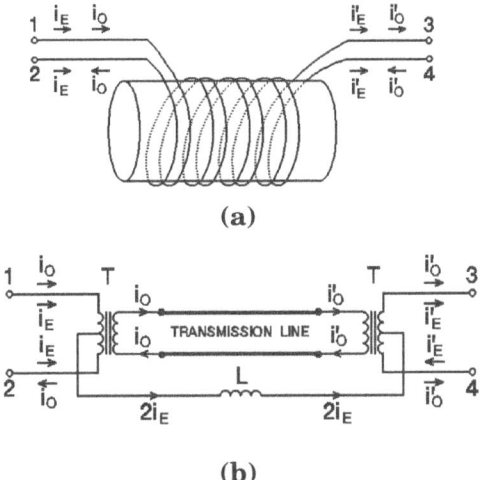

(a)

(b)

Figure 2-77 *Bifilar coil and its equivalent circuit.*

For an intuitive explanation, it is convenient to divide the currents which circulate in a bifilar coil into odd-mode currents i_0, and even-mode currents i_E, as illustrated in Figure 2-77. The odd-mode currents generate a negligible external magnetic field (because they generate opposite, yet equal, magnetic fields). Consequently, there is no magnetic coupling between the adjacent turns and, for the odd-mode currents, the bifilar coil is equivalent to a transmission line with the same length wires. The even-mode currents generate in-phase equal magnetic fields, creating a strong magnetic field and strong coupling between two adjacent turns. Thus, the bifilar coil is equivalent to a standard coil (usually having a large inductance) for the even-mode currents, L in Figure 2-77(b). The odd- and even-mode currents can be separated using two ideal transformers, T, with median taps in one of their windings to obtain the equivalent circuit of the bifilar coil presented in Figure 2-77 (b). If inductance L is large enough, the even-mode currents are negligible and the bifilar coil becomes a transmission line that allows only odd-mode currents.

Several basic circuits can be built around just one bifilar coil, as shown in Figure 2-78. In all cases, optimum performance is obtained if the impedance matching conditions are achieved. If $Z_0 = Z_L$, Equation 2.92 shows that the input impedance of the transmission line is $Z = Z_0 = Z_L$ for any length of the line.

$$R = Z_0 = R_L \qquad\qquad 2.107$$

where Z_0 is the characteristic impedance of the transmission line used for the bifilar coil, and R is the input impedance of the transmission line (the impedance seen by the signal source).

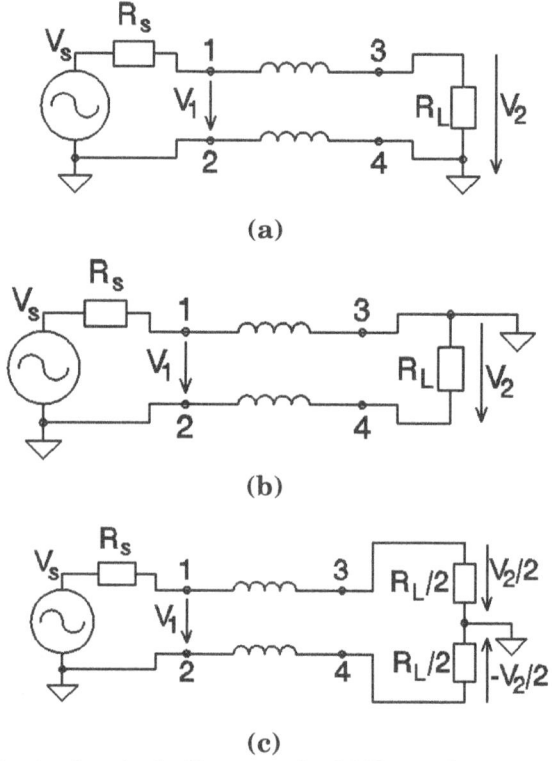

(a)

(b)

(c)

Figure 2-78 *Basic circuits built around a bifilar coil.*

a. In the circuit of Figure 2-78 (a), the bifilar coil behaves as a delay line. The coiling of the transmission line around a ferrite core and the ferrite core itself play no role.
b. The circuit in Figure 2-78 (b) is a phase inverter. If the reactance of L (Figure 2-77) is much greater than R_L, then only odd-mode currents flow in the circuit, and V_1 and V_2 are out-of-phase. If the reactance of L is insufficient, a shunting current will flow through it (from the signal source to ground), resulting in a decrease of the input impedance and the presence of magnetic flux in the ferrite core.
c. A balun (balanced-to-unbalanced) configuration is shown in Figure 2-78 (c). If the center of R_L is left off (floating load), the currents in the two windings will be equal and opposite. If the center of R_L is connected to ground, as shown in Figure 2-78 (c), the reactance of L must be far greater than R_L to assure that the even-mode currents are negligible and the load is balanced to ground. A similar configuration can be used to match a balanced signal source (for example, a push-pull PA) to an unbalanced load (for example, a coaxial feeder).

By combining bifilar coils in parallel-series arrangements, it is possible to obtain baluns (balanced-to-unbalanced transformers) or ununs (unbalanced-to-unbalanced transformers) with impedance transforming ratios of $1:n^2$, where n is the number of bifilar coils. This technique is based on summing in-phase voltages at the high-impedance side and in-phase currents at the low-impedance side. Figure 2-79 shows a Guanella's 1:4 transmission-line transformer as an example.

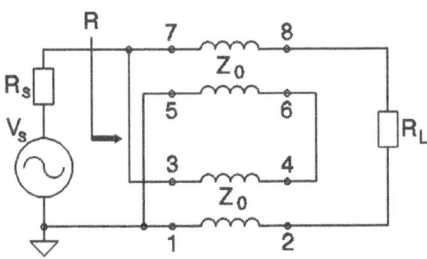

Figure 2-79 *Guanella's 1:4 transmission-line transformer.*

The two bifilar coils are in parallel at the low-impedance side (the signal source) and in series at the high-impedance side (the load). With the ground connected to terminal 1 (and 5), this transmission-line transformer is used to match an unbalanced generator with a floating load. According to the placement of two ground connections, other configurations such as ununs, phase-reversals, and hybrids are possible.

Because each transmission line sees one half of the load, the optimum value of the characteristic impedance is $Z_0 = R_L/2$. With this value, $R = R_L/4$. If n bifilar coils are used (Figure 2-80), the optimum value of Z_0 is $Z_0 = R_L/n$ and $R = R_L/n^2$. If the bifilar coils are in parallel at the load side and in series at the signal source side (Figure 2-81), then $Z_0 = nR_L$ and $R = n^2R_L$.

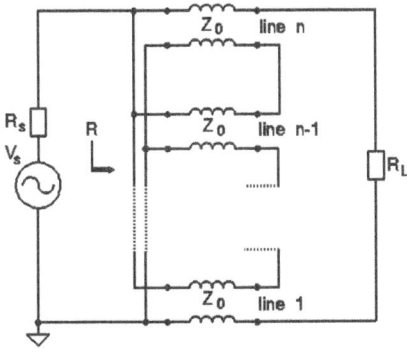

Figure 2-80 *Guanella's $1:n^2$ transmission-line transformer.*

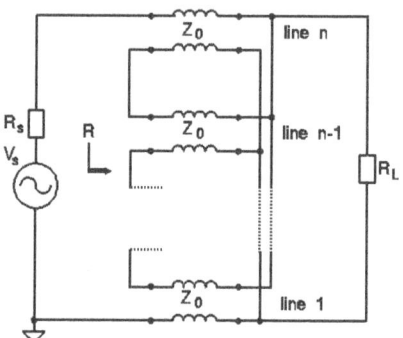

Figure 2-81 *Guanella's $n^2{:}1$ transmission-line transformer.*

The high-frequency response of Guanella's transmission-line transformer is limited by parasitics not absorbed into the characteristic impedance of the transmission line (for example, the parasitic inter-winding capacitance between adjacent bifilar turns), change of Z_0 with frequency, or increasing power loss in the transmission line with increasing frequency. The low frequency response depends on several factors and requires a more detailed discussion. By way of example, a low-frequency model of the Guanella's 1:4 transmission-line transformer is presented in Figure 2-82.

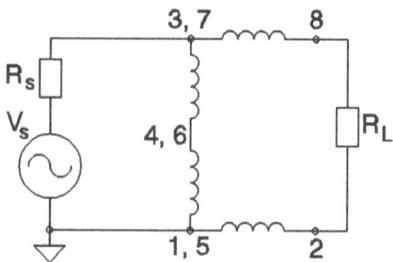

Figure 2-82 *Low-frequency model of Guanella's 1:4 transmission-line transformer.*

There are several distinct cases, depending on the ground connections and the number of magnetic cores used.

1. The circuit is used as a balun with floating load, i.e., only terminal (1,5) is grounded. The low-frequency response is given by the magnetizing inductance (that shunts the signal source), comprised of winding 3-4 in series with winding 6-5. Its value is calculated as follows:

a. If the two transmission lines are coiled on separate cores, the magnetizing inductance is the sum of the two inductances (3-4 and 6-5).
b. If the two transmission lines are coiled on the same core, the magnetic coupling between them must be considered. Assuming a unity coupling factor, the total magnetizing inductance would be greater than in the previous case by a factor of two.

2. The circuit is used as an unun transformer, with terminals (1, 5) and 2 connected to ground. Consequently, winding 1-2 is shorted, as is winding 3-4.
    a. If two cores are used, windings 5-6 and 7-8 are not affected by shorted windings 1-2 and 3-4. The low-frequency response is provided by the inductances of windings 5-6 and 7-8 and the magnetic coupling between them. In this case, the magnetic core for windings 1-2 and 3-4 is no longer needed and can be replaced by a nonmagnetic form for mechanical purposes.
    b. If only one core is used, windings 5-6 and 7-8 are also shorted, resulting in a very poor low-frequency response. If a good low-frequency unun transformer response is needed, it is best (as suggested by the results above) to connect a 1:4 balun in series with a 1:1 balun for isolation (see Figure 2-83).

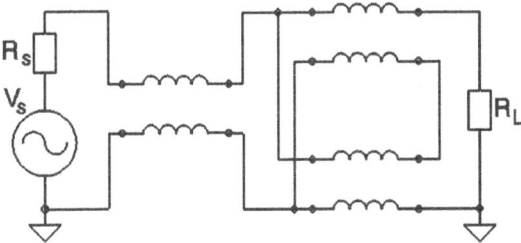

Figure 2-83 *Wideband Guanella 1:4 unun using a 1:1 balun back-to-back with a 1:4 balun.*

3. The circuit is used as a balun with terminal (1, 5) and the center of R_L connected to ground. In this case, the voltage across winding 7-8 (and across winding 5-6) is zero. The situation is similar to that of the unun, except that now the top transmission line is shorted instead of the bottom transmission line.

Ruthroff's Transmission-Line Transformer

Reference [42] presents the basic principles of Ruthroff-type transmission-line transformers for a circuit configuration with 1:4 impedance ratio (see Figure 2-84). The same circuit appears in more detail in this section to illustrate a simple mathematical approach applied to these transmission-line transformers.

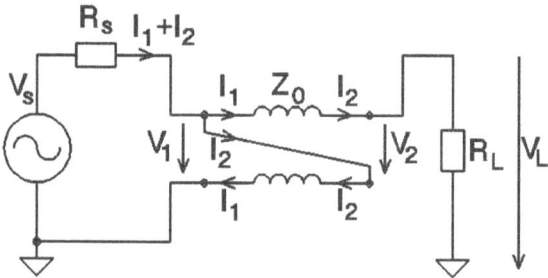

Figure 2-84 *Ruthroff's 1:4 transmission-line transformer (unun).*

Ruthroff's transmission-line transformer mainly sums a direct voltage (or current) with a delayed voltage (or current), which traverses a transmission line. In the Ruthroff 1:4 transformer shown in Figure 2-84, the direct voltage, V_1, is added to the delayed voltage, V_2, to obtain the voltage across the load.

The simple mathematical approach presented below is based on the following assumptions:

a.   The windings have sufficient longitudinal reactance so only odd-mode currents (I_1 and I_2 in Figure 2-84) flow through the bifilar coil.

b.   The main purpose of the circuit is to obtain maximum transfer of power from signal source V_s to load R_L over a wide frequency band. Although this is not usually the case in RF PAs, it simplifies the mathematical approach without affecting the generality of the results. Consequently, for optimum operating conditions, the input impedance of the transformer should be purely resistive and equal to R_s. For RF power applications (when the signal-source impedance is not equal to the load impedance), R_s is the input impedance seen at one end of the transformer with the opposite end terminated in R_L.

c.   A lossless transmission line, with characteristic impedance Z_0 and physical length l is used in the circuit.

Based on these assumptions, the circuit in Figure 2-84 is described by the following equations [10, 12, 15]:

$$V_s = V_1 + R_s(I_1 + I_2) \qquad V_L = V_1 + V_2 = R_L I_2 \qquad \text{2.108}$$

$$V_2 = V_1 \cos \beta l - j Z_0 I_1 \sin \beta l \qquad I_2 = -j \frac{V_1}{Z_0} \sin \beta l + I_1 \cos \beta l$$

where β is given by Equation 2.93.

This set of equations is solved for the output power P_0.

$$P_0 = |I_2|^2 R_L =$$

$$= \frac{V_s^2 R_L \left(1 + \cos \beta l\right)^2}{\left[2R_s \left(1 + \cos \beta l\right) + R_L \cos \beta l\right]^2 + \left(Z_0 + \dfrac{R_S R_L}{Z_0}\right)^2 \sin^2 \beta l} \qquad 2.109$$

From this expression, the conditions for maximum power transmission are obtained by setting

$$\frac{dP_0}{dZ_0} = 0 \qquad 2.110$$

The maximum value of P_0 is obtained for

$$Z_0 = \sqrt{R_s R_L} \qquad 2.111$$

and is given by

$$P_{0,max} = \frac{V_s^2 R_L \left(1 + \cos \beta l\right)^2}{\left[2R_s \left(1 + \cos \beta l\right) + R_L \cos \beta l\right]^2 + 4R_s R_L \sin^2 \beta l} \qquad 2.112$$

$P_{0,max}$ can be further maximized by setting $l = 0$ and

$$\frac{dP_{0,max}}{dR_L} = 0 \qquad 2.113$$

Next

$$R_s = \frac{R_L}{4} \qquad 2.114$$

and, for these values, the signal source delivers its available power into the load resistance

$$P_A = \frac{V_s^2}{4R_s} \qquad 2.115$$

As a result, the optimum operating conditions of the Ruthroff 1:4 transformer are given by Equations 2.111 and 2.114. Using these equations in Equation 2.109,

$$\frac{P_0}{P_A} = \frac{(1+\cos\beta l)^2}{\frac{5}{4}\left(1 + \frac{6}{5}\cos\beta l + \cos^2\beta l\right)} \qquad 2.116$$

The variation of P_0/P_A with βl is depicted in Figure 2-85. The high-frequency response of the Ruthroff transmission-line transformer is limited by the allowed insertion loss at the high end of the frequency band, and can be improved using a short transmission line. For example, if at the maximum operating frequency f_{max} the electrical length of the line is $\beta l = 45°$ (corresponding to $\lambda/8$), then

$$Z_{in} = Z_0\frac{R_L\cos\beta l + jZ_0\sin\beta l}{2Z_0(1+\cos\beta l) + jR_L\sin\beta l} \qquad 2.117$$

The insertion loss is approximately 1 dB when the line is a quarter-wavelength ($\beta l = 90°$) and infinite when it is a half-wavelength ($\beta l = 180°$).

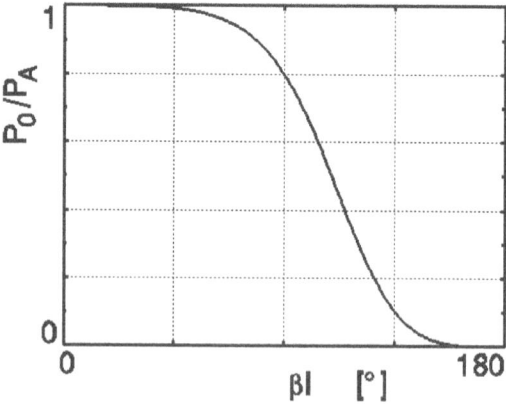

Figure 2-85 *Theoretical P_0/P_A versus electrical length of the transmission line for a Ruthroff 1:4 transformer.*

Using Equation 2.108, the input impedance of the transmission-line transformer is

$$Z_{in} = Z_0\frac{R_L\cos\beta l + jZ_0\sin\beta l}{2Z_0(1+\cos\beta l) + jR_L\sin\beta l} \qquad 2.118$$

If matching conditions 2.111 and 2.114 are satisfied, Equation 2.118 becomes

$$Z_{in} = R_{in} + jX_{in} = R_s \frac{2\cos\beta l + j\sin\beta l}{1 + \cos\beta l + j\sin\beta l}$$

$$\frac{R_{in}}{R_s} = \frac{1 + \cos\beta l}{2} \quad \frac{X_{in}}{R_s} = \frac{\sin\beta l(1 - \cos\beta l)}{2(1 + \cos\beta l)}$$

2.119

The variation of R_{in}/R_s and X_{in}/R_s with the electrical length of the transmission line βl is depicted in Figure 2-86. For a $\lambda/8$ line (at the maximum operating frequency), Equation 2.119 yields $R_{in}/R_s \approx 0.8536$ and $X_{in}/R_s \approx 0.0607$.

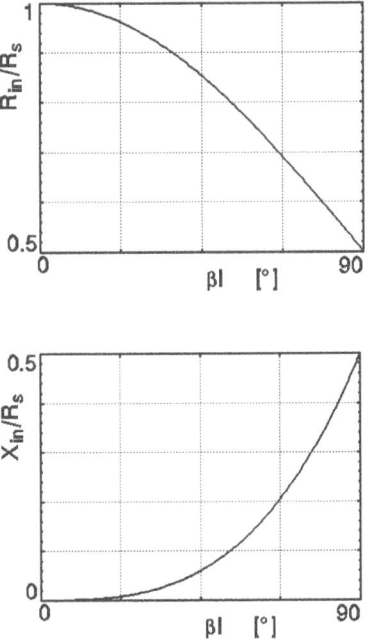

Figure 2-86 *Normalized input resistance and input reactance versus electrical length of the transmission line for a Ruthroff 1:4 transformer.*

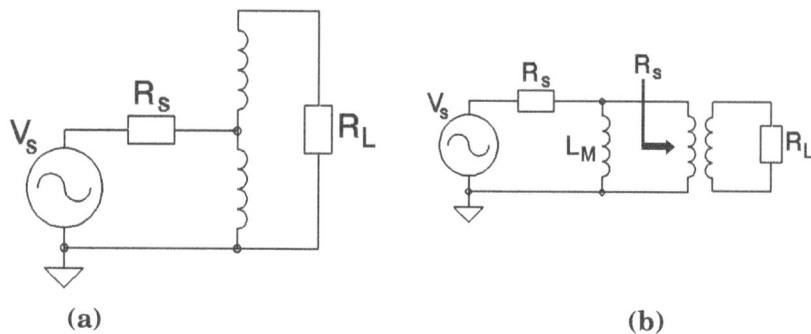

(a) (b)

Figure 2-87 *Low-frequency equivalent circuits of the Ruthroff 1:4 transformer (unun).*

The low-frequency model of the Ruthroff 1:4 transformer is shown in Figure 2-87 (a). This schematic is quite similar to that of a 1:4 autotransformer. A further simplified equivalent circuit is presented in Figure 2-87 (b). The autotransformer is replaced by an ideal transformer shunted by the core magnetizing inductance, L_M. Denoting the reactance of L_M with $X_M = 2\pi f L_M$, the output power (dissipated in R_L) is given by

$$P_0 = \frac{V_s^2}{R_s} \frac{X_M^2}{R_s^2 + 4X_M^2} \qquad \frac{P_0}{P_A} = \frac{4X_M^2}{R_s^2 + 4X_M^2} \qquad 2.120$$

where P_A is the available power from the source V_s.

A practical figure for the low-frequency response of transmission-line transformers is the –0.45 dB frequency. This represents a power loss of about 10 percent and corresponds to an SWR of about 2:1. According to Equation 2.120, the –0.45 dB frequency is

$$2\pi f_{min} = \frac{3R_s}{2L_M} \qquad 2.121$$

Thus, there is only one way to improve the low-frequency response of the transmission-line transformer: increasing L_M. This can be done by a) increasing the number of turns of the bifilar coil, b) increasing the permeability of the magnetic core, c) increasing the effective cross-sectional area of the core, and d) decreasing the average magnetic path length in the core. Note that a) and c) yield an increase of the transmission-line length, degrading the high-frequency response.

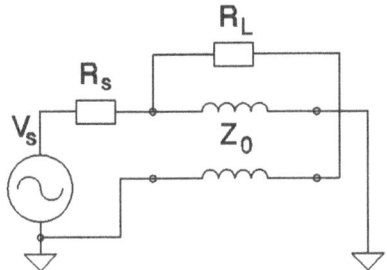

Figure 2-88 *Ruthroff's 1:4 transmission-line transformer (balun).*

The circuit in Figure 2-84 can be used as a 4:1 unun if the signal source and the load change places. The load resistor must be connected to obtain a Ruthroff 1:4 balun, as shown in Figure 2-88.

Other Types of Transmission-Line Transformers

The transmission-line transformers presented in the preceding sections have limited broadband impedance matching capabilities. The matched impedances must be in a ratio of $1:n^2$, where n is an integer. Other impedance ratios are useful in practical applications and can be obtained by a) adequate connection of Guanella's and/or Ruthroff's transmission-line transformers, and b) using multifilar windings and/or tapped transmission-line transformers. Both techniques are described briefly below.

A general synthesis procedure for an arbitrary $m^2:n^2$ impedance ratio (where m and n are integers) is presented in Reference [48]. This technique may be considered an extension of Guanella's circuit for more complex transmission-line transformer configurations. The basic idea is illustrated in Figure 2-89. Suppose a certain transmission-line diport behaves as a transformer with a voltage ratio of $m:n$. A new transmission line is connected to the terminals of that diport: in parallel at one end and in series at the other. The result is a new transformer with a voltage ratio of $m:(m + n)$.

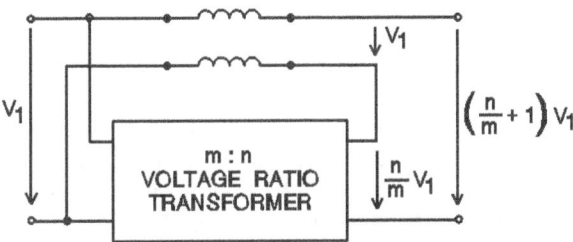

Figure 2-89 *A m:n voltage ratio transformer and one more transmission line yield a m:(m + n) voltage ratio transformer.*

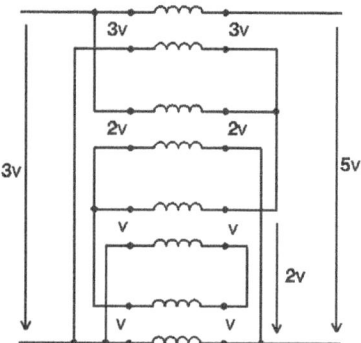

Figure 2-90 *A 3:5 voltage ratio transmission-line transformer.*

Figure 2-90 shows a 3:5 voltage ratio transformer (that is, a 9:25 impedance ratio). The voltages across each transmission line are shown. Another technique based on parallel-series arrangements is illustrated in Figure 2-91 [39]. The bottom transmission line is connected as a 1:4 Ruthroff unun, the top transmission line is connected as a Guanella 1:1 balun. The result is a 1:9 impedance transformer.

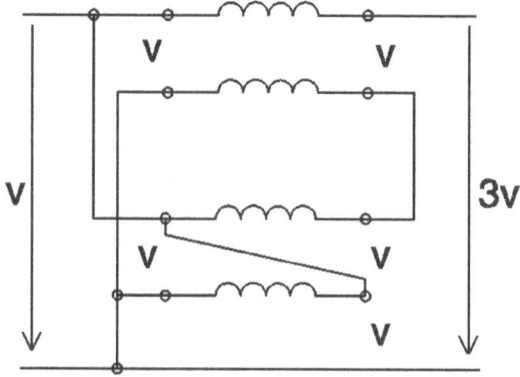

Figure 2-91 *A 1:9 Ruthroff-Guanella transformer.*

Circuits using multifilar and/or tapped windings are Ruthroff-type transformers; that is, one or more delayed voltages (or currents) are added to a direct voltage (or current). This technique is illustrated in Figure 2-92 for a trifilar 1:2.25 impedance ratio unun [5, 39].

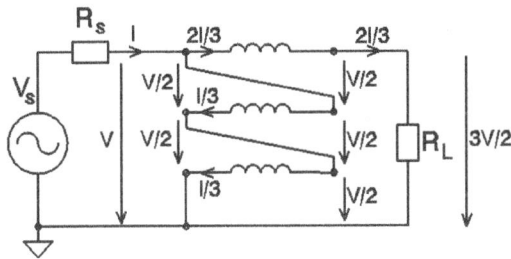

Figure 2-92 *A trifilar 1:2.25 impedance ratio unun.*

The circuit uses a trifilar winding, similar to the bifilar type shown in Figure 2-77 (a), coiled on a single ferrite core. The trifilar winding behaves as two transmission lines, the center winding being common. The output voltage is the sum of two direct voltages of $V/2$ and one delayed voltage of $V/2$. Consequently, the voltage ratio of this transformer is 1:1.5 and the impedance ratio is 1:2.25. Ignoring the even-mode currents, the current flowing through the top winding is $2I/3$, while the currents flowing through the other two windings is $I/3$. Thus, the flux in the magnetic core is canceled out and the power loss in the core is negligible. The transformer in Figure 2-92 has a better frequency response than a 1:4 Ruthroff transformer using a transmission line of equal length. The low-frequency response is improved because the bottom two windings are series-aiding (unity mutual coupling) in a low-frequency equivalent circuit of the trifilar transformer. The high-frequency response is improved because a delayed voltage of only $V/2$ is added to a direct voltage of V.

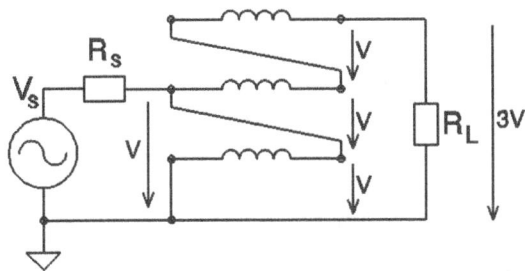

Figure 2-93 *A trifilar 1:9 impedance ratio unun.*

A similar circuit can be used as a 1:9 trifilar unun (see Figure 2-93). The output voltage is the sum of a direct voltage of V, a delayed voltage of V, and a double-delayed voltage of V. This causes significant degradation of the circuit's high-frequency response.

The two examples in Figures 2-92 and 2-93 suggest that a n-filar winding coiled on a single core can be used to obtain a transformer with any voltage ratio between 1:n and $(n-1):n$; that is, any impedance ratio between 1:n^2 and $(n-1)^2:n^2$. The $(n-1):n$ configuration yields the best frequency response because (n-1) windings are series aiding in a low-frequency equivalent circuit, improving the low-frequency response. At the same time, a delayed voltage of only $V/(n-1)$ is added to a direct voltage of V, extending the high-frequency response. The 1:n configuration yields the worst-case frequency response, at both low and high frequencies. In the low frequency equivalent circuit, only one winding shunts the signal source. At high-frequency, the output voltage is the sum of a direct voltage of V, a delayed voltage of V, a double-delayed voltage of V, . . ., and a $(n-1)$ times delayed voltage of V.

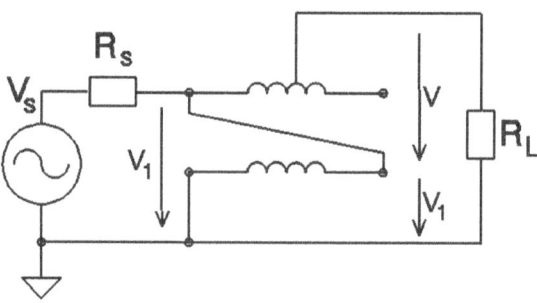

Figure 2-94 *A tapped bifilar transmission-line transformer.*

It is possible to use tapped windings in transmission-line transformers because of the longitudinal potential gradient from the input to the output of the bifilar coil. The circuit configuration shown in Figure 2-94 is similar to a 1:4 Ruthroff unun, except the output voltage is obtained from a tap in the top winding [39]. When the longitudinal reactance is high enough, the even-mode currents are negligible, and a potential gradient of V_1 exists along the windings. Depending on the position of the tap, voltage V varies between 0 and V_1. This configuration allows for any voltage ratio between 1:1 and 1:2 and impedance ratios between 1:1 and 1:4. The same idea can be used with multifilar transformers [39, 44]. For example, by tapping the top winding of the transformer shown in Figure 2-92, it is possible to obtain impedance ratios of 1:1 to 1:2.25. By tapping the top winding of the transformer shown in Figure 2-93, it is possible to obtain impedance ratios of 1:4 to 1:9.

A Design Example
The 1:4 Ruthroff unun provides a simple example. Given $R_s = 25\ \Omega$, $R_L = 100\ \Omega$, $f_{max} = 100$ MHz, and $f_{min} = 1$ MHz, the design procedure is outlined below.[10]

1. Equation 2.111 yields $Z_0 = 50\ \Omega$, which makes it convenient to use a standard coaxial cable such as RG-58/U. Depending on the required power handling capabilities, a more appropriate coaxial cable may be chosen.

2. The dielectric of the coaxial cable results in a relative phase velocity of 0.79 (i.e., $\lambda = 0.79\ \lambda_0$, where λ is the wavelength in the coaxial cable and λ_0 is the wavelength in free space). According to the results in section *Ruthroff's Transmission-Line Transformer*, the cable length must be equal to $\lambda/8$ at 100 MHz.

$$l = \frac{\lambda}{8} = \frac{0.79\lambda_0}{8} \approx 29.6\ \text{cm} \qquad 2.122$$

3. According to Equation 2.121, the core magnetizing inductance L_M must have a minimum value.

$$L_M = \frac{3R_S}{4\pi f_{min}} \approx 5.97\ \mu\text{H} \qquad 2.123$$

4. Taking into account the power level, the operating frequency range, and the physical and mechanical characteristics of the transmission line, a ferrite core (most often, a toroidal one) is used. For a toroidal core, the magnetizing inductance is given in References [9, 37, 39, 44].

$$L_M = \mu_0 \mu_r N^2 \frac{A_E}{l_E} \qquad 2.124$$

where $\mu_0 = 4\pi \times 10^{-7}$ H/m, μ_r is the relative permeability of the core, N is the number of turns, A_E is the effective cross-sectional area of the core, and l_E is the average magnetic path length in the core. A composite parameter may be provided by the manufacturer to simplify the inductance calculations

$$L_M = A_L N^2 \qquad 2.125$$

Equations 2.124 and 2.125 give the minimum number of turns needed to obtain the desired low-frequency response. Considering that a toroidal core with $A_L = 250$ nH/turn^2 is used in this design, Equation 2.125 yields $N \approx 4.9$. Thus, the desired frequency response is obtained for $N = 5$. Whether the coil can be realized or not can be decided once the physical dimensions of the ferrite core and the transmission line are known. Several different cores may be tried to obtain the best design.

5. The peak value of the magnetic induction in the core is given by [9, 37]

$$B_{max} = \frac{V}{2\pi f_{min} NA_E} \qquad 2.126$$

where V is the peak value of the voltage across the transmission line. The magnetic induction must be kept below the saturation level to avoid nonlinearities, generation of harmonics, and inefficiency. With B_{max} known, it is possible to estimate (from the manufacturer's data sheet) the maximum power loss density in the ferrite core, followed by the power loss in the core. Excessive power dissipation may be a problem in high-power transmission-line transformers, because the thermal conductivity of ferrites is low. As a result, the core may overheat and crack.

Some Practical Considerations

Transmission-line transformers used in the HF and VHF bands are usually made by coiling a transmission line on a ferrite core. However, at very high frequencies and/or when a good low-frequency response is not important, they can be built without a ferrite. At high power levels, when heavy-gauge conductors must be used, it may be useful to use a single-turn construction with stacked ferrite cores. (More practical details are found in [37, 39, 51, 52]). Ferrite toroids and rods can be used as magnetic cores for transmission-line transformers. Although ferrite rods have the advantage of a simpler form, toroids are preferred because

a. A better low-frequency response can be obtained with a toroid. (For a given ferrite type and transmission-line length, a much higher inductance is obtained using a toroid.)

b. The two ends of the transmission line are physically close, which is advantageous for some transmission-line transformer configurations.

Another important matter for practical implementation of transmission-line transformers is the characteristic impedance of the transmission line. Standard coaxial lines and strip lines are usually the best choice for transmission-line transformers. Their characteristic impedance value is not affected by the particular characteristics of a certain design. Twisted pairs or pairs of untwisted wires are also often used because they are very simple to make. By varying the conductor diameter, the electrical insulation material and its thickness, and the number of twists per unit of length, it is possible to obtain characteristic impedances from 10 to 200 ohms [37, 39, 44].

Although transmission-line transformers have been used widely for over 50 years, their operation is still not well understood. The classic and intuitive approach presented in most of the literature uses two different equivalent circuits to describe the operation in the low- and high-frequency range: a conventional "wire-wound" transformer or autotransformer and a transmission line. This classic theory provides pertinent explanations for the frequency and power limitations of the transmission-line transformer and also offers simple and accurate design equations. However, the boundary between the two equivalent circuits is rather unclear and the model for the mid-frequency range appears to be missing, both of which are serious drawbacks. On the other hand, some experimental results cannot be explained with this theory. (For example, theory says that at high frequencies, the energy is transmitted by the transmission-line mode and the core is not needed. Experiments show that the magnetic core plays a major role throughout the entire passband, even at high frequencies.)

A different, more complicated approach uses the theory of coupled transmission lines to describe the operation of the transmission-line transformer [41, 49]. A single equivalent circuit is used over the entire frequency band of the circuit, and the practical benefits seem to be very promising. Accurate CAD models of the transmission-line transformer can be built and used for simulating more complex RF circuits.

2.8   Gain Leveling and VSWR Correction

Broadband RF power amplifiers often create two additional problems:
a. The overall gain must be relatively flat over the operating frequency range $f_{min} \cdots f_{max}$. The required flatness usually varies from tenths of a dB to 1 to 2 dB.
b. The input VSWR of the amplifier must be kept within a specified range (for example, 2:1 or lower) over the operating frequency range.

Although these two problems are sometimes discussed separately, they cannot be completely divorced from each other. Both problems can be solved by including various R, C and L components at the amplifier input or in a feedback path. A circuit designed for input VSWR correction will, to some extent, affect the overall gain of the amplifier and vice versa. For example, if some of the driving power is lost in the compensating input network, the amplifier gain will decrease. In fact, this could be helpful at low frequencies, where the amplifier gain is usually higher.

This discussion intends only to present a qualitative approach and some basic design ideas. More detailed explanations and design equations can be found in the literature. Since some of the design equations are only approximate, some of the design parameters are difficult to estimate from the manufacturers' data sheets, and it is almost impossible to consider all the parasitic elements in the circuits, numerical optimization and experimental adjustments are often required.

The gain disparity in a broadband amplifier can be greater than 10 to 15 dB from the low to the high end of the operating frequency range. For example, BJT gain falls off at high frequencies with a slope of 6 dB per octave.

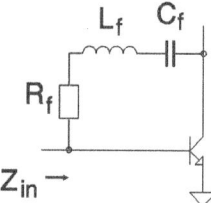

Figure 2-95 *Frequency-dependent shunt feedback.*

A possible solution for gain leveling is to use a frequency-dependent shunt feedback, as suggested in Figure 2-95 [4, 7, 53, 54]. C_f is a DC-blocking capacitor. L_f tends to increase the impedance of the feedback path with the operating frequency, preserving amplifier gain at higher frequencies. At lower frequencies, the feedback becomes stronger, decreasing amplifier gain as well as the input impedance Z_{in}.

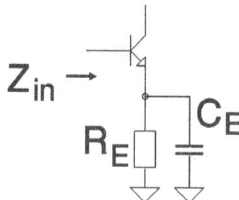

Figure 2-96 *Frequency-dependent degenerative feedback.*

Somewhat similar results can be obtained using degenerative feedback, as shown in Figure 2-96 [4, 7]. C_E reduces the gain at low frequencies. This solution is rarely used because it is difficult to avoid unwanted parasitic inductances in the emitter circuit.

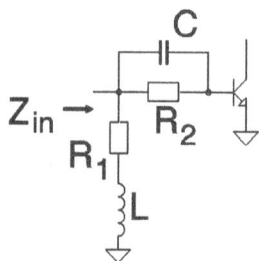

Figure 2-97 *An RLC network for gain leveling and VSWR correction.*

Gain leveling can also be accomplished using an RLC network, as suggested in Figure 2-97 [7]. At low frequencies, L has a low reactance while C has a high reactance. Consequently, the amount of power driven into the base-emitter junction is reduced, as is the overall gain. At higher frequencies, L and C tend to eliminate the effect of R_1-and R_2-preserving power gain. Obviously, the input impedance also changes with frequency; thus, this circuit can be used for both gain leveling and VSWR correction [54, 55].

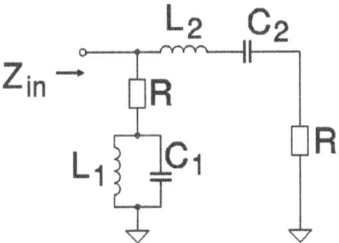

Figure 2-98 *An RLC network (with constant input impedance) for gain leveling.*

Figure 2-98 presents another gain leveling circuit [27, 32]. If its load resistance, R, is constant over the entire frequency range, this circuit is able to provide a constant input impedance $Z_{in} = R$ and compensate for a 6 dB per octave gain fall-off. The two resonant circuits are tuned to the highest frequency of the operating range to minimize losses.

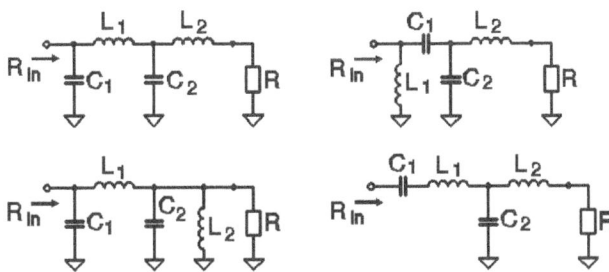

Figure 2-99 *Four-reactance networks for broadband impedance matching.*

Four-reactance networks can also be used for broadband matching (see Figure 2-99). A four-reactance network is able to transform load impedance R into R_{in}, at two frequencies. An approximate transformation is achieved over a reasonable frequency band (see [60] and its accompanying references). If a very large bandwidth is required and/or the active device has a high Q, band-pass Chebyshev filters are often used [27, 32]. Such a circuit can be designed to take into account the gain roll-off [43].

2.9   Amplitude Modulation

Three basic procedures can be used to obtain a high-power AM signal: base-bias modulation, AM signal amplification, and collector modulation [1, 2, 4, 5, 9, 25]. Combined modulation techniques can also be used to improve circuit performance.

In base-bias modulation, the low-frequency modulating signal is used to control the base-bias of a Class C operated transistor. A change in the base-bias voltage affects the conduction angle, and thus the amplitude of the output signal, generating an AM signal. Base-modulation is not generally used due to several disadvantages, such as low efficiency (a typical maximum value is about 40 percent), significant nonlinearity of the modulation characteristics, and adverse operating conditions for the amplifier providing the modulating signal. (It operates into the base-emitter capacitance whose value varies with the signal level.) Although somewhat better results can be obtained using FETs instead of BJTs, this method is used only in very low-cost, simple transmitters. Base-bias modulation is also sometimes used in collector-modulated stages to improve the linearity of the modulation characteristics [4, 5].

AM signal amplification is widely used in SSB transmitters. The SSB signal is generated easily at low power levels and must be amplified linearly to the desired power level. This is usually accomplished with Class AB push-pull linear amplifiers [4, 5].

The main form of amplitude modulation used in solid-state RF amplifiers is collector modulation [2, 4, 5, 9, 25, 56]. This technique can be used directly to obtain an AM *double sideband* (DSB) signal. In an envelope *elimination and restoration* (EER) system, collector modulation can be used to generate any type of AM signal. The basic circuit of a collector-modulated RF amplifier and the corresponding waveforms are shown in Figure 2-100.

The modulating signal

$$v_m(t) = V_m \cos \omega_m t \qquad 2.127$$

is used to produce a time-varying collector-supply voltage for the RF amplifier

$$v_C(t) = V_{dc} + V_m \cos \omega_m t \qquad 2.128$$

The voltage across the load is an AM DSB signal

$$v_0(t) = V_0 \cos \omega_0 t = V_{dc}(1 + m \cos \omega_m t) \cos \omega_0 t \qquad 2.129$$

where

$$m = \frac{V_m}{V_{dc}} \leq 1 \qquad\qquad 2.130$$

is the modulation depth.

Ignoring V_{cEsat} and taking into account that the collector voltage must be positive, $v_C(t) \geq 0$, $m \leq 1$. Under the peak modulation condition, the maximum collector voltage is $2V_{dc}$. AM signals with $m > 1$ cannot be obtained using collector modulation.

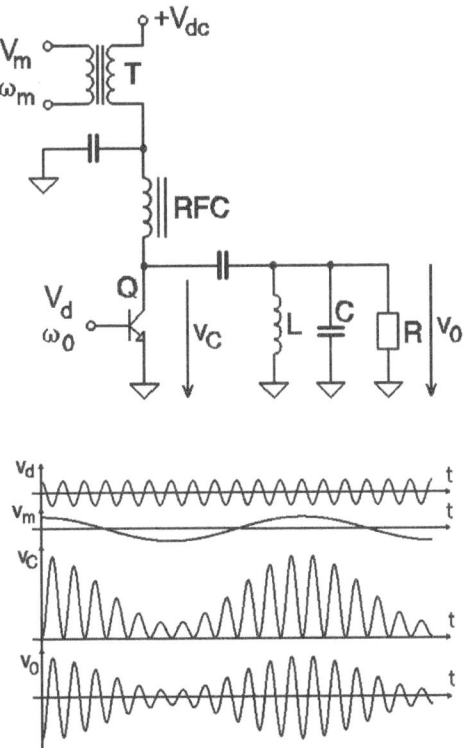

Figure 2-100 *Circuit of a collector-modulated RF amplifier using a modulation transformer and corresponding waveforms.*

An important advantage of this type of modulation is that the RF power amplifier can be designed to operate with high efficiency. The efficiency being almost constant with the modulation depth. The main disadvantage is that the modulating signal must be amplified to a high power level. For example, if the carrier power is P_0 and the collector efficiency of the RF stage is η_{RF}, the amplifier for the modulating signal must be able to provide

$$P = \frac{\frac{m^2}{2}P_0}{\eta_{RF}} \qquad P_{max} = \frac{P_0}{2\eta_{RF}} \text{ if } m_{max} = 1 \qquad 2.131$$

As a result, the efficiency of this amplifier is also important for the overall efficiency of the transmitter. The modulation transformer must be able to pass low-frequency, high-power signals with low distortion and high-efficiency; thus, it is usually a bulky, heavy, and expensive component. An excellent solution is a series-coupled Class S modulator (see Chapter 6) because it does not need a modulation transformer and because its theoretical efficiency is 100 percent.

To obtain a linear modulation characteristic, the output voltage must be linearly related to the collector supply voltage. This may be accomplished by overdriving the Class C RF amplifier (see section *Saturated Class C Amplifiers*). The linearity of the modulation characteristic (see Figure 2-101) is a function of several parameters: drive level and drive feedthrough to the output via C_{CB}, the saturation characteristics of the RF transistor, and the amount of voltage-variable capacitance in the collector circuit.

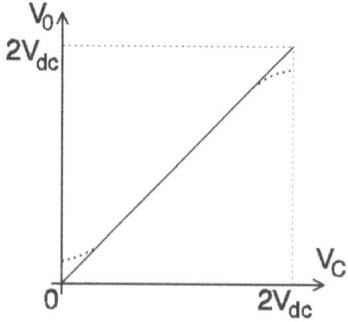

— *ideal characteristic*
..... *real characteristic*

Figure 2-101 *Amplitude of the output voltage V_0 versus the collector voltage V_C.*

The drive must be large enough to ensure that the transistor is overdriven (i.e., it saturates during a portion of the RF cycle) — even for large values of V_C, approaching V_{dc}. Otherwise, at high V_C levels, the saturation angle will vary with V_C (see Figure 2-17) and the output voltage will increase nonlinearly with V_C (the dotted line in Figure 2-101). The capacitive coupling between input and output via C_{CB} [11] produces a signal across the load for low values of V_C (even when $V_C = 0$), as shown in Figure 2-101. The amount of unwanted signal across the load depends on the drive level, the value of C_{CB}, and the collector load impedance. The reactance of C_{CB} is usually larger than the collector load impedance and so the feedthrough signal is almost in phase-quadrature with the amplified output signal. This determines a phase variation of the carrier at low V_C levels, producing unwanted interference with signals in adjacent channels.[12] Thus, too much drive results in nonlinearity at low output levels and too little drive results in nonlinearity at high output levels.

A reasonable tradeoff can be found in practical applications; however, good linearity of the modulation characteristic requires restricting the maximum value of the modulation depth, which reduces the overall performance of the circuit. A practical solution is to decrease the drive level at low V_C and to increase it at high V_C, i.e., use an AM signal to drive the circuit. This technique is called double-collector modulation because the final stage and its driver are both collector modulated. A similar effect (i.e., an improved linearity of the modulation characteristic) can be obtained using a base-bias modulation of the final stage, although the double-collector modulation method usually gives better results. At low V_C, the base-bias is adjusted to drive the transistor toward cutoff; at high V_C, the base-bias tends to drive the transistor toward saturation.

2.10   Class C Frequency Multipliers

Frequency multipliers are often used to multiply the frequency of the master oscillator or to increase the modulation index in the case of phase or frequency modulation [4, 5, 9, 16, 22]. The Class C frequency multiplier has the same schematic as the Class C power amplifier (see Figure 2-1) and operates in much the same way. The only difference is that the collector resonant circuit is tuned to the desired harmonic, suppressing all other harmonics. Assuming that the parallel LC output circuit is ideal, tuned to the nth harmonic, a sinusoidal output voltage is obtained.

$$v_0(\theta) = V_0 \cos\theta = RI_M \alpha_n(\theta_c)\cos\theta \qquad 2.132$$

The output power is given by

$$P_0 = \frac{V_0^2}{2R} = \frac{1}{2}RI_M^2\alpha_n^2(\theta_c) \qquad 2.133$$

and taking into account that

$$P_{dc} = V_{dc}I_{dc} = V_{dc}I_M\alpha_0(\theta_c) \qquad 2.134$$

the collector efficiency is

$$\eta = \frac{P_0}{P_{dc}} = \frac{V_0}{V_{dc}} \frac{|\alpha_n(\theta_c)|}{2\alpha_0(\theta_c)}$$

2.135

The collector efficiency is highest if $V_0 = V_{dc}$.

$$\eta_{max} = \frac{|\alpha_n(\theta_c)|}{2\alpha_0(\theta_c)}$$

2.136

Finally, the power output capability (for $V_0 = V_{dc}$) is given by

$$C_P = \frac{P_0}{v_{max}I_M} = \frac{|\alpha_n(\theta_c)|}{4}$$

2.137

The variation of the maximum collector efficiency η_{max} with the conduction angle θ_c, for a Class C amplifier ($n = 1$), a doubler ($n = 2$), and a tripler ($n = 3$), is shown in Figure 2-102. Note that the collector efficiency decreases as the multiplying order n increases. Also note that a Class B circuit cannot be used as a frequency tripler — η_{max} ($\theta_c = 90°$) = 0 — because a half-wave sinusoidal waveform does not contain the third harmonic (see Figure 2-9).

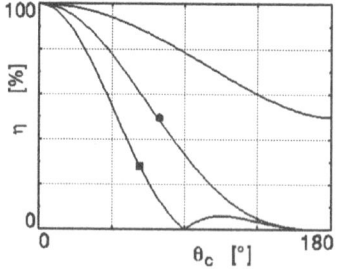

— amplifier ($n = 1$)
—•— doubler ($n = 2$)
—■— tripler ($n = 3$)

Figure 2-102 *Maximum theoretical collector efficiency of Class C circuits.*

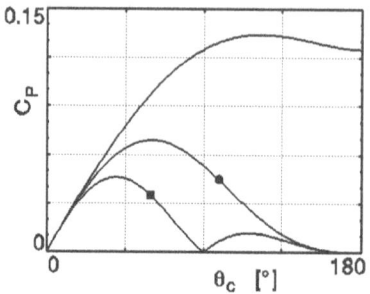

— amplifier ($n = 1$)
—•— doubler ($n = 2$)
—■— tripler ($n = 3$)

Figure 2-103 *Maximum theoretical power output capability of Class C circuits.*

Figure 2-103 shows the variation of power output capability C_P with the conduction angle θ_c. Optimum performance of frequency multipliers (i.e., maximum C_P) is obtained for

a. frequency doubler: $\theta_c = 60°$, $C_P = 0.06892$, $\eta_{max} = 63.23\%$
b. frequency tripler: $\theta_c = 39.86°$, $C_P = 0.04613$, $\eta_{max} = 63.01\%$

EXAMPLE 2.20

A Class C frequency tripler must deliver $P_0 = 1$ W at $V_{dc} = 12$ V. For optimum performance, choose $\theta_c = 39.86°$ and $V_0 = V_{dc}$. The load resistance must be $R = 72$ Ω and the maximum value of the collector current is $I_M = V_0/[R\alpha_3(\theta_c)] = 0.903$A. Finally, $\eta = 63.01\%$ and $C_P = 0.04613$.

The graphs of Figure 2-102 and 2-103 and the equations above suggest that as multiplication factor n increases, the output power (and also the power gain of the stage), the collector efficiency, and the power output capability decrease. On the other hand, if n increases, it becomes more difficult to filter out adjacent harmonics $n - 1$ and $n + 1$ because they lie closer to the desired harmonic, and the relative bandwidth becomes narrower. As a result, Class C frequency multipliers are not recommended for use at high power levels or for a multiplication factor exceeding $n = 3$.

A problem common to any type of frequency multiplier is the output voltage damping (see Figure 2-104). This effect becomes stronger as multiplication factor n increases and output Q decreases. A quantitative evaluation is possible, considering that the envelope of the output voltage varies as $\exp(-\omega t/2Q)$; this would allow the determination of the amplitude variation of the output signal. If this variation exceeds the allowable limit, an amplitude limiter should be required as the next stage in the amplifying chain.

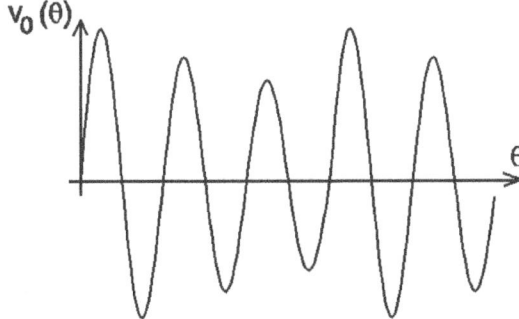

Figure 2-104 *Waveform of the output voltage in a practical frequency tripler circuit.*

The characteristics of the real active device (V_{CEsat} and R_{ON}) can be considered in the same light as those for the Class C amplifier. However, Class C frequency multipliers are much more difficult to design using the large-signal impedance method because RF power transistor manufacturers usually do not present large-signal parameters for the frequency multiplication mode. The parameters provided in the data sheets for Class C amplification could be used to estimate roughly the multiplier circuit parameters, but considerable experimental adjustments would be necessary to obtain optimum performance.

2.11   Stability of RF Power Amplifiers

RF power amplifier instability manifests itself in spurious oscillations. These oscillations can occur at frequencies related or unrelated to the operating frequency (for example, harmonics or subharmonics). They can have complex waveforms and amplitude with a wide range of variations. Depending on their characteristics, spurious oscillations may lead to various unwanted effects: interference with the useful signal, spurious modulations, noise, and often overloading and damage to the active device.

Several of the many oscillation mechanisms that exist in RF power amplifiers are described below along with suggestions to prevent and/or cure spurious oscillations.

* Spurious oscillation can occur only under a particular set of operating conditions, such as frequency, bias, signal level, load impedance, and temperature.

* Several different oscillation mechanisms may coexist in the same circuit, either simultaneously or under different sets of operating conditions.

* It is often very difficult to construct a model of the RF power amplifier that allows an accurate (and reasonably complete) mathematical description of the oscillation mechanism.

* Different types of oscillation can lead to similar external symptoms.

* An attempt to prevent or cure one form of oscillation may generate another.

* Achieving stability may require some sacrifice of gain, selectivity, or overall efficiency (and a reasonable tradeoff may be difficult to obtain).

Types of Instability

Linear Feedback

It is well known that a sufficiently large amount of positive feedback in an amplifier may give rise to oscillations. Positive linear feedback can be caused by transistor interelectrode capacitances and lead inductances (for example, C_{CB} in a common-emitter stage or L_B in a common-base stage) or by capacitive or magnetic coupling between the input and output circuits. This latter may take place directly (for example, magnetic coupling between inductors) or through common circuits, such as DC power-supply circuits or ground return.

Linear feedback is conceptually simple and can be analyzed using the classic techniques of linear system analysis [5]. To apply such a method, it is necessary to find a complete equivalent amplifier circuit. This circuit must include the significant parasitic resistances, inductances, and capacitances associated with the components and the circuit layout. Accurate parasitic values are often difficult to estimate, especially if the magnetic couplings between inductors or impedances associated with the DC power-supply or ground circuits must be considered. Furthermore, an RF power amplifier is always a nonlinear system. Even a Class A amplifier is nonlinear when it operates at high signal levels, although the degree of nonlinearity involved is significantly less than that of a Class C amplifier. In these cases, the quasi-linear approach must be used, and the nonlinear elements of the circuit must be replaced by their linear equivalents. The main problem is the active device: The transistor parameters change during the RF cycle — they change dramatically if the transistor goes into the cutoff or the saturation region — and should be described by an averaged set of parameters. Finding reasonably accurate values of these averaged parameters is, in most cases, an extremely difficult task. This is a very serious limitation of the practical use of the classic linear feedback theory.

An oscillation caused by linear feedback may exist without a signal being applied, or it may start when the RF input signal is added. In some cases, the oscillation may continue after the signal is removed; in other cases, it may stop. An oscillation may occur only within a certain range of DC power-supply voltage and/or a certain level of RF input signal. The oscillation may be at any frequency, and its amplitude may vary from small to large values. Often, the oscillation is seen as a signal arithmetically added to the desired signal. The behavior of the RF power amplifier is determined by the variations of the transistor(s) parameters with the instantaneous voltage and current values. As a result, the conditions required to start or sustain oscillation could be fulfilled only for certain values of DC-supply voltage and RF signal level.

In most RF power amplifiers it is almost impossible to separate linear feedback from other feedback mechanisms. It is also difficult to describe it in mathematical terms and predict with accuracy the magnitude and phase of the loop gain. Consequently, it is desirable to avoid or to reduce linear feedback as much as possible. This can be done through physical construction and layout considerations [5, 7, 57]. Several practical recommendations are presented below.

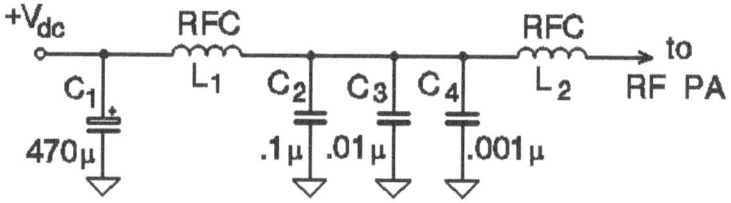

Figure 2-105 *DC power-supply decoupling.*

a. DC power-supply decoupling must be effective from very low to very high frequencies. A good approach is presented in Figure 2-105. C_1 to C_4 ensure an effective bypass over the whole frequency range. Special attention must be paid to the lead lengths of the capacitors, and the series parasitic inductances must be kept as low as possible. Ceramic chip capacitors are the best choice for C_2 to C_4. L_1 and L_2 are RF chokes. The arrangement shown in Figure 2-105 is effective for preventing RF currents from migrating along the V_{dc} power bus to other stages of the RF amplifier.
b. Avoid undesired magnetic coupling between inductors by leaving space between them, placing them perpendicular to each other, and/or using magnetic shielding.
c. Avoid undesired capacitive coupling between the input and output circuit. This can be achieved with careful circuit layout and/or use of electrostatic shielding.
d. Reduce the parasitic inductances and capacitances associated with the leads and the circuit layout as much as possible.
e. Pay attention to the ground circuit. A double-sided circuit board (with one of its sides a ground plane) is an excellent solution.
f. Use resistive damps (for example, lossy ferrite beads) to attenuate undesired high-frequency resonances.

Thermal Feedback

RF power transistors often dissipate a significant amount of power that is converted to heat. The operating temperature of the junctions is determined by a) the dissipated power, and b) the thermal resistances of the transistor, heatsink, and mounting hardware. Their thermal capacitances (together with the thermal resistances) determine the junction temperature variation with time. However, it is well known that transistor characteristics are strongly affected by temperature, and this may create the conditions for thermally induced oscillation [58]. The feedback loop (that in some cases may conduct to oscillation) is as follows: a change of the transistor voltage or current tends to modify the transistor temperature and this, in turn, affects the transistor electrical characteristics, changing the voltage across and/or the current through the transistor.

Possible ways to minimize this type of oscillation include

a. Reducing the operating junction temperature by using a more effective cooling system (i.e., reducing the junction-to-ambient thermal resistance) or reducing the amount of dissipated power.

b. Using a high-efficiency power amplifier if possible. A switching-mode power amplifier is less susceptible to the thermal induced oscillation because its performance values are less affected by transistor characteristics.

Oscillation Caused by the Transistor Bias Network

Experience shows that if the RF chokes and their damping effects are improperly chosen, self-oscillation may appear in RF power amplifiers. An explanation of this behavior is based on a nonlinear oscillation mechanism in the transistor bias network [57, 59]. Depending on the impedance seen in the collector of the RF transistor, the impedance of the base-bias circuit, and the transistor parameters (especially, the capacitances) it is possible to establish a positive feedback path in the circuit.

Suppose the collector current changes. This would affect the base current, changing its DC component (more exactly, its average value over one period of the operating frequency) [47]. Depending on the impedance of the transistor base circuit, the change in the base current DC component may affect the base bias and, eventually, the collector current. As a result, a possible positive feedback path may be established in the circuit. The oscillation appears at a frequency lower than the intended operating frequency.

To avoid this type of oscillation, it is best to provide resistive damping in the base-bias circuit (Figure 2-106). A low enough value for R_B (tens of ohms) can prevent the oscillations at any load, DC-supply voltage, or drive power conditions [59]. However, a low value for R_B means a low amplifier stage power gain, because of the power lost in this resistor; consequently, the value of R_B should ensure a reasonable compromise between stability and power gain.

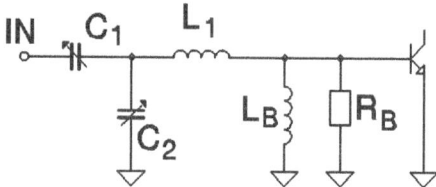

Figure 2-106 *Resistive damping in the base-bias circuit.*

Parametric Oscillation

The theory of circuits with variable parameters shows that oscillations can be generated by periodically varying the parameter (capacitance or inductance) of one of the energy-storing elements of a tuned circuit.[13] Because the capacitances of RF power transistors vary strongly with the voltage across them, their nonlinear behavior may cause parametric oscillations [57, 60-63]. The most significant capacitance change of the collector junction occurs near zero collector voltage. As a result, the circuits operating over the entire cycle at high collector-to-emitter voltages (for example, linear Class A amplifiers) are much less likely to oscillate due to this mechanism. On the contrary, if the active device reaches saturation (for example, overdriven Class C, mixed-mode, or switching mode amplifiers) the tendency toward parametric oscillation is considerably higher.

Parametric oscillation usually occurs at a subharmonic of the circuit operating frequency (f/k, where f is the operating frequency and k is an integer), although it is not necessary that the oscillation frequency is an integer submultiple of f. Most often $k = 2$ or $k = 3$; that is, the frequency of oscillation is one half or one third the operating frequency. On the oscilloscope, the parametric oscillation is seen as a sequence (repeating every k cycles) of different size and/or different shape waveforms of individual RF cycles (occurring at the operating frequency f) [60, 61, 63].

Recommendations for curing this type of oscillation include

a. Reducing the nonlinearity of the capacitance by adding a linear capacitance in parallel. This solution can be applied only at low frequencies; at high frequencies, transistor capacitances are high enough to be supplementally increased with parallel capacitances.

b. Avoiding the low voltage region of the $C = C(v)$ characteristic (where the nonlinearity is very strong). This means decreasing the collector efficiency of the RF power amplifier.

c. Adding some resistive damping in the circuit; for example, a small inductance-free series resistor or a lossy ferrite bead in the collector line. This also involves a decrease of the collector efficiency.[14]

Load Current Drawn through Base-Collector Junction in Forward Conduction

Several authors have touched on this mode of oscillation [57, 64] but have not presented a detailed explanation or a mathematical analysis. Such oscillation may occur if the base-collector junction is forced into forward conduction part of the RF cycle. The direct connection between the input and output tuned circuits may cause oscillation. Another unwanted effect of a negative collector voltage is to exceed the V_{EBX} rating of the transistor [26]. A possible solution is to connect an anti-parallel commutating diode across the transistor. Such a diode carries the load current if the collector voltage becomes negative.

Oscillation Caused by the Base Stored Charge

Oscillation caused by the base stored charge may occur in amplifiers in which the transistor saturates during the RF cycle (for instance, overdriven Class C, mixed-mode, or switching-mode amplifiers) [57]. The transistor remains saturated during the transistor storage time, which depends on the base and collector currents [65]. Suppose that the collector current during the ON time increases. This will decrease the storage time and reduce the total ON time of the transistor. Depending on the collector circuit configuration, the reduction of the transistor ON time may cause a reduction of the collector current, leading to a larger storage time, and increasing the total ON time of the transistor. This sequence suggests that there is a possibility of subharmonic oscillations. Reference [57] provides a more detailed description of this phenomenon in a Class E amplifier along with an approximate condition for oscillation.

Other Practical Considerations

A chain of RF power amplifiers is always more susceptible to spurious oscillations. This susceptibility is due to the increased complexity of the overall feedback paths (magnetic or capacitive coupling, common DC power-supply circuits or ground return and such) and to the possible interactions between different oscillation mechanisms occurring in subsequent stages [5, 57, 63].

It is possible to increase the output power in an RF power amplifier by using several power transistors connected in parallel. Unfortunately, these simple circuits are less stable than those using only one active device [66]. Possible explanations for this are an imbalance between the paralleled transistors and the increased complexity of the practical circuit with paralleled transistors, where the possible types of oscillation can occur in common mode or in various differential modes. The solution presented in Reference [66] assures better isolation between transistors, improving the circuit stability.

2.12   Thermal Calculation and Mounting Considerations

Because RF power transistors dissipate a significant amount of power, it is often necessary to provide them with heatsinks. Data sheets specify a maximum junction temperature T_{jmax} that should not be exceeded. Moreover, it is desirable to maintain as low a junction temperature as possible because many of the semiconductor failure mechanisms are temperature-dependent. A general rule is: the lower the junction temperature, the higher the amplifier reliability.

It is very difficult to develop an exact mathematical description of the heat dissipation process because a large number of parameters affecting the process that cannot be predetermined, such as the extent to which air can flow unhindered, the screening effect of nearby components and heating from these components. A detailed description of the heat transfer process is beyond the scope of this book; only basic principles and practical recommendations are presented here [4, 7, 12, 67-70].

Heat Dissipation Process

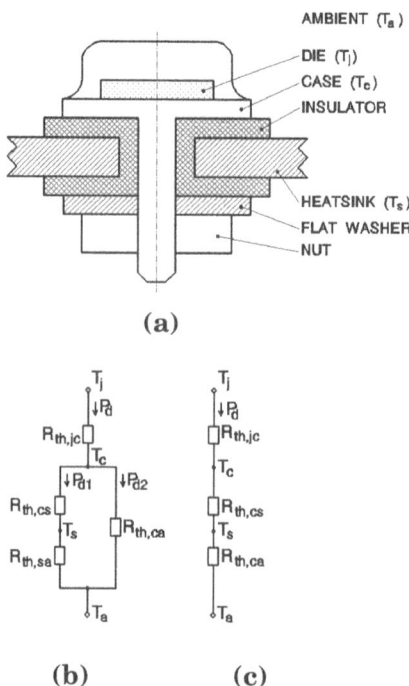

(a)

(b)   **(c)**

Figure 2-107 *Thermal resistance model for a power transistor mounted on a heatsink.*

The heat generated in the power transistor chip flows from the junction (with a temperature T_j) to the mounting base, which is in close contact with the case (T_c) [see Figure 2-107 (a)]. Via the contact pressure, the heat flows from the transistor case to the heatsink (T_s), and the heatsink loses heat to its surroundings (T_a) by conduction, convection, and radiation. Some heat, usually a negligible amount, flows directly from the transistor case to the ambient air. Generally, natural air-cooling is used so the ambient air referred to in Figure 2-107 is the surrounding air. The heat mainly flows from the heatsink to the ambient air by natural convection and radiation. In some applications, forced air cooling (by fans) or forced water cooling may be provided.

Table 2-3 *Equivalent thermal and electrical parameters.*

Thermal parameter	Electrical parameter
Temperature T [°C]	Potential U [V]
Temperature difference ΔT [°C] $\Delta T = T_1 - T_2$	Voltage V [V] $V = U_1 - U_2$
Dissipated power P_d [W]	Current I [A]
Thermal resistance R_{th} [°C/W] $R_{th} = \Delta T/P_d$	Resistance R [Ω] $R = V/I$
Quantity of heat Q [J] $Q = P_d t$ (t = time)	Charge q [C] $Q = I \cdot t$ (t = time)
Heat capacity C_h [J/°C] $C_h = Q/\Delta T$	Capacity C [F] $C = q/V$
Thermal conductivity K [W/(m°C)]	Conductivity σ [S/m]
Thermal time constant $C_h R_{th}$ [sec]	Time constant RC [sec]

From a practical point of view, the concept and rules for heat flow can be better understood using the analogy of thermal and electrical parameters (see Table 2-3). This analogy allows description and analysis of a heat transfer process using a thermal model similar to an electrical circuit. Rules similar to the ones known from circuit theory can be applied to the thermal model.

Figure 2-107 (b) presents the thermal resistance model for a power transistor mounted on a heatsink. The thermal resistances involved in the heat flow process are

- $R_{th,jc}$ = thermal resistance between junction and case
- $R_{th,cs}$ = thermal resistance between case and heatsink
- $R_{th,sa}$ = thermal resistance between heatsink and ambient
- $R_{th,ca}$ = direct thermal resistance between case and ambient

For most applications, the direct heat loss from the transistor to the surroundings through $R_{th,ca}$ is negligible, i.e.,

$$R_{th,ca} \gg R_{th,cs} + R_{th,sa} \qquad\qquad 2.138$$

(A notable exception to this rule applies to small heatsinks in cases where a considerable part of the heat is dissipated directly from the transistor case to the ambient.)

Consequently, it is possible to obtain the simplified equivalent circuit given in Figure 2-107(c). The total thermal resistance between junction and ambient is given by

$$R_{th,ja} = R_{th,jc} + R_{th,cs} + R_{th,sa} \qquad\qquad 2.139$$

The junction-to-case thermal resistance is determined by the path between junction and mounting base; therefore, it depends on package materials and physical layout. To obtain a low junction-to-case thermal resistance, the package body of today's RF power transistors is a beryllium oxide (BeO) disc with high-thermal conductivity. Attached to the bottom of the BeO disc is a copper stud used for heat transfer and mechanical mounting. The junction-to-case thermal resistance can be obtained from the data sheets for any given device, because its value is fixed by the transistor type used.

The case-to-heatsink thermal resistance depends on the quality and size of the contact areas, the type of any intermediate plates used (for example, insulators), and the contact pressure. Care must be taken to ensure that the heatsink mounting surface is flat and fine finished, the mounting hole(s) are burr-free and perpendicular to the mounting surface, and the nuts are properly torqued. A significant improvement in thermal transfer is achieved using a silicone- (or another synthetic oil) based heatsink compound. The quantity of thermal compound used must be kept to the minimum required to fill in any air gaps between the transistor mounting surface and the heatsink surface. The thermal conductivity of silicone grease is significantly higher than that of air. This explains why the case-to-heatsink thermal resistance can be lowered several times using a thermal grease [70]. The rougher the surfaces, the more valuable the silicone grease becomes in lowering the thermal resistance. When an insulating washer or bushing is used, thermal grease is of great importance because the number of interfaces in the heat transfer flow is increased, and some insulating materials (for example, mica) have a markedly uneven surface. The case-to-heatsink thermal resistance is typically a few tenths of a degree C/W. There is no simple rule to calculate its value, but extensive measurements [70] and practical experience allow accurate estimation of this thermal resistance.

The heatsink-to-ambient thermal resistance depends on factors, such as material, size, heatsink finish, construction, and position. For example, a vertically mounted mat-black heatsink is more efficient than a horizontally mounted bright heatsink due to improved convection and radiation conditions. The material normally used for heatsink construction is aluminum and, in most cases, ready-made extruded finned heatsinks are preferred. They are more economical to use and have well-known thermal characteristics (experimentally measured).

Heatsink Design

The following steps are required to design a heatsink:

1. Estimate a maximum value (worst case) for the ambient temperature $T_{a,max}$ (considering the characteristic working conditions of the circuit).

2. Adopt a suitable value for the junction temperature, T_j. T_j must be less than $T_{j,max}$. T_j could be lowered, taking into account reliability considerations.

3. Calculate/estimate the maximum dissipated power, P_d. (This value could vary with frequency, voltage or signal level. The worst case scenario should be considered.)

4. Calculate the required value for the junction-to-ambient thermal resistance.

$$R_{th,ja\,(req)} = \frac{T_j - T_{a,max}}{P_d}$$ 2.140

5. Obtain the value of $R_{th,jc}$ from the data sheets.

6. Estimate a value for $R_{th,cs}$ (considering the concrete mounting details).

7. Calculate the required value for the heatsink-to-ambient thermal resistance

$$R_{th,sa\,(req)} = R_{th,ja\,(req)} - R_{th,jc} - R_{th,cs}$$ 2.141

8. Choose an appropriately sized heatsink using heatsink nomograms provided by manufacturers.

Note that

1. If the value of $R_{th,sa(req)}$ given by Equation 2.141 is negative, choose another

transistor with lower $R_{th,jc}$ and/or reduce $R_{th,cs}$ as much as possible.

2. If the value of $R_{th,sa(req)}$ is very low, forced air cooling may be required.

3. The nomograms do not provide exact design values because the practical conditions always deviate to some extent from those under which the nomograms were obtained. Thus, check the design by measuring the important temperatures in the finished equipment under the worst possible operating conditions. This is especially important in light of today's trend toward higher frequencies, higher power devices, and equipment miniaturization.

Mounting Considerations

Most of the mounting considerations for RF power transistors are designed to prevent tension at the metal-ceramic package interfaces because the application of such stresses can result in device destruction. Faulty mounting can not only cause mechanical damage to the package, but can also lead to a much higher junction temperature than expected and as well as to low reliability. In fact, proper mounting minimizes the temperature difference between the semiconductor case and the heatsink.

Many practical details and recommendations for mounting RF power transistors are discussed in [67, 69, 70]. Several basic considerations are given below.

- The transistor should not be mounted in such a way as to place ceramic to metal joints in tension, apply force on the strip leads in a vertical direction towards the cap, or apply shear forces to the leads.

- The leads must not be used to prevent device rotation during stud torque application.

- The device leads must be soldered into the circuit only after the device is properly secured into the heatsink.

- The recommended maximum torque must not be exceeded. If repeated assembly/disassembly operation is expected, a lesser torque must be used.

2.13 Notes

1. If a conventional (wire-wound) transformer is used, the m/n turns ratio can have almost any desired value. However, in practice, wideband transmission line transformers must be used rather than wire-wound transformers (see Section 2.7). As a result, the m/n turns ratio is the ratio of two small integer numbers (2, 3, 4,. . . , 1/2, 1/3, 1/4, . . .).

2. The single-ended Class B amplifier can be considered a particular example of the Class C amplifier. The analysis presented in this section is valid for any conduction angle $2\theta_c$.

3. For instance, a parallel-tuned circuit with $Q = 5$ used in a 100 MHz PA operating in a 50-ohm load needs $L \approx 16$ nH and $C \approx 159$ pF. Suppose that this amplifier must provide $P_0 = 10$ W at $V_{dc} = 12$ V. The equivalent load resistance must be $R = 7.2$ W, and if a pi matching network with $Q = 3$ is used to provide the impedance transformation, then $L \approx 12.9$ nH, $C_1 \approx 663$ pF, and $C_2 \approx 263$ pF (see Section 2.6, Figure 2-56).

4. A series-tuned circuit with $Q = 5$ used in a 100 MHz PA operating in a 50-ohm load needs L ≈ 0.4 mH and C ≈ 6.4 pF. If a T network with $Q = 3$ is used to transform the 50-ohm load into 7.2 ohms, then $L \approx 34$ nH, $C_1 \approx 52$ pF, and $C_2 \approx 48$ pF (see Section 2.5, Figure 2-59).

5. A more detailed example and comparison data are provided in Wood and Davidson [31], for a Motorola MRF873 BJT.

6. The design of the test fixture and the performance measurement is a complicated process. It is described in more detail in Wood and Davidson [31].

7. The large-signal impedances given in the data sheets are usually determined for maximum power gain at a given output power. However, there is no standard procedure, and the published impedance data can be determined for any other optimum operating conditions (for example, maximum output power or efficiency), as specified by the manufacturer.

8. If the source has an internal impedance $Z_s = R + jX$, the maximum power is delivered to a complex conjugate load $Z_s^* = R - jX$. This maximum power is called *available power of the source.*

9. For a microstrip, the highest value of Z_0 is limited by how narrow the strip can be made and by how high its series resistance can become and still be considered a lossless line. The lowest value of Z_0 is limited by the space available for a wide

strip, and the transverse resonance that excites high-order propagation modes when the strip width approaches $\lambda/4$.

10. Additional details and more practical information regarding the design procedure of transmission-line transformers can be found in [9, 37, 39, 44, 50, and 52].

11. Similar effects also occur in circuits using FETs due to the C_{RSS} capacitance.

12. A similar phase modulation effect can be caused by the voltage-variable collector capacitance.

13. Note that a parametric oscillator is a circuit presenting a unique feature among standard types of oscillators. It needs a pumping oscillator that is the primary energy source in the circuit. In a standard oscillator (either self-oscillating or externally excited), the DC energy is converted into AC energy.

14. Note that active devices with a low resistance in the ON state (for example, switching transistors) are more sensitive to parametric oscillations than those having a higher resistance (for example, RF power transistors).

2.14 References

1. Terman, F. E. *Radio Engineering.* New York: McGraw-Hill, 1947.
2. Clarke, K. K. and D. T. Hess. *Communication Circuits: Analysis and Design,* Boston: Addison-Wesley, 1971.
3. Norris, B. *MOS and Special-Purpose Bipolar Integrated Circuits & RF Power Transistor Circuit Design.* New York: McGraw-Hill, 1976.
4. Krauss, H. L., C. V. Bostian, and F. H. Raab. *Solid-State Radio Engineering.* New York: John Wiley & Sons, 1980.
5. Shakhgildyan, V. V. *Radio Transmitter.* Moscow: Mir, 1981.
6. Bowick, C. *RF Circuit Design.* Indianapolis: Howard W. Sams, 1982.
7. DeMaw, D. *Practical RF Design Manual.* Upper Saddle River, NJ: Prentice-Hall, 1982.
8. Smith, J. *Modern Communication Circuits.* New York: McGraw-Hill, 1986.
9. Shakhgildyan, V. V. *Radio Transmitter Design,* Moscow: Mir, 1987.
10. Sokal, N. O., I. Novak, and J. Donohue. "Classes of RF Power Amplifiers A Through S, How They Operate, and When to Use Them." *Proceedings of RF Expo West 1995,* (San Diego, CA) (1995): 131–138.
11. MOTOROLA RF DEVICE DATA, DL110/D, Rev 8, Motorola Inc., Phoenix, AZ, 1997.
12. Ostroff, E. D., M. Borkowski, H. Thomas, and J. Curtis. *Solid-State Radar Transmitters.* Boston: Artech House, 1985.
13. Furlan, J. "Preliminary Analysis of the Transistor Tuned Power Amplifier." *Proceedings of the IEEE,* vol. 52 (1964): 311.
14. Scott, T. M. "Tuned Power Amplifiers." *IEEE Transactions Circuit Theory,* vol. CT-11 (September 1964): 385–389.
15. Slatter, J. A. G. "An Approach to the Design of Transistor-Tuned Power Amplifiers." *IEEE Transactions Circuit Theory,* vol. CT-12 (June 1965): 206–211.
16. Harrison, R. G. "A Nonlinear Theory of Class C Transistor Amplifiers and Frequency Multipliers." *IEEE Journal Solid-State Circuits,* vol. SC-2 (September 1967): 93–102.
17. Snider, D. M. "A Theoretical Analysis and Experimental Confirmation of the Optimally Loaded and Overdriven RF Power Amplifier." *IEEE Transactions Electron Devices,* vol. ED-14 (December 1967): 851–857.
18. Rose, B. E. "Notes on Class-D Transistor Amplifier." *IEEE Journal Solid-State Circuits,* vol. SC-4 (June 1969): 178–179.
19. Bailey, R. L. "Large-Signal Nonlinear Analysis of High-Power High-Frequency Junction Transistor." *IEEE Transactions Electron Devices,* vol. ED-17 (February 1970): 108–119.
20. El-Said, M. A. H. "Analysis of Tuned Junction-Transistor Circuits Under Large Sinusoidal Voltages in the Normal Domain." *IEEE*

Transactions Circuit Theory, vol. CT-17 (February 1970): 8–18.

21. Peden, R. D. "Charge-Driven HF Transistor-Tuned Power Amplifier." *IEEE Journal Solid-State Circuits*, vol. SC-5 (April 1970): 55–63.

22. Johnston R. H. and A. R. Boothroyd. "High-Frequency Transistor Frequency Multipliers and Power-Amplifiers." *IEEE Journal Solid-State Circuits*, vol. SC-7 (February 1972) 81–89.

23. Vidkjaer, J. "A Computerized Study of the Class-C - Biased RF - Power Amplifier." *IEEE Journal Solid-State Circuits*, vol. SC-13 (April 1978): 247–258.

24. Vidkjaer, J. "A Described Function Approach to Bipolar RF-Power Amplifier Simulation." *IEEE Transactions Circuits Systems*, vol. CAS-28 (August 1981): 758–767.

25. Krauss, H. L. and J. F. Stanford. "Collector Modulation of Transistor Amplifiers." *IEEE Transactions Circuit Theory*, vol. CT-12 (September 1965): 426–428.

26. Sokal, N. O. "Unseen Emitter-Base Breakdown in RF Power Amplifiers — A Possible Hazard." *IEEE Journal Solid-State Circuits*, vol. SC-12 (June 1977): 319–321.

27. Horwitz, J. H. "Design Wideband UHF Power Amplifiers with These Techniques for Broadband Matching, Gain Compensation and Parasitic Inductance Reduction." *Electronic Design*, no. 11 (1969).

28. Granberg, H. O. "Biasing Solid State Amplifiers to Linear Operation." *Motorola Semiconductor Products, Inc.*, Phoenix, AZ, Article Reprint AR511/D.

29. Hejhall, R. "Systemizing RF Power Amplifier Design," *Motorola Semiconductor Products, Inc.*, Phoenix, AZ, Application Note AN282A.

30. Houselander, L. S. et al. "Transistor Characterization by Effective Large-Signal Two-Port Parameters." *IEEE Journal Solid-State Circuits*, vol. SC-5 (April 1970): 77–79.

31. Wood, A. and B. Davidson. "RF Power Device Impedances: Practical Considerations." *Motorola Semiconductor Products, Inc.*, Phoenix, AZ, Application Note AN1526.

32. Becciolini, B. "Impedance Matching Networks Applied to RF Power Transistors." *Motorola Semiconductor Products, Inc.*, Phoenix, AZ, Application Note AN-721.

33. Gilbert, E. N. "Impedance Matching with Lossy Components." *IEEE Transactions Circuits Systems*, vol. CAS-22 (February 1975): 96–100.

34. Albulet, M. "A New CAD Method for Narrowband Matching Circuits." *RF Design* 17, no. 5 (1994): 52–57.

35. Sun, Y. and J. K. Fidler, "Design of P Impedance Matching Networks." *IEEE International Symposium Circuits Systems*, (London, UK), vol. 5 (1994): 5–8.

36. Davis, F. "Matching Network Designs with Computer Solutions," *Motorola Semiconductor Products, Inc.*, Phoenix, AZ, Application Note AN-267.

37. Grossner, N. R. and I. S. Grossner. *Transformers for Electronic Circuits*. New York: McGraw-Hill, 1983.

38. Hilbers, A. H. "Design of HF

Wideband Power Transformers (Part II)." *Philips Semiconductors,* Application Report ECO 7213.

39. Sevick, J. *Transmission Line Transformers.* Newnington: ARRL, 1990.

40. Guanella, G. "Nouveau Transformateur d'Adaptation pour Haute Frequence." *Brown Boveri Review,* vol. 31 (September 1994): 327–329.

41. Talkin, A. I. and J. V. Cuneo. "Wide-Band Balun Transformer." *Review of Science Institute,* vol. 28 (October 1957): 808–815.

42. Ruthroff, C. L. "Some Broad-Band Transformers." *Proceedings of IRE,* vol. 47 (August 1959): 1337–1342.

43. Pitzalis, O. and R. A. Gilson. "Tables of Impedance Matching Networks which Approximate Prescribed Attenuation Versus Frequency Slopes." *IEEE Transactions MTT,* vol. MTT-19 (April 1971).

44. Krauss, H. L. and C. W. Allen. "Designing Toroidal Transformers to Optimize Wideband Performance." *Electronics* (August 1973): 113–116.

45. Blocker, W. "Extended Impedance Matching Capabilities in the Wide-Band Transmission Line Transformer." *Proceedings of the IEEE,* vol. 65 (September 1977): 1405–1406.

46. Blocker, W. "The Behavior of the Wide-Band Transmission-Line Transformer for Nonoptimum Line Impedance." *Proceedings of the IEEE,* vol. 66 (April 1978): 518–519.

47. Pitzalis, O. and T. P. M. Couse. "Practical Design Information for Broadband Transmission Line Transformers." *Proceedings of the IEEE,* vol. 67 (April 1979): 738–739.

48. Rotholz, E. "Transmission-Line Transformers." *IEEE Transactions MTT,* vol. MTT-29, (April 1981): 327–331.

49. Radu, S. and M. Albulet, "A SPICE Model of Transmission Line Transformers (TLT)." *Proceedings of the 4th Biennial International Baltic Electronics Conference,* (Tallinn, Estonia) vol. II (1994): 487–498.

50. Granberg, H. O. "Broadband Transformers and Power Combining Techniques for RF." *Motorola Semiconductor Products, Inc.,* Phoenix, AZ, Application Note AN749.

51. Hilbers, A. H. "HF Wideband Power Transformers." *Philips Semiconductors,* Application Report ECO 6907.

52. Hilbers, A. H. "Power Transformers for the Frequency Range of 30–80 MHz." *Philips Semiconductors,* Application Report ECO 7703.

53. Granberg, H. O. "Broadband Linear Power Amplifiers Using Push-Pull Transistors." *Motorola Semiconductor Products, Inc.,* Phoenix, AZ, Application Note AN593.

54. Granberg, H. O. "A Two-Stage 1 kW Solid-State Linear Amplifier." *Motorola Semiconductor Products, Inc.,* Phoenix, AZ, Application Note AN758.

55. Koppen, M. J. "A Single Stage Wideband (1.6–28MHz) Linear Power Amplifier for 300 Watts PEP Using 2xBLX15." *Philips Semiconductors,* Application Report ECO 7308.

56. Hollander, D. "A 15-Watt AM Aircraft Transmitter Power Amplifier Using Low Cost Plastic Transistors." *Motorola Semiconductor Products, Inc.,* Phoenix, AZ, Application Note AN793.

57. Sokal, N. O. "Parasitic Oscillation in Solid-State RF Power Amplifiers — Causes and Cures Demystified." *RF Design* 3, no. 9 (1980): 32–36.

58. Muller, O. "Thermal Feedback in Power Semiconductor Devices," *IEEE Transactions Electron Devices*, vol. ED-17 (September 1970): 770–782.

59. Vidkjaer, J. "Instabilities in RF-Power Amplifiers Caused by a Self-Oscillation in the Transistor Bias Network." *IEEE Journal Solid-State Circuits*, vol. SC-11 (October 1976): 703–712.

60. Lohrmann, D. R. "Parametric Oscillations in VHF Transistor Power Amplifier." *Proceedings of the IEEE*, vol. 54 (March 1966): 409–410.

61. Lohrmann, D. R. "Amplifier Has 85% Efficiency While Providing up to 10 Watts Power Over a Wide Frequency Band." *Electronic Design*, vol. 14 (1966): 38–43.

62. Muller, O. and W. G. Figel. "Stability Problems in Transistor Power Amplifiers." *Proceedings of the IEEE*, vol. 55 (1967): 1458–1466.

63. Lohrmann, D. R. "Exotic Effects?" *Proceedings of the IEEE*, vol. 56 (March 1968): 332–333.

64. Sokal, N. O. and A. D. Sokal. "Class E - A New Class of High-Efficiency Tuned Single-Ended Switching Power Amplifiers." *IEEE Journal Solid-State Circuits*, vol. SC-10 (June 1975): 168–176.

65. Sokal, N. O. "RF Power Transistor Storage Time: Theory and Measurements." *IEEE Journal Solid-State Circuits*, vol. SC-11 (April 1976): 344–346.

66. Kazimierczuk, M. and N. O. Sokal. "Cause of Instability of Power Amplifier with Parallel-Connected Power Transistors." *IEEE Journal Solid-State Circuits*, vol. SC-19 (August 1984): 541–542.

67. Danley, L. "Mounting Stripline-Opposed-Emitter (SOE) Transistors." *Motorola Semiconductor Products, Inc.*, Phoenix, AZ, Application Note AN555.

68. Johnson, R. J. "Thermal Rating of RF Power Transistors." *Motorola Semiconductor Products, Inc.*, Phoenix, AZ Application Note AN790.

69. Swanson, H. "Mounting Techniques for PowerMacro Transistors," *Motorola Semiconductor Products, Inc.*, Phoenix, AZ, Application Note AN938.

70. Roehr, B. "Mounting Considerations for Power Semiconductors," *Motorola Semiconductor Products, Inc.*, Phoenix, AZ, Application Note AN1040.

$$3$$

Class D RF Power Amplifiers

A Class D amplifier is a switching-mode amplifier that uses two active devices driven in a way that they are alternately switched ON and OFF. The active devices form a two-pole switch that defines either a rectangular voltage or rectangular current waveform at the input of a tuned circuit that includes the load. The load circuit contains a band- or low-pass filter that removes the harmonics of the rectangular waveform and results in a sinusoidal output.

In its simpler form, the load circuit can be a series or parallel resonant circuit tuned to the switching frequency; such a load circuit will be considered here. In practical applications, this circuit can be replaced by narrowband pi- or T-matching circuits, or by band- or low-pass filters (in wideband amplifiers).

3.1   Idealized Operation of the Class D Amplifier

Complementary Voltage Switching (CVS) Circuit

The CVS circuit is presented in Figure 3-1(a) [1-11]. Input transformer T_1 applies the drive signal to the bases of Q_1 and Q_2 in opposite polarities. If the drive is sufficient for the transistors to act as switches, Q_1 and Q_2 switch alternately between cut-off (OFF state) and saturation (ON state). The transistor pair forms a two-pole switch that connects the series-tuned circuit alternately to ground and V_{dc}.

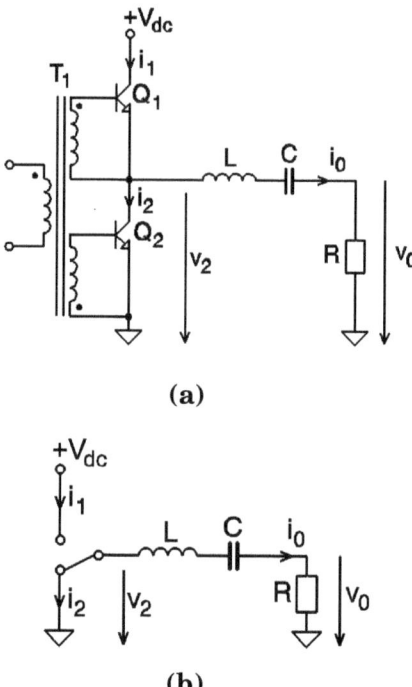

(a)

(b)

Figure 3-1 CVS Class D circuit; (a) basic circuit, (b) equivalent idealized circuit.

The analysis below is based on the following assumptions:

1. The series resonant circuit, tuned to the switching frequency, f, is ideal, resulting in a sinusoidal load current. The CVS circuit requires a series-tuned circuit or an equivalent (that imposes a sinusoidal current), such as a T-network. A parallel-tuned circuit (or an equivalent, such as a pi-network) cannot be used in the CVS circuit because it imposes a sinusoidal voltage and the two-pole switch imposes a rectangular voltage waveform.

2. The active devices act as ideal switches: zero saturation voltage, zero saturation resistance, and infinite OFF resistance. The switching action is instantaneous and lossless.

3. The active devices have null output capacitance.

4. All components are ideal. (The possible parasitic resistances of L and C can be included in the load resistance R; the possible parasitic reactance of the load can be included in either L or C.)

Based on the assumptions above, the equivalent circuit of Figure 3-1(b) is obtained. Assuming a 50 percent duty cycle (that is, 180 degrees of saturation and 180 degrees of cut off for each transistor), voltage $v_2(\theta)$ applied to the output circuit is a periodical square wave (see Figure 3-2).

$$v_2(\theta) = \begin{cases} V_{dc}, & 0 \leq \theta \leq \pi \\ 0, & \pi \leq \theta \leq 2\pi \end{cases} \qquad 3.1$$

where $\theta = \omega t = 2\pi f t$.

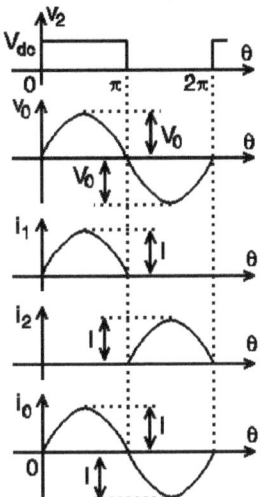

Figure 3-2 *Waveforms in CVS Class D circuit.*

Decomposing $v_2(\theta)$ into a Fourier series yields

$$v_2(\theta) = V_{dc}\left(\frac{1}{2} + \frac{2}{\pi}\sum_{n=1}^{\infty}\frac{\sin(2n-1)\theta}{2n-1}\right) \qquad 3.2$$

Because the series-tuned circuit is ideal, the output current is sinusoidal

$$i_0(\theta) = I\sin\theta = \frac{2}{\pi}\frac{V_{dc}}{R}\sin\theta \qquad 3.3$$

giving a sinusoidal output voltage

$$v_0(\theta) = V_0 \sin\theta = \frac{2}{\pi} V_{dc} \sin\theta \qquad 3.4$$

At one moment, the sinusoidal output current flows through either Q_1 or Q_2, depending on which device is ON. As a result, collector currents $i_1(\theta)$ and $i_2(\theta)$ are half sinusoid with the amplitude

$$I = \frac{2}{\pi} \frac{V_{dc}}{R} \qquad 3.5$$

The output power (dissipated in the load resistance R) is given by

$$P_0 = \frac{I^2}{2} R = \frac{2}{\pi^2} \frac{V_{dc}^2}{R} \approx 0.2026 \frac{V_{dc}^2}{R} \qquad 3.6$$

The DC input current is the average value of $i_1(\theta)$ (see Figure 3-2).

$$I_{dc} = \overline{i_1(\theta)} = \frac{1}{2\pi} \int_0^{2\pi} i_1(\theta)\,d\theta = \frac{I}{\pi} = \frac{2}{\pi^2} \frac{V_{dc}}{R} \qquad 3.7$$

The current drawn from the DC power supply takes the form of a half-sinusoid pulse train; therefore, a local bypass capacitor is required. In practice, the use of an additional filter in the power supply line is recommended (see Figure 3-3). C_1 and C_2 are often 1 µF + 0.1 µF + 0.01 µF (+470 pF...1 nF) parallel combinations and provide efficient bypassing for a wide range of frequencies. C_2 must be able to store enough energy to supply the required current pulses without a significant voltage drop in the Q_1 collector.

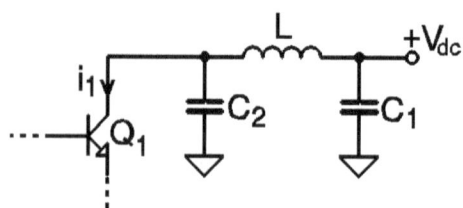

Figure 3-3 *DC supply filtering in CVS circuits.*

The DC input power is given by

$$P_{dc} = V_{dc} I_{dc} = \frac{2}{\pi^2} \frac{V_{dc}^2}{R} = P_0 \qquad 3.8$$

and the collector efficiency (for the idealized operation) is 100 percent.

$$\eta = \frac{P_0}{P_{dc}} = 1 \qquad\qquad 3.9$$

The power output capability is obtained by normalizing the output power (P_0) by the number of active devices (two), the peak collector voltage (V_{dc}), and the peak collector current (I)

$$C_P = \frac{P_0}{2V_{dc}I} = \frac{1}{2\pi} \approx 0.1592 \qquad\qquad 3.10$$

EXAMPLE 3.1

A CVS circuit delivers an output power of $P_0 = 100$ W in an $R = 10$ Ω load. To obtain this output power, the required DC supply voltage is taken from Equation 3.6, or $V_{dc} = 70.25$ V. The peak collector current is determined from Equation 3.5, $I = 4.47$ A, and the DC input current is given by Equation 3.7, $I_{dc} = 1.42$ A. The required device ratings are V_{dc} and I.

The circuit of Figure 3-1 is called a *quasi-complementary circuit* [8, 10] because it uses two identical transistors (npn BJTs or n-channel MOSFETs). A true-complementary configuration (as used in audio-frequency amplifiers) requires npn and pnp BJTs, or n- and p-channel MOSFETs.

Examples of several true-complementary Class D circuits are shown in Figure 3-4 [7, 8, 12, 13]. Those of Figures 3-4 (c-e) are particularly interesting because they do not need transformers. The circuit of Figure 3-4 (c) requires two DC supply voltages; however, the transistors must not be mounted on the same heatsink without electrical insulation. The circuits of Figures 3-4(d) and (e) use only one DC supply voltage and both transistors may be mounted on a single heatsink (without electrical insulation). Both transistors are in common-emitter configuration and provide high-power gain. In contrast to the CVS circuit of Figure 3-1, the true-complementary circuit of Figure 3-4 (d) allows biasing of the MOSFET gates, reducing drive power requirements. The true-complementary configurations cannot be used in practical circuits because there are no pnp BJTs or p-channel MOSFETs (intended for power switching or RF applications) that are electrically complementary in all respects to their npn or n-channel counterparts [8, 10, 14]. The pnp BJTs and the p-channel MOSFETs for these applications are more expensive and have significantly reduced performance. Experiments with approximate complementary pairs of power switching MOSFETs showed that they cannot be used at frequencies higher than several megahertz [10]. In this context, further development of truly complementary pairs of MOSFETs or BJTs would offer new possibilities for the Class D circuit, including the ability to integrate the amplifier and its drive circuit in a monolithic or hybrid structure.

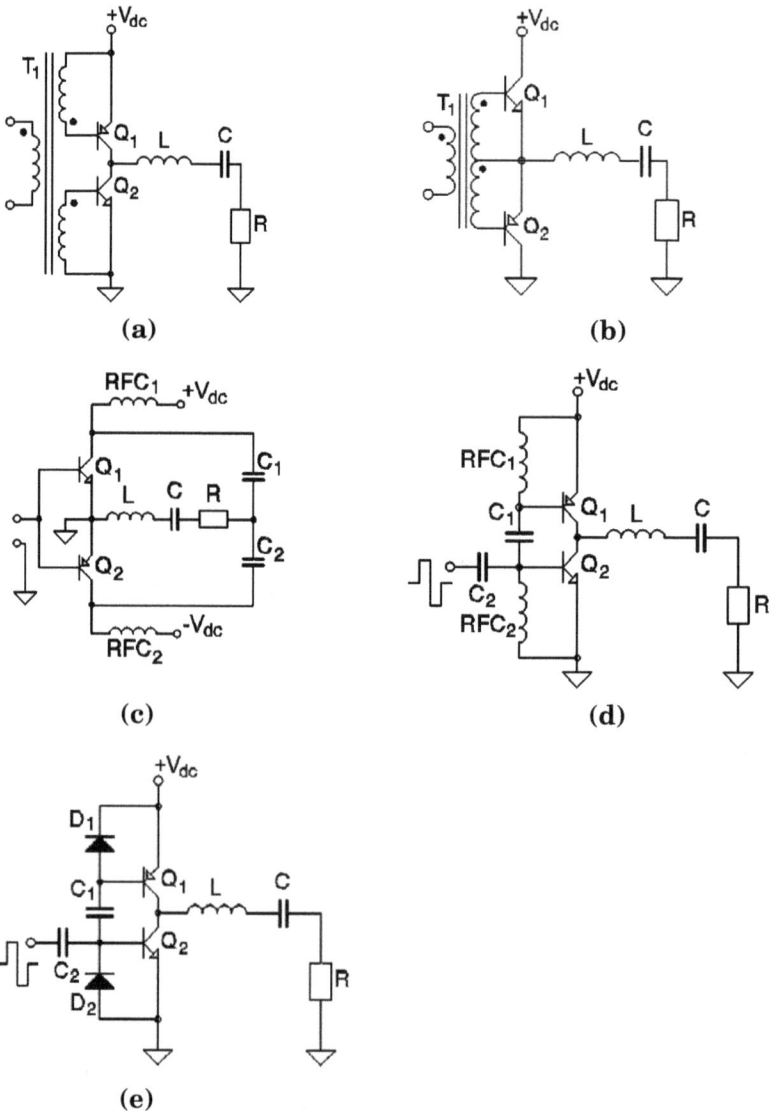

Figure 3-4 *True-complementary Class D circuits.*

Transformer-Coupled Voltage Switching (TCVS) Circuit

The TCVS circuit is presented in Figure 3-5 [2, 5, 9, 11, 15, 16, 17].

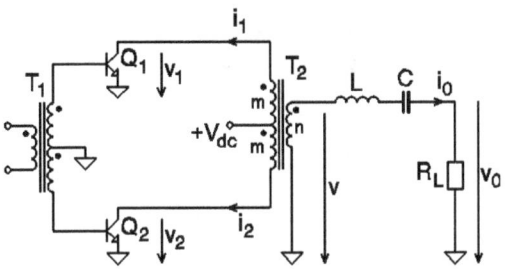

Figure 3-5 *TCVS Class D circuit.*

As in the CVS circuit, input transformer T_1 allows Q_1 and Q_2 to switch ON and OFF. The circuit analysis is based on the simplifier assumptions provided in section ***Complementary Voltage Switching (CVS) Circuit***; a 50 percent duty cycle is assumed. Output transformer T_2 is ideal, having m turns in each half of the primary winding and n turns in the secondary winding. For $0 \leq \theta \leq \pi$, $v_2(\theta) = 0$ if Q_2 is ON and Q_1 is OFF. In this case, the voltage across the lower half of the primary winding is V_{dc} (see Figure 3-6). Because the transformer is ideal, the voltage across the upper half of the primary winding is also V_{dc}. Consequently, for $0 \leq \theta \leq \pi$, $v_1(\theta) = 2\,V_{dc}$. For $\pi \leq \theta \leq 2\pi$, Q_1 is ON, Q_2 is OFF, thus $v_1(\theta) = 0$ and $v_2(\theta) = 2\,V_{dc}$. The voltage across each half of the primary winding is V_{dc}, but it travels in the opposite direction of that shown in Figure 3-6. Therefore, the voltage across each half of the primary winding is a square wave with levels of $\pm V_{dc}$. The voltage across the secondary winding is also a square wave with levels of $\pm(n/m)V_{dc}$.

$$v(\theta) = \begin{cases} \dfrac{n}{m}V_{dc}\,, & 0 \leq \theta \leq \pi \\[2mm] -\dfrac{n}{m}V_{dc}\,, & \pi \leq \theta \leq 2\pi \end{cases} \qquad 3.11$$

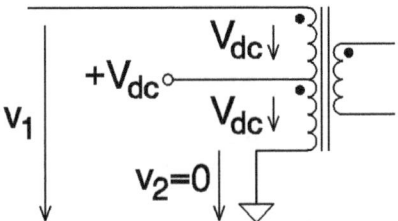

Figure 3-6 *Voltages across the primary winding of T_2 ($0 \leq \theta \leq \pi$).*

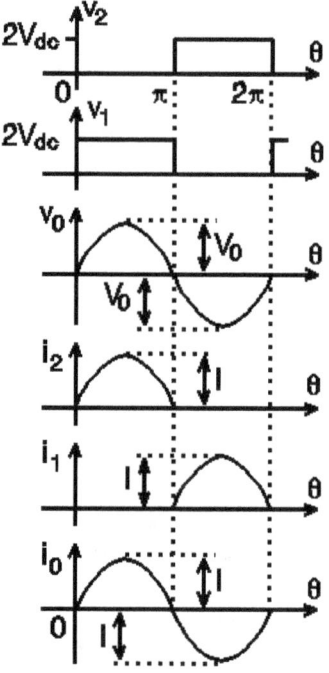

Figure 3-7 *Waveforms in TCVS Class D circuit.*

The output voltage (see Figure 3-7) is the fundamental frequency component of $v(\theta)$, with an amplitude

$$V_0 = \frac{4}{\pi}\frac{n}{m}V_{dc} \qquad\qquad 3.12$$

Thus, the output power is given by

$$P_0 = \frac{V_0^2}{2R_L} = \frac{8}{\pi^2}\frac{V_{dc}^2}{\left(\dfrac{m}{n}\right)^2 R_L} = \frac{8}{\pi^2}\frac{V_{dc}^2}{R} \approx 0.8106\frac{V_{dc}^2}{R} \qquad\qquad 3.13$$

where

$$R = \left(\frac{m}{n}\right)^2 R_L \qquad\qquad 3.14$$

is the resistance seen across half of the primary winding of T_2 with the other half of the primary winding open (in other words, the equivalent resistance presented to each transistor). For the same DC supply voltage V_{dc} and the same equivalent load resistance, R, the output power obtained in the TCVS circuit is four times that obtained in the CVS circuit. The collector currents of Q_1 and Q_2 are half sinusoid with the amplitude

$$I = \frac{4}{\pi} \frac{V_{dc}}{R}$$ 3.15

double in comparison with the case of the CVS circuit.

The DC input current may be calculated as the average value of the current drawn into the center tap of the output transformer T_2.

$$I_{dc} = \overline{i_1(\theta) + i_2(\theta)} = \frac{1}{2\pi} \int_0^{2\pi} [i_1(\theta) + i_2(\theta)] \, d\theta = \frac{2}{\pi} I = \frac{8}{\pi^2} \frac{V_{dc}}{R}$$ 3.16

Note that, as in the case of the CVS circuit, a local bypass capacitor is required. The solution suggested in Figure 3-3 can also be applied to the TCVS circuit.

With I_{dc} given by Equation 3.16, the DC input power is

$$P_{dc} = V_{dc} I_{dc} = \frac{8}{\pi^2} \frac{V_{dc}^2}{R} = P_0$$ 3.17

and the collector efficiency (for the idealized operation) is 100 percent. The power output capability is the same as in the previous case.

$$C_P = \frac{P_0}{2(2V_{dc})I} = \frac{1}{2\pi} \approx 0.1592$$ 3.18

EXAMPLE 3.2

A TCVS circuit using MOSFETs delivers an output power of $P_0 = 100$ W in an $R_L = 50\ \Omega$ load. If the output transformer were to have $m = n$, then $V_{dc} = 78.54$ V and $I = 2$ A would result from Equations 3.13 and 3.15. As a result, the transistor ratings would be $2\,V_{dc} = 157$ V and $I = 2$ A, values not convenient in practice. If 100-volt MOSFETs are used, then V_{dc} must be less than 50 volts, and from Equation 3.13, $R \leq 20.26$ W. These results suggest[1] the use of an output transformer with $m/n = 1/2$, so $R = 12.5\ \Omega$. This yields $V_{dc} = 39.27$ V, $I = 4$ A, and $I_{dc} = 2.55$ A. The required devices ratings are about 80 volts and 4 amps.

Transformer-Coupled Current Switching (TCCS) Circuit

The TCCS circuit appears in Figure 3-8 [2, 5, 7-9, 11]. As in the previous configurations, input transformer T_1 determines that Q_1 and Q_2 will switch alternately ON and OFF (a 50 percent duty cycle is assumed).

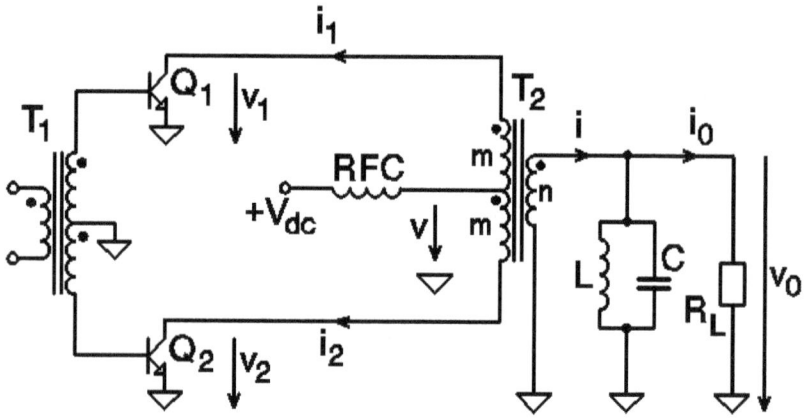

Figure 3-8 *TCCS Class D circuit.*

The following analysis is based on the simplifier assumptions provided in section ***Complementary Voltage Switching (CVS) Circuit***, with one exception: The TCCS circuit requires a parallel resonant circuit instead of a series resonant type. As in the previous case, output transformer T_2 is ideal, having m turns in each half of the primary winding, and n turns in the secondary winding. The RF choke (RFC) is also ideal because it has no series resistance, and its reactance at the switching frequency is infinite. Therefore, the RF choke[2] allows only a constant (DC) input current, I_{dc}. The RF choke forces input current I_{dc} into the center tap of the primary winding. This current is directed to ground through the active device that is in the ON state. As a result, collector currents $i_1(\theta)$ and $i_2(\theta)$ are square waves with levels of 0 and I_{dc} (see Figure 3-9). Transformation of these currents in the secondary winding of T_2 determines a square-wave current.

$$i(\theta) = \begin{cases} \dfrac{m}{n} I_{dc}, & 0 \le \theta \le \pi \\[2mm] -\dfrac{m}{n} I_{dc}, & \pi \le \theta \le 2\pi \end{cases} \qquad 3.19$$

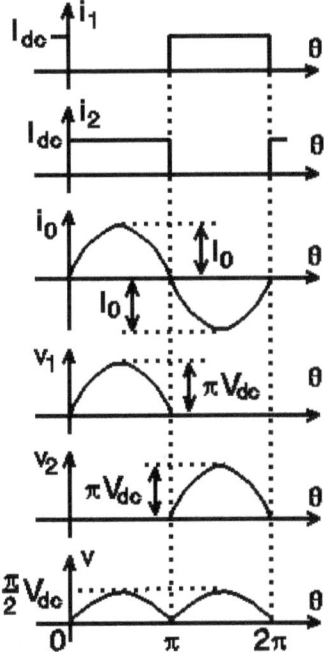

Figure 3-9 *Waveforms in TCCS Class D circuit.*

Decomposing $i(\theta)$ into a Fourier series yields

$$i(\theta) = \frac{4}{\pi}\frac{m}{n}I_{dc}\sum_{n=1}^{\infty}\frac{\sin(2n-1)\theta}{2n-1} \qquad 3.20$$

Because the parallel-tuned circuit is ideal, it provides a zero reactance path to ground for the harmonic currents in $i(\theta)$. On the other hand, the parallel-tuned LC circuit provides an infinite reactance to the fundamental frequency component of $i(\theta)$, forcing it entirely into the load

$$i_0(\theta) = I\sin\theta = \frac{4}{\pi}\frac{m}{n}I_{dc}\sin\theta \qquad 3.21$$

where it generates a sinusoidal output voltage

$$v_0(\theta) = V_0\sin\theta = \frac{4}{\pi}\frac{m}{n}I_{dc}R_L\sin\theta \qquad 3.22$$

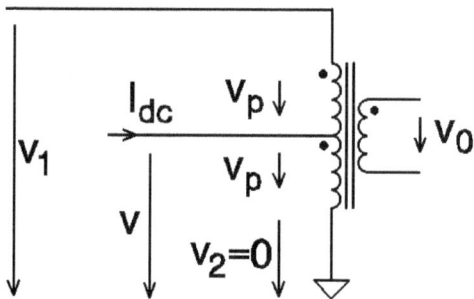

Figure 3-10 *Voltages across the windings of T_2 ($0 \le \theta \le \pi$).*

For $0 \le \theta \le \pi$, suppose that Q_2 is ON and Q_1 is OFF, yielding $v_2(\theta) = 0$. Because the voltage across the secondary winding is sinusoidal, the voltage across each half of the primary winding is also sinusoidal (see Figure 3-10).

$$v_p(\theta) = V_p \sin\theta = \frac{m}{n} V_0 \sin\theta \qquad 3.23$$

Thus, for $0 \le \theta \le \pi$

$$v_1(\theta) = 2v_p(\theta) = 2\frac{m}{n} V_0 \sin\theta = \frac{8}{\pi} I_{dc} R \sin\theta \qquad 3.24$$

where

$$R = \left(\frac{m}{n}\right)^2 R_L \qquad 3.25$$

is the resistance seen across each half of the primary winding of T_2 with the other half open (the equivalent load resistance of each transistor).

For $\pi \le \theta \le 2\pi$, Q_1 is ON and Q_2 is OFF. Thus, $v_1(\theta) = 0$ and

$$v_2(\theta) = -2v_p(\theta) = -\frac{8}{\pi} I_{dc} R \sin\theta \qquad 3.26$$

At the center tap of the primary winding, the voltage with respect to ground is given by

$$v(\theta) = \begin{cases} \dfrac{1}{2} v_1(\theta) & 0 \le \theta \le \pi \\[2mm] \dfrac{1}{2} v_2(\theta) & \pi \le \theta \le 2\pi \end{cases} \qquad 3.27$$

and is plotted in Figure 3-9. Because the RF choke is ideal, there is no DC voltage drop across it. Consequently, the average value of the voltage at the center tap of the primary winding of T_2 must be V_{dc}

$$V_{dc} = \overline{v(\theta)} = \frac{1}{2\pi} \int_0^{2\pi} \left| \frac{4}{\pi} I_{dc} R \sin\theta \right| d\theta = \frac{8}{\pi^2} I_{dc} R \qquad 3.28$$

and this condition gives

$$I_{dc} = \frac{\pi^2}{8} \frac{V_{dc}}{R} \qquad 3.29$$

With I_{dc} given by Equation 3.29 and using Equations 3.24, 3.26, and 3.27, the peak values of $v_1(\theta)$, $v_2(\theta)$, and $v(\theta)$ shown in Figure 3-9 are obtained. The output power (dissipated in the load resistance R_L) is

$$P_0 = \frac{V_0^2}{2R_L} = \frac{\pi^2}{8} \frac{V_{dc}^2}{R} \approx 1.2337 \frac{V_{dc}^2}{R} \qquad 3.30$$

Using Equation 3.29, the DC input power is

$$P_{dc} = V_{dc} I_{dc} = \frac{\pi^2}{8} \frac{V_{dc}^2}{R} = P_0 \qquad 3.31$$

and the collector efficiency (for the idealized operation) is 100 percent. The power output capability is the same as in the previous cases.

$$C_P = \frac{P_0}{2(\pi V_{dc}) I_{dc}} = \frac{1}{2\pi} \approx 0.1582 \qquad 3.32$$

Finally, note that the TCCS circuit is the "dual" of the TCVS circuit, because the voltage and current waveforms are interchanged. In contrast with the TCVS circuit that uses a series-tuned circuit, the TCCS circuit uses a parallel-tuned circuit (or an equivalent, such as a pi-network) that imposes a sinusoidal output voltage across the load.

EXAMPLE 3.3

A TCCS circuit using 100-volt MOSFETs delivers an output power of $P_0 = 100$ W in an $R_L = 50\ \Omega$ load. Because the maximum drain voltage is πV_{dc}, $V_{dc} \leq 31.83$ V, and from Equation 3.30, $R \leq 12.5\ \Omega$. Using an output transformer with $m/n = 1/2$, $V_{dc} = 31.83$ V and $I_{dc} = 3.14$ A. The required device ratings are about 100 volts and 3.14 amps.

3.2 Practical Considerations

In practical circuits, the simplifier assumptions given in section *Complementary Voltage Switching (CVS) Circuit* usually cannot be accepted. For example:

1. The real transistors have nonzero switching times, parasitic inductances and capacitances, nonzero ON resistance and, possibly, a saturation voltage.

2. The capacitor and inductor from the output-tuned circuit have parasitic resistances (that is, finite unloaded quality factors).

3. The output-tuned RLC circuit has a finite-loaded quality factor.

4. The load circuit can have a nonzero net reactance at the operating frequency caused by mistuning, for example.

5. It is possible for frequency variations or changes to appear in the transistors' timing.

The effects of all these factors cannot be estimated analytically because a quasi-complete circuit model would yield very complicated equations. For this reason, the analysis below evaluates the effects of each circuit imperfection on the operation of the Class D power amplifier. The results provide a more accurate estimation of the output power, collector efficiency, and active device stresses, allowing more accurate circuit and thermal designs. The total effects of more than one of these factors can be estimated by a convenient combination of the individual effects. This estimation is only an approximation; an exact analysis requires numerical simulations.

Reactive Loads

As a jumping off point, consider a CVS Class D circuit with a reactive load (see Figure 3-11). C_{p1} and C_{p2} include the output capacitances of the transistors and the stray circuit capacitances. The series net reactance X (at the switching frequency f) can model a circuit mistuning or a change of the operating frequency.

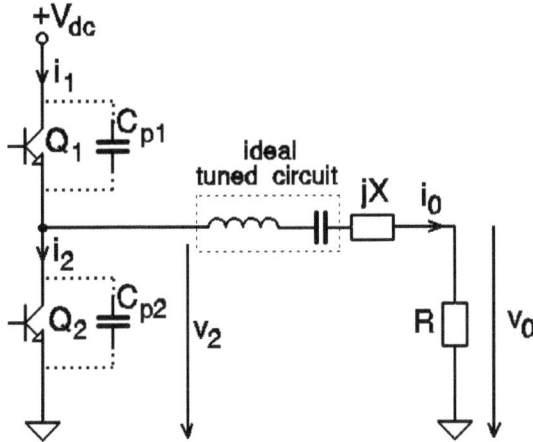

Figure 3-11 *CVS Class D circuit with reactive load.*

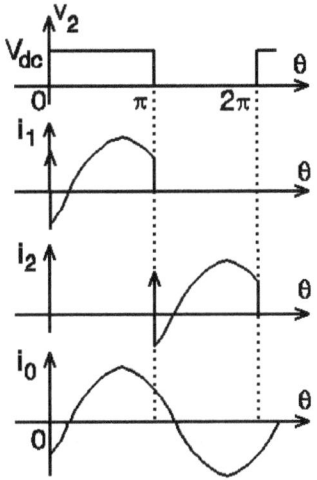

Figure 3-12 *Waveforms in CVS Class D circuit with reactive load.*

With the exception of these specifications, the simplifier assumptions presented in section ***Complementary Voltage Switching (CVS) Circuit*** remain valid. The circuit waveforms are presented in Figure 3-12 and assume a 50 percent duty cycle. Because of the series reactance, the output current $i_0(\theta)$ (and the output voltage $v_0(\theta)$) is phase shifted relative to the $v_2(\theta)$ voltage waveform, and both $i_1(\theta)$ and $i_2(\theta)$ are negative during a portion of each period. If Q_1 and Q_2 are MOSFETs, the negative currents pass without difficulty. If Q_1 and Q_2 are BJTs, the negative currents (which cannot pass through the transistors) charge capacitances C_{p1} and C_{p2}, producing large voltage spikes that can damage the transistors.

A suitable path for the negative collector currents is provided by diodes D_1 and D_2 in Figure 3-13. The output current $i_0(\theta)$ is directed through one of the four devices (Q_1, D_1, Q_2, or D_2), preventing the collector voltage spikes (see Figure 3-14).

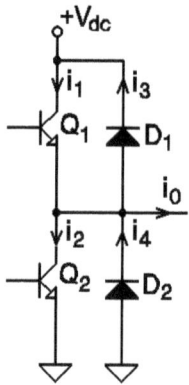

Figure 3-13 *Anti-parallel protection diodes in Class D amplifiers using BJTs.*

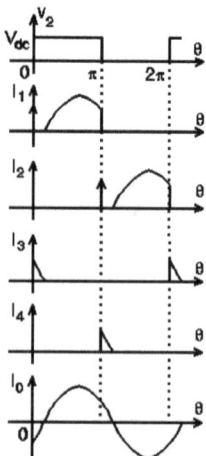

Figure 3-14 *Waveforms in CVS Class D circuit with protection diodes.*

Note that the current pulses[3] through Q_1 and Q_2, which charge/discharge C_{p1} and C_{p2} when switching occurs, are shown for accuracy in Figures 3-12 and 3-14. The power losses due to the parasitic capacitances will be calculated in the next section.

The bipolar transistors used in TCVS Class D circuits can be protected in a similar fashion, using anti-parallel diodes connected from the collectors to the emitters. The active devices used in TCCS Class D amplifiers must withstand negative collector voltages when they are OFF. Because the construction of a power MOSFET is such that a parasitic diode will conduct when the drain voltage is negative (regardless of gate-to-source voltage), a diode in series with each drain is required. Bipolar transistors generally have a somewhat limited capability for negative collector voltages[4] (in the OFF state). However, they can usually be protected from negative collector voltages by placing diodes in series with the collectors.

The series reactance reduces the amplitude of the output current and the output power. Denoting $Z = R + jX$, the output current of a CVS circuit is given by

$$i_0(\theta) = \frac{2}{\pi} \frac{V_{dc}}{|Z|} \sin\left[\theta + \arctan\left(\frac{X}{R}\right)\right]$$ 3.33

resulting in

$$P_0 = \frac{2}{\pi^2} \frac{V_{dc}^2}{|Z|^2} R = \frac{2}{\pi^2} \frac{V_{dc}^2}{R} \rho^2 \quad \text{where} \quad \rho = \frac{R}{|Z|} < 1$$

$$P_{dc} = V_{dc} I_{dc} = V_{dc}\left[\overline{i_1(\theta) - i_3(\theta)}\right] =$$ 3.34

$$\frac{V_{dc}}{2\pi} \int_0^\pi \frac{2}{\pi} \frac{V_{dc}}{|Z|} \sin\left[\theta + \arctan\left(\frac{X}{R}\right)\right] d\theta = \frac{2}{\pi^2} \frac{V_{dc}^2}{R} \rho^2$$

$$C_P = \frac{\rho}{2\pi} \approx 0.1592\rho$$

The efficiency of the Class D circuit is unaffected, but the power output capability decreases. Note that, in this case, the currents through the active devices jump when switching occurs. This causes
a. longer transition times, because the actual active devices switches have limited di/dt capability,
b. higher power dissipation during the transition times,
c losses due to the lead inductances associated with the active devices.

EXAMPLE 3.4
The output circuit in Example 3.1 is detuned, resulting in a series reactance, $X = 10\ \Omega$, at the switching frequency. In accordance with Equations 3.33 and 3.34, the amplitudes of the output current decreases to $I = 3.16$ A and the output power decreases to $P_0 = 50$ W.

The DC input current decreases to $I_{dc} = 0.71$ A (the collector efficiency remains at 100 percent), and the power output capability decreases to $C_P = 0.1125$.

Shunt Capacitances or Series Inductances at Switches

Linear Output Capacitances

The parasitic capacitance shunting an active device (C_{p1} and C_{p2}, Figure 3-11) includes the output capacitance of the transistor and the stray capacitance. In the following analysis, the assumptions made in section **Complementary Voltage Switching (CVS) Circuits** are accepted, except the one regarding the null output capacitance of the active devices. The parasitic capacitance is assumed to be independent of the collector voltage.

If Q_1 is ON and Q_2 is OFF, then C_{p1} is discharged and C_{p2} is charged to V_{dc}. When Q_1 turns OFF and Q_2 turns ON, C_{p2} is discharged instantaneously through the zero ON resistance of Q_2. C_{p1} is charged (also instantaneously through Q_2) to V_{dc}. When C_{p2} is discharged, the energy stored in it is dissipated (in Q_2).

$$E_{dis,2} = \frac{1}{2} C_{p2} V_{dc}^2 \qquad 3.35$$

Simultaneously, energy is dissipated (also in Q_2) to charge C_{p1} from zero to V_{dc}.

$$E_{dis,1} = \frac{1}{2} C_{p1} V_{dc}^2 \qquad 3.36$$

When Q_1 turns ON and Q_2 turns OFF, C_{p2} is charged to V_{dc} and C_{p1} is discharged. The energy dissipated in Q_1 (in this process) is $E_{dis1} + E_{dis2}$, equal to the energy dissipated in Q_2 at the instant of the other commutation. Denoting the switching frequency as f, the power losses due to the parasitic capacitances are given by

$$P_{dis} = 2\left(E_{dis,1} + E_{dis,2}\right)f = \left(C_{p1} + C_{p2}\right)V_{dc}^2 f = 2C_p V_{dc}^2 f \qquad 3.37$$

where $C_{p1} = C_{p2} = C_p$ (because the transistors are always identical).

Some observations and practical considerations

1. The power losses due to this mechanism do not cause a decrease of output power. This is because the currents that charge and discharge capacitances C_{p1} and C_{p2} flow only through Q_1, Q_2, and the power supply, and do not circulate through the load. As a result, the equations given in Section 3.1 for the output power are not affected by the para-

sitic capacitances. The DC input power is the sum of the output power and the power losses given by Equation 3.37.

2. In a real amplifier, the charge or discharge of a parasitic capacitance requires a nonzero length of time, during which the collector voltage and current are simultaneously nonzero. Because the energy dissipated to charge or discharge a capacitor does not depend on the series resistance or on the current or voltage waveforms, Equation 3.37 is valid in any real amplifier.

3. For thermal calculations, it must be taken into account that Q_1 and Q_2 each dissipate half of the power given by Equation 3.37. The results presented in References [18] and [19] show that the parasitic series resistance associated with the output capacitance of a MOSFET is (at least at frequencies that are not too high) about three to ten times greater than the ON resistance. The presence of this resistance (whose value is never specified by transistor manufacturers) does not affect the power losses calculated above, but limits the amplitude of the current pulses that appear during switching.

4. If Q_1 and Q_2 are bipolar transistors, the parasitic capacitances are charged and discharged between V_{dc} and V_{CEsat}. Thus, in Equation 3.37, V_{dc} must be replaced by $V_{dc} - V_{CEsat}$. If Q_1 and Q_2 are *field-effect transistors* (FETs), Equation 3.37 can be used as it stands.

5. Equation 3.37 shows that power losses increase with the parasitic capacitance C_p, the switching frequency f, and the square of the DC supply voltage V_{dc}. This loss mechanism easily can be the dominant one, especially in cases where the DC supply voltage is large and/or the frequency is high. These losses limit the usability of the Class D power amplifier at high frequencies. For the numerical example presented below, the different power losses are calculated in the next sections. The power losses due to the parasitic capacitances are approximately one magnitude order greater than the power losses due to other causes.

EXAMPLE 3.5

A CVS Class D amplifier operates at $V_{dc} = 75$ V and $f = 13.56$ MHz in an $R = 3.8\ \Omega$ load. The output power for the idealized operation in Equation 3.6 is $P_0 = 300$ W. A suitable transistor for this amplifier is IRF540 [20], a 100 V/28 A MOS device (the peak drain current is, in accordance with Equation 3.5, $I = 12.56$ A). The output capacitance of this device is $C_{OSS} = 560$ pF (at $V_{DS} = 25$ V).

Ignoring the stray capacitances, the charge/discharge losses are obtained from Equation 3.37, P_{dis} = 85.43 W. These losses are about one quarter of the output power. The efficiency is reduced significantly to below 80 percent because the parasitic capacitances do not affect the output power, the drain efficiency is given by $\eta = P_0 / (P_0 + P_{dis}) = 300 / 385.43 = 77.84\%$.

In the TCVS Class D circuit, the parasitic capacitances of Q_1 and Q_2 are charged and discharged between 0 and 2 V_{dc} (see Figure 3-7). Therefore, the power losses are given by

$$P_{dis} = 2\left[\frac{1}{2}C_{p1}\left(2V_{dc}\right)^2\right]f + 2\left[\frac{1}{2}C_{p2}\left(2V_{dc}\right)^2\right]f =$$
$$\left(C_{p1} + C_{p2}\right)4V_{dc}^2 f = 8C_p V_{dc}^2 f$$

3.38

where $C_{p1} = C_{p2} = C_p$. A comparison of Equation 3.37 with Equation 3.38 shows that the charge/discharge losses in the TCVS circuit are four times greater than the losses in the CVS circuit, for the same C_p, V_{dc}, and f. However, for a fair comparison, it must be observed that if a TCVS circuit operates at half of the supply voltage used by the CVS circuit, the output powers and the transistor ratings are the same for both configurations (see Equations 3.13, 3.15, 3.5, and 3.6).

In this case, the charge/discharge losses are the same for both Class D circuits. Note that in the TCVS circuit, only one half of power loss P_{dis} given by Equation 3.38 is dissipated in Q_1 and Q_2 (one quarter of P_{dis} each). This is because (in contrast with the CVS configuration) parasitic capacitances C_{p1} and C_{p2} are charged through the primary winding of the output transformer (dissipating half of the power loss in its series resistance) and are only discharged through Q_1 and Q_2.

If Q_1 and Q_2 are bipolar transistors, the parasitic capacitances are charged and discharged between 2 $V_{dc} - V_{CEsat}$ and V_{CEsat}. In Equation 3.38, V_{dc} must be replaced by $V_{dc} - V_{CEsat}$.

Voltage Dependent Output Capacitances

Most of the parasitic capacitance, C_p (in both BJTs and MOSFETs amplifiers) is an abrupt-junction capacitance. As a result, the variation of C_p with voltage is given by [21].

$$C_p(v) \approx \frac{C_{p0}}{\sqrt{1 + \dfrac{v}{V_B}}} \quad (v > 0)$$

3.39

where C_{p0} is the zero-voltage capacitance and V_B is the barrier potential ($V_B = 0.5...1$ V). A typical variation of $C_p = C_p(v)$ is shown in Figure 3-15. For a given device, C_{p0} and V_B can be determined easily using either the data sheets curve $C_p = C_p(v)$, or two small-signal impedance measurements with different bias voltages. For increased accuracy, it is recommended that several measurements be performed with various bias voltages followed by a standard fitting technique to determine C_{p0} and V_B. In practical applications, a linear capacitance term is always present in the structure of the parasitic capacitance $C_p(v)$. This linear term could be considered in Equation 3.39 to increase the accuracy.

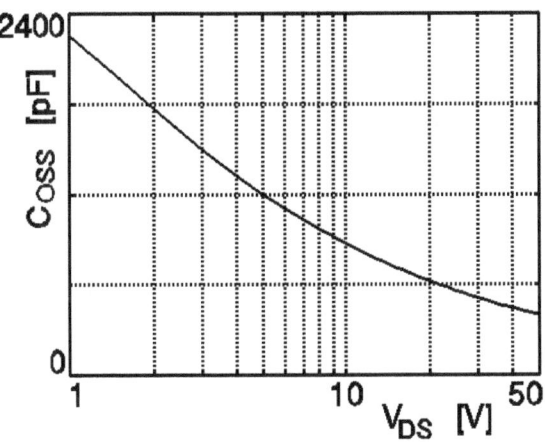

Figure 3-15 *Output capacitance C_{OSS} versus drain-to-source voltage V_{DS} (IRF540 MOSFET, f = 1 MHz, V_{GS} = 0 V) [20].*

EXAMPLE 3.6

For an IRF540 MOSFET (see Reference [20] and Figure 3-15) at a bias voltage of $v_1 = V_{DS1} = 25$ V, the output capacitance is $C_p(v_1) = C_{OSS}(V_{DS1}) = 560$ pF. At a bias voltage of $v_2 = V_{DS2} = 5$ V, the output capacitance is $C_p(v_2) = C_{OSS}(V_{DS2}) = 1200$ pF. Substituting these numerical values in Equation 3.39 and solving the equation system obtained, $C_{p0} = 3757$ pF and $V_B = 0.568$ V.

A CVS Class D circuit that uses two identical MOS transistors with output capacitances $C_{p1}(v) = C_{p2}(v) = C_p(v)$ given by Equation 3.39 is considered below. When capacitance C_p is charged between 0 and V_{dc}, the energy dissipated in its series resistance is

$$P_{dis} = 4E_{dis}f = 4V_{dc}C_{p0}V_B\left(\sqrt{1+\frac{V_{dc}}{V_B}}-1\right)f \qquad 3.40$$

An equal amount of energy is dissipated in the series resistance when C_p is discharged. If the switching frequency is f, the power losses due to this mechanism are given by

$$E_{dis} = \frac{1}{2}V_{dc}\int_0^{V_{dc}} C_p(v)\, dv = V_{dc}C_{p0}V_B\left(\sqrt{1+\frac{V_{dc}}{V_B}}-1\right) \qquad 3.41$$

EXAMPLE 3.7

Considering a voltage-dependent capacitance in Example 3.5 ($C_{p0} = 3757$ pF and $V_B = 0.568$ V), $P_{dis} = 91.45$ W is obtained, which is roughly the same as the previously obtained value assuming linear C_{OSS}. Figure 3-16 presents the variation of P_{dis} with V_{dc} for linear $C_{OSS} = C_{OSS}$ (25 V) = 560 pF and, respectively, for nonlinear C_{OSS}. These curves suggest that, for Class D amplifiers using 100-volt MOSFETs and operating at $V_{dc} > 45$ to 50 V, it is possible to make satisfactory practical predictions of the circuit operation using Equation 3.37 and the C_{OSS} value measured at 25 volts (and given in the data sheets).

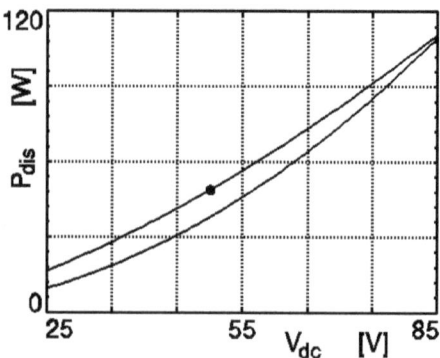

Figure 3-16 *Variation of power losses with V_{dc}, for linear C_{OSS} (—) and nonlinear C_{OSS} (—•—), respectively; IRF540 MOSFET, f = 13.56 MHz.*

In the TCVS Class D circuit, the parasitic capacitances are charged and discharged between 0 and 2 V_{dc}. Thus, Equation 3.41 becomes

$$P_{dis} = 8V_{dc}C_{p0}V_B\left(\sqrt{1 + \frac{2V_{dc}}{V_B}} - 1\right)f \qquad 3.42$$

Series Inductances

In the TCCS Class D circuit, the parasitic capacitances do not have charge/discharge losses due to the mechanism described.[5] However, in the TCCS circuit, there is another power loss mechanism. The currents through the active devices jump when switching occurs (Figure 3-9), and this causes losses due to the series inductances at the switch. The total series inductance L_s (with each transistor) is the sum of the lead inductance (inside and outside the package) and of the leakage inductance of the output transformer. During each period, the current in L_s is changed instantaneously from 0 to I_{dc} and vice versa. The power losses due to the series inductance are given by

$$P_{dis} = 2\left[2\left(\frac{1}{2}L_s I_{dc}^2\right)\right]f = 2L_s I_{dc}^2 f \qquad 3.43$$

Similar power losses also appear in the Class D voltage switching amplifier, if the collector currents jump (for example, if the series circuit is detuned or if the duty cycle is not 50 percent).

EXAMPLE 3.8

A TCVS Class D circuit operating at $f = 13.56$ MHz delivers an output power of $P_0 = 300$ W in an $R_L = 3.8\ \Omega$ load. Using 100-volt MOSFETs, $V_{dc} \leq 31.83$ V (because the peak collector voltage is πV_{dc}). The equivalent load resistance $R \leq 4.167\ \Omega$ is obtained from Equation 3.30. This suggests the choice of an output transformer having $m = n$, so $R = R_L = 3.8\ \Omega$. Thus, from Equation 3.30, $V_{dc} = 30.40$ V is obtained and the required device ratings are $\pi V_{dc} = 95.5$ V and $I_{dc} = 9.87$ A. Supposing that each transistor has a series inductance of $L_s = 20$ nH, Equation 3.43 yields $P_{dis} = 52.84$ W.

Saturation Voltage of Transistors

If the active devices from the Class D amplifier are bipolar transistors with a saturation voltage V_{CEsat}, the analysis presented in Section 3.1 must be modified as follows. The saturation voltage is of the order of 0.1 volt at low frequencies (DC to of the order of $f_T/1000$), and up to several volts at high frequencies (of the order of $f_T/10$).

CVS Circuit

The voltage $v_2(\theta)$ varies between V_{CEsat} and $V_{dc} - V_{CEsat}$.

$$P_0 = \frac{2}{\pi^2} \frac{\left(V_{dc} - 2V_{CEsat}\right)^2}{R} \tag{3.44}$$

$$P_{dc} = V_{dc}I_{dc} = \frac{2}{\pi^2} \frac{V_{dc}\left(V_{dc} - 2V_{CEsat}\right)}{R} \tag{3.45}$$

$$\eta = \frac{P_0}{P_{dc}} = 1 - \frac{2V_{CEsat}}{V_{dc}} \tag{3.46}$$

$$C_P = \frac{1}{2\pi} \frac{V_{dc} - 2V_{CEsat}}{V_{dc} - V_{CEsat}} \tag{3.47}$$

TCVS Circuit

In this case, voltages $v_1(\theta)$ and $v_2(\theta)$ vary between V_{CEsat} and $2\,V_{dc} - V_{CEsat}$.

$$P_0 = \frac{8}{\pi^2} \frac{\left(V_{dc} - V_{CEsat}\right)^2}{R} \tag{3.48}$$

$$P_{dc} = V_{dc}I_{dc} = \frac{8}{\pi^2} \frac{V_{dc}\left(V_{dc} - V_{CEsat}\right)}{R} \tag{3.49}$$

$$\eta = \frac{P_0}{P_{dc}} = 1 - \frac{V_{CEsat}}{V_{dc}} \tag{3.50}$$

$$C_P = \frac{V_{dc} - V_{CEsat}}{\pi\left(2V_{dc} - V_{CEsat}\right)} \tag{3.51}$$

TCCS Circuit

The currents $i_1(\theta)$ and $i_2(\theta)$ vary between 0 and I_{dc}, where

$$I_{dc} = \frac{\pi^2}{8} \frac{V_{dc} - V_{CEsat}}{R} \tag{3.52}$$

while the voltages $v_1(\theta)$ and $v_2(\theta)$ vary between V_{CEsat} and $\pi(V_{dc} - V_{CEsat}) + V_{CEsat}$.

$$P_0 = \frac{\pi^2}{8} \frac{(V_{dc} - V_{CEsat})^2}{R} \qquad 3.53$$

$$P_{dc} = V_{dc}I_{dc} = \frac{\pi^2}{8} \frac{V_{dc}(V_{dc} - V_{CEsat})}{R} \qquad 3.54$$

$$C_P = \frac{1}{2\pi + \dfrac{2V_{CEsat}}{V_{dc} - V_{CEsat}}} \qquad 3.55$$

and the collector efficiency is given by Equation 3.50.

Some observations and practical considerations
1. In all cases, P_0, P_{dc}, η, and C_P decrease in comparison with the values corresponding to the ideal case (Section 3.1).

2. At first sight, the CVS circuit appears to have disadvantages in comparison with the TCVS circuit (see Equations 3.46 and 3.50). As was shown earlier, if a TCVS circuit operates at half the supply voltage used by a CVS circuit, the output power and the transistor ratings are the same for both configurations. In this case, the power losses due to the parasitic capacitances are the same in both circuits and the effects of V_{CEsat} on the circuit performance are similar.

3. The "effective" DC supply voltage for the CVS circuit is denoted as

$$V_{eff} = V_{dc} - 2V_{CEsat} \qquad 3.56$$

and with

$$V_{eff} = V_{dc} - V_{CEsat} \qquad 3.57$$

the "effective" DC supply voltage for the TCVS and TCCS circuits, the following general relation is obtained for the collector efficiency

$$\eta = \frac{V_{eff}}{V_{dc}} \qquad 3.58$$

The "effective" supply voltage V_{eff} can be used instead of supply voltage V_{dc} to calculate (using the equations from Section 3.1) the current through the load, the output power, and the DC input current. The DC input power is calculated using the supply voltage V_{dc}.

4. Assuming equal peak collector currents, the saturation power losses of the TCCS circuit are greater than those of the voltage switching circuit. This is because, for a number of reasons, V_{CEsat} is always higher for the TCCS circuit. The transistors in a voltage switching circuit conduct a high collector current for just a small portion of the ON state (the current being a sinusoidal waveform); the transistors in a TCCS circuit conduct the peak collector current during the entire ON state (the current being a rectangular waveform). The saturation voltage for a rectangular current waveform is always higher than that for a sinusoidal current with the same peak value and the same frequency. Although this explanation is complex, it is related to the increase of V_{CEsat} with frequency [7].

ON Resistance of Transistors

During saturation, bipolar transistors are characterized by a roughly constant saturation voltage, V_{CEsat}, and a roughly constant saturation resistance, R_{CEsat}. Field-effect transistors are characterized (during the ON state) by a roughly constant drain-to-source resistance, $R_{DS,ON}$. Generally, the active device ON resistance is denoted below by R_{ON}. It is assumed constant during the ON state and independent of the current through (or the voltage across) it. The other assumptions from Section 3.1 remain valid. The analysis presented in Section 3.1 must be modified as follows.

CVS Circuit

The ON resistance, R_{ON}, is placed in series with the load resistance, R [see the equivalent circuit of Figure 3-1(b)]. Thus,

$$P_0 = \frac{2}{\pi^2} \left(\frac{V_{dc}}{R + R_{ON}} \right)^2 R \qquad 3.59$$

$$P_{dc} = V_{dc} I_{dc} = \frac{2}{\pi^2} \frac{V_{dc}^2}{R + R_{ON}} \qquad 3.60$$

$$\eta = \frac{P_0}{P_{dc}} = \frac{R}{R + R_{ON}} \qquad 3.61$$

$$C_P = \frac{1}{2\pi} \frac{R}{R + R_{ON}} \qquad 3.62$$

The relations above show that the output power, the DC input power, the collector efficiency, and the power output capability decrease with increasing R_{ON}.

EXAMPLE 3.9

ON resistance losses in the CVS circuit of Example 3.5. The IRF540 MOSFET has an ON resistance of $R_{ON} = 0.077\ \Omega$ [20]. It affects the output power and the DC input power. Equations 3.59 and 3.60, with $R = 3.8\ \Omega$ and $R_{ON} = 0.077\ \Omega$, yield $P_0 = 288.2$ W and $P_{dc} = 294$ W. To obtain $P_0 = 300$ W, Equation 3.59 gives the necessary load resistance, $R = 3.644\ \Omega$. With this value, $P_{dc} = 306.34$ W and the power losses (in both transistors) are $P_{dis} = 6.34$ W. These losses are significantly lower (about thirteen times) than those due to the parasitic capacitances. However, each transistor dissipates a considerable power (about 50 watts, including the gate drive losses), which can determine that the transistors will operate at a high junction temperature. Thus, the ON resistance increases, increasing the conduction losses. For example, if the junction temperature is 100°C, the ON resistance of IRF540 is about 1.5 times the value measured at 25°C (0.077 Ω) [20].

TCVS Circuit

The ON resistance, R_{ON}, is placed in series with the equivalent load resistance of each transistor.

$$P_0 = \frac{8}{\pi^2}\left(\frac{V_{dc}}{R + R_{ON}}\right)^2 R \qquad\qquad 3.63$$

$$P_{dc} = V_{dc}I_{dc} = \frac{8}{\pi^2}\frac{V_{dc}^2}{R + R_{ON}} \qquad\qquad 3.64$$

The collector efficiency is given by Equation 3.61; the power output capability is given by Equation 3.62.

TCCS Circuit

The ON resistance is placed in series in the path of I_{dc} (see Figure 3-8). The currents $i_1(\theta)$ and $i_2(\theta)$ vary between 0 and I_{dc} where

$$I_{dc} = \frac{V_{dc}}{\dfrac{8R}{\pi^2} + R_{ON}} = \frac{V_{dc}}{R_{dc} + R_{ON}} \qquad\qquad 3.65$$

and $R_{dc} = 8R/\pi^2$ is the resistance (load) that an ideal TCCS circuit shows to the power supply. The collector voltages $v_1(\theta)$ and $v_2(\theta)$ vary between $I_{dc}R_{ON}$ and $(8R/\pi + R_{ON})I_{dc}$.

$$P_0 = \frac{V_{dc}^2 R_{dc}}{\left(R_{dc} + R_{ON}\right)^2} \qquad 3.66$$

$$P_{dc} = V_{dc} I_{dc} = \frac{V_{dc}^2}{R_{dc} + R_{ON}} \qquad 3.67$$

$$\eta = \frac{P_0}{P_{dc}} = \frac{R_{dc}}{R_{dc} + R_{ON}} \qquad 3.68$$

$$C_P = \frac{1}{2\pi + 2R_{ON} / R_{dc}} \qquad 3.69$$

For all Class D configurations, it is obvious that the waveforms are affected by the nonzero ON resistance R_{ON}. For example, the waveform of $v_2(\theta)$ in a CVS circuit is presented in Figure 3-17; the other waveforms are only quantitatively affected.

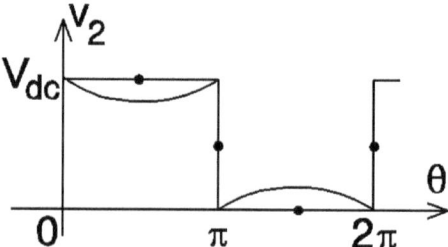

Figure 3-17 *Waveform of $v_2(\theta)$ in a CVS circuit — $R_{ON} \neq 0$; —•— $R_{ON} = 0$.*

The effects of R_{ON} on Class D circuits can be evaluated by introducing an "effective" DC supply voltage V_{eff},

$$V_{eff} = \frac{R}{R + R_{ON}} V_{dc} \qquad 3.70$$

in the voltage switching circuits and

$$V_{eff} = \frac{R_{dc}}{R_{dc} + R_{ON}} V_{dc} \qquad 3.71$$

in the current switching circuit. V_{eff} can be used instead of the supply voltage V_{dc} in all calculations except the DC input power. The collector efficiency is given by Equation 3.58.

Optimum Load Resistance for Maximum Efficiency

In many practical applications, the desired output power can be obtained with different combinations of DC supply voltage V_{dc} and load resistance R. In the analysis below, a CVS Class D circuit using MOSFETs is considered, but a similar analysis can be made for a TCVS circuit using MOSFETs. In the Class D voltage switching amplifier using BJTs, the conduction losses due to V_{CEsat} can be considered instead of the conduction losses due to R_{ON}. In the TCCS Class D circuit, a similar analysis could consider the switching losses due to the series inductances at the time of the switch, instead of the power losses due to the parasitic capacitances.

Given specified active devices, characterized by ON resistance R_{ON} and output capacitance C_p, the conduction losses (due to R_{ON}) increase as R decreases (see Equation 3.61). On the other hand, the power losses due to the parasitic capacitances, C_p, increase as V_{dc} increases (see Equation 3.37). Equation 3.59 for the output power shows there is a value of the load resistance (and of the DC supply voltage) that minimizes the total power losses[6] and thus maximizes the collector efficiency. Given the specified output power P_0, DC supply voltage V_{dc}, and ON resistance R_{ON}, the necessary load resistance R is found from Equation 3.59.

$$R = \frac{V_{dc}^2}{\pi^2 P_0} - R_{ON} + \sqrt{\frac{V_{dc}^2}{\pi^2 P_0}\left(\frac{V_{dc}^2}{\pi^2 R_{ON}} - 2R_{ON}\right)} \qquad 3.72$$

and the power dissipation due to R_{ON} is given by

$$P_{dRON} = \frac{R_{ON}}{R} P_0 \qquad 3.73$$

Assuming linear parasitic capacitances (independent of the collector voltage), the power losses due to their charge and discharge are obtained from Equations 3.37 and 3.59

$$P_{dCp} = \pi^2 R \left(1 + \frac{R_{ON}}{R}\right)^2 C_p f P_0 \qquad 3.74$$

EXAMPLE 3.10

For the circuit presented in Example 3.5, $P_0 = 300$ W, IRF540 MOSFET ($C_{OSS} = 560$ pF, $R_{ON} = 0.077$ Ω), $f = 13.56$ MHz, the variation of the necessary load resistance, R, with the DC supply voltage V_{dc} is depicted in Figure 3-18(a). Figures 3-18 (b, c, d) present the peak drain current $I = 2 V_{dc}/[\pi(R + R_{ON})]$, the power losses P_{dRON} and P_{dCp}, the total power dissipation $P_d = P_{dRON} + P_{dCp}$, and the collector efficiency $\eta = P_0/(P_0 + P_d)$. These curves show there is an optimum operating point (minimum power losses, therefore maximum efficiency) for $V_{dc} = 41.73$ V; with this value, $R = 1.02$ Ω and $\eta = 85.92\%$. Note that this optimum operating point can be calculated easily by setting the derivative of the total dissipated power with respect to the supply voltage equal to zero.

$$R = \sqrt{R_{ON}^2 + \frac{R_{ON}}{\pi^2 C_p f}} \qquad\qquad 3.75$$

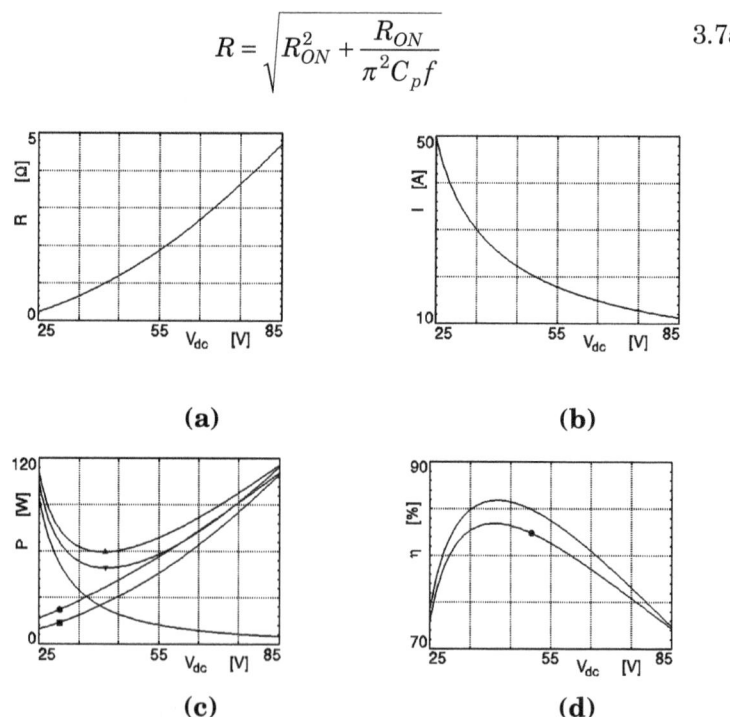

(a)   **(b)**

(c)   **(d)**

Figure 3-18 (a) Load resistance R, and (b) peak drain current I, versus DC supply voltage V_{dc}; (c) power losses P_{dRON} (—), P_{dCp} (—■—), $P_d = P_{dRON} + P_{dCp}$ (—▼—), P_{dCpn} (—•—), $P_{dn} = P_{dRON} + P_{dCpn}$ (—▲—) and (d) collector efficiency η, with linear C_p (—) and with nonlinear Cp (—•—), respectively. CVS Class D amplifier; IRF540 MOSFETs, $f = 13.56$ MHz, $P_0 = 300$ W.

Now consider nonlinear parasitic capacitances (see Equation 3.39). The power losses due to their charge and discharge are given by Equation 3.41. This power (denoted by P_{dCpn}) and the total power dissipation $P_{dn} = P_{dRON} + P_{dCpn}$ are shown in Figure 3-18 (c). The optimum operating point is $V_{dc} = 40.93$ V, $R = 0.97$ Ω, $\eta = 83.44$ %, in agreement with that previously calculated for linear parasitic capacitances. This suggests that satisfactory practical predictions of circuit operations can be made using a circuit model with linear C_p.

Some observations and practical considerations

1. For nonlinear parasitic capacitances, the maximum efficiency point cannot be determined analytically; the solution above is obtained numerically. An approximate solution is as follows [21]:

   a. ignoring R_{ON} in comparison with R, from Equations 3.59 and 3.73

$$R \approx \frac{2V_{dc}^2}{\pi^2 P_0} \qquad P_{dRON} \approx \frac{\pi^2}{2} \frac{R_{ON}P_0^2}{V_{dc}^2} \qquad \text{3.76}$$

   b. assuming $V_{dc} >> V_B$, Equation 3.41 yields

$$P_{dCpn} \approx 4C_{p0}fV_B^{1/2}V_{dc}^{3/2} \qquad \text{3.77}$$

   c. the total power losses are given by

$$P_{dn} = P_{dRON} + P_{dCpn} \approx \frac{\pi^2}{2} \frac{R_{ON}P_0^2}{V_{dc}^2} + 4C_{p0}fV_B^{1/2}V_{dc}^{3/2} \qquad \text{3.78}$$

and setting the derivative of P_{dn} with respect to V_{dc} equal to zero, the maximum efficiency point is determined

$$V_{dc} \approx \left(\frac{\pi^2 R_{ON}P_0^2}{6C_{p0}fV_B^{1/2}} \right)^{2/7} \qquad \text{3.79}$$

For the example considered above, these approximate equations yield $V_{dc} = 36.61$ V, $R = 0.91\ \Omega$, and $\eta = 83.44\ \%$, in a satisfactory agreement with the exact values previously obtained.

2. In practical applications, the parasitic capacitances, C_p, are primarily abrupt-junction capacitances, but some linear capacitances are also present. Therefore, a convenient combination of the two models presented above can provide a very precise estimation of the maximum efficiency operating point.

Nonideal LC Components

CVS Circuit

Assume that the series-tuned RLC circuit resonates at the switching frequency, f, and has a loaded quality factor Q given by

$$Q = \frac{2\pi fL}{R} = \frac{1}{2\pi fCR} \qquad \text{3.80}$$

If Q_L is the unloaded quality factor values of series inductor L and Q_C is the the unloaded quality factor values of series capacitor C, then the equivalent series resistance of the LC circuit is

$$r = \frac{2\pi fL}{Q_L} + \frac{1}{2\pi fCQ_C} = RQ\left(\frac{1}{Q_L} + \frac{1}{Q_C}\right) \qquad 3.81$$

This parasitic resistance is placed in series with the load resistance, R. Consequently, the equivalent load presented to each transistor is $R + r$, and the output power dissipated in load resistance R is given by (see Equation 3.6)

$$P_0 = \frac{2}{\pi^2}\frac{V_{dc}^2}{R+r}\frac{R}{R+r} = \frac{2}{\pi^2}\frac{V_{dc}^2 R}{(R+r)^2} \qquad 3.82$$

Finally, the collector efficiency becomes

$$\eta = \frac{P_0}{P_{dc}} = \frac{R}{R+r} = \frac{1}{1 + Q\left(\dfrac{1}{Q_L} + \dfrac{1}{Q_C}\right)} \qquad 3.83$$

Some observations and practical considerations

1. Equations 3.82 and 3.83 are only approximations because P_0 and η are evaluated when the harmonic components of the load current are ignored.

2. If the loaded Q increases, the reactances of L and C and their parasitic resistances also increase. As a result, the output power (at the fundamental frequency) decreases with increasing loaded Q (assuming a constant load resistance, R).

3. To obtain a high collector efficiency, the values of Q_L and Q_C must be as high as possible. These values depend on the physical volume allocated to the LC components and on the technological possibilities. Q_C is usually at least five to ten times higher than Q_L. Thus, the power losses due to the inductor dominate the power losses due to the capacitor. Therefore, a good approximation for the collector efficiency is given by

$$\eta \approx \frac{1}{1 + \dfrac{Q}{Q_L}} \qquad 3.84$$

4. The choice of the loaded Q of the series-tuned circuit is a tradeoff among several factors.
   a. Power losses in the parasitic series resistance of L and C. These losses increase as loaded Q increases.

b. Harmonic content of the power delivered to the load (or at the input port of the harmonic-suppression filter, if one is used in circuit). The harmonic content decreases as loaded Q increases.

c. Operating frequency range with fixed component values (without variable tuning). The amplifier bandwidth decreases as loaded Q increases.

Because the unloaded quality factor of L is usually about 100...200, reasonable values of loaded Q are below 10. This assures about 2 to 5 percent power loss in the parasitic series resistance. Thus, if the harmonic content of the output power must be low (for example, typical specifications for radio transmitters require the harmonics to be attenuated to at least 60 dB below the carrier) a band- or low-pass filter must be inserted in the circuit.

5. In many practical applications, a maximum limit for the harmonic output at load is specified. It is necessary to choose a suitable configuration of the harmonic-suppression filter and a suitable value of the loaded Q to minimize the total losses in the load network and the subsequent filter (and to achieve the desired harmonic suppression).

6. A sinusoidal current is assumed through L and C in the analysis above. However, the harmonic components cannot be ignored, especially if the loaded Q is low. A more precise estimation of power losses due to the series-parasitic resistance of L and C must take into account the following considerations:

a. The capacitance and the inductance do not vary with frequency.

b. Q_C varies approximately inverse proportional with frequency because the power losses (other than dissipation in the electrodes resistance) vary with frequency as f^2 (for a given AC voltage across the capacitor).

c. Q_L varies with frequency as $f^{1/2}$ because the power losses (other than dissipation in the winding resistance) vary with frequency as $f^{1/2}$ (for a given AC current through the inductor).

EXAMPLE 3.11

For the circuit presented in Example 3.9 ($P_0 = 300$ W, $V_{dc} = 75$ V, $R = 3.644$ Ω, IRF540 MOSFET with $R_{ON} = 0.077$ Ω and $C_{OSS} = 560$ pF), the series-tuned circuit has a loaded quality factor of $Q = 5$. Estimating $Q_L = 150$ and $Q_C = 1000$, Equation 3.81 yields an equivalent series resistance of $r = 0.140$ Ω. The output power (dissipated in the load resistance R) decreases to

$$P_0 = \frac{2}{\pi^2} \frac{V_{dc}^2 R}{(R + R_{ON} + r)^2} = 278.63 \text{ W} \qquad 3.85$$

One way to maintain $P_0 = 300$ W is to reduce load resistance R. Substituting r from Equation 3.81 in Equation 3.85 and solving the resulting equation with respect to R, $R = 3.376$ Ω and then $r = 0.129$ Ω are obtained. Taking into account the power losses due to the parasitic capacitances, the DC input power is given by

$$P_{dc} = \frac{2}{\pi^2} \frac{V_{dc}^2}{R + R_{ON} + r} + 2C_{OSS}V_{dc}^2 f \approx 403.6 \text{ W} \qquad 3.86$$

resulting in the collector efficiency $\eta = 74.32\%$. The total power losses are 103.6 watts. The 85.4 watt losses are due to the parasitic capacitances; 11.4 watts are losses due to the parasitic-series resistance, and 6.8 watts are conduction losses in R_{ON}. The peak drain current is $I = 13.33$ A.

TCVS Circuit

All considerations above, as well as Equations 3.80, 3.81, 3.83, and 3.84 remain valid (substituting R_L for R).

TCCS Circuit

Assuming a loaded quality factor Q for the parallel-tuned RLC circuit (see Figure 3-8).

$$Q = \frac{R_L}{2\pi f L} = 2\pi f C R_L \qquad 3.87$$

The equivalent loss resistance of the parallel LC circuit is given by

$$r = 2\pi f L Q_L \left\| \frac{Q_C}{2\pi f C} \right. = \frac{R_L}{Q} \frac{Q_L Q_C}{Q_L + Q_C} \qquad 3.88$$

This parasitic resistance is connected in parallel with load resistance R_L. The load presented to the amplifier is $R_L \| r$. The output power dissipated in load resistance R_L is given by Equation 3.30. The collector efficiency is

$$\eta = \frac{R_L \| r}{R_L} = \frac{r}{r + R_L} = \frac{1}{1 + Q\left(\dfrac{1}{Q_L} + \dfrac{1}{Q_C}\right)} \qquad 3.89$$

the same as in the previous cases. All considerations presented for the CVS circuit remain valid, with one exception: the output power (dissipated in the load, at the fundamental frequency) is not affected by Q.

Loaded Q of the Tuned Circuit

The following analysis refers to the CVS Class D circuit (see Figure 3-1). However, the results obtained can be applied, with minor modifications, to the other configurations. The analytical description of a CVS Class D amplifier with finite output Q is based on the assumptions given in section *Complementary Voltage Switching (CVS) Circuit*, except the one regarding the output network Q.

Because the waveform of $v_2(\theta)$ is not affected by the loaded Q, Equations 3.2 and 3.80 yield the amplitudes of the odd harmonic components of the output current

$$I_{2n-1} = \frac{2}{\pi} \frac{V_{dc}}{R} \frac{1}{(2n-1)\sqrt{1+Q^2\left(2n-1-\dfrac{1}{2n-1}\right)^2}} \; , \; n = 1, 2, \dots \qquad 3.90$$

The even harmonics are zero. Using Equation 3.90, the output power is

$$P_0 = \frac{R}{2} \sum_{n=1}^{\infty} I_{2n-1}^2 = \frac{2}{\pi^2} \frac{V_{dc}^2}{R} \sum_{n=1}^{\infty} \frac{1}{(2n-1)^2\left[1+Q^2\left(2n-1-\dfrac{1}{2n-1}\right)^2\right]} \qquad 3.91$$

The total harmonic distortion is

$$THD = \frac{\sqrt{\sum\limits_{n=2}^{\infty} I_{2n-1}^2}}{I_1} = \sqrt{\frac{\pi^2}{2} \frac{P_0 R}{V_{dc}^2} - 1} \qquad 3.92$$

Assuming that for $0 \le \theta \le \pi$, Q_1 is OFF and Q_2 is ON, the current through Q_2 (in the ON state) is given by

$$i_2(\theta) = \sum_{n=1}^{\infty} I_{2n-1} \sin\left\{(2n-1)\theta - \arctan\left[Q\left(2n-1-\frac{1}{2n-1}\right)\right]\right\} \qquad 3.93$$

and the peak current through switches I_{SM} can be numerically obtained by setting the derivative of $i_2(\theta)$ equal to zero.

Finally, the power output capability is given by

$$C_P = \frac{P_0}{2V_{dc}I_{SM}} \qquad 3.94$$

The variation of these performance values of a Class D amplifier with the output Q is depicted in Figure 3-19 (for standard normalized values of

$V_{dc} = 1$ V and $R = 1$ Ω).

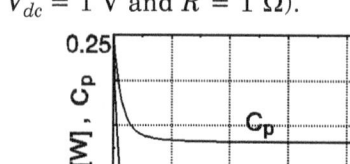
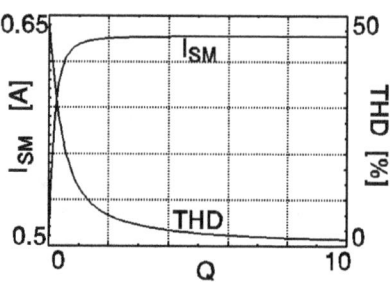

Figure 3-19 *Performance of CVS Class D amplifier versus the output Q* ($V_{dc} = 1$ V, $R = 1$ Ω, $D = 0.5$).

Equation 3.90 may be used to estimate the output power for each harmonic and the required attenuation for the harmonic-suppression filter. The output power at the fundamental frequency is not directly affected by the loaded Q. However, because the harmonic components of the output current flow through the transistors, the peak collector current and the power output capability are directly affected by the loaded Q.

EXAMPLE 3.12

For a CVS Class D amplifier operating with $V_{dc} = 75$ V, $R = 1$ Ω, and $Q = 1$, the equations presented above give $P_0 = 231.6$ W, $I_{SM} = 9.47$ A, $C_P = 0.16308$, and THD = 12.69%. The amplitude of the fundamental frequency component of the output current is $I_1 = 9.55$A; the amplitudes of the harmonic components are $I_3 = 1.12$ A (–18.6 dBc) and $I_5 = 0.39$ A (–27.8 dBc). Figure 3-20 depicts the waveform of the output current. It is interesting to note that $i_0(0) = -i_0(\pi) = -2.22$ A. This is caused by the harmonic components of the output current and, as a result, the currents through the active devices jump at the moment switching occurs, with well-known drawbacks (see **Reactive Loads**). The magnitude of these jumps decreases as Q increases and, only for infinite Q, is zero current switching rigorously obtained.

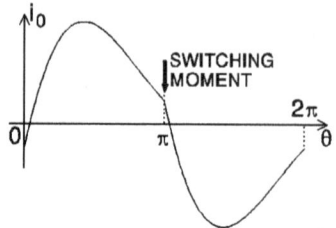

Figure 3-20 *Output current for the circuit in Example 3.12.*

Transition Time

The idealized operation of the Class D amplifier (presented in Section 3.1) assumes instantaneous switching of the active devices between the ON and OFF states. Real devices take a finite time to switch from ON to OFF and vice versa.[7] An exact analysis of waveforms and power losses during transitions is very complicated [7] and therefore only a few simple considerations are presented below.

Consider a CVS configuration and assume that the ON-to-OFF and OFF-to-ON switching times are equal. Denoting the switching time with t_s and the switching frequency with f, the angular time required for transition is $\tau = 2\pi f t_s$. During the transition, it is assumed that the transistor conductance varies linearly with time between zero and $G = 1/R_{ON}$ (see Figure 3-21).

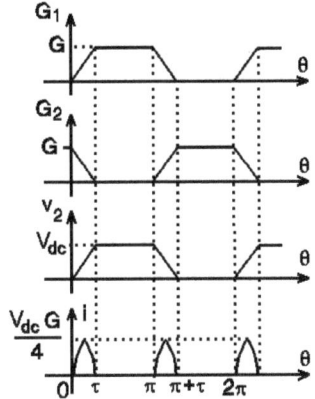

Figure 3-21 *Waveforms in CVS Class D circuit with nonzero transition times.*

For $0 \leq \theta \leq \tau$,

$$G_1(\theta) = G\frac{\theta}{\tau} \; ; \; G_2(\theta) = G\left(1 - \frac{\theta}{\tau}\right) \qquad 3.95$$

Without considering the output current that flows alternately through the two transistors,[8] the current through G_1 and G_2 during the transition is

$$i(\theta) = V_{dc}G\frac{\theta}{\tau}\left(1 - \frac{\theta}{\tau}\right) \qquad 3.96$$

resulting in (for $0 \leq \theta \leq \tau$)

$$v_2(\theta) = \frac{i(\theta)}{G_2(\theta)} = V_{dc} \frac{\theta}{\tau}$$

3.97

Decomposing $v_2(\theta)$ into a Fourier series, the amplitude of the output current is

$$I = \frac{4}{\pi} \frac{V_{dc}}{R} \frac{\sin(\tau/2)}{\tau}$$

3.98

and the output power is given by

$$P_0 = \frac{I^2}{2} R = \frac{8}{\pi^2} \frac{V_{dc}^2}{R} \frac{\sin^2(\tau/2)}{\tau^2}$$

3.99

The DC input current is the sum between the average value of $i(\theta)$ and the average value of the current through Q_1 due to the output current

$$I_{dc} = \frac{1}{2\pi} \int_0^{2\pi} i(\theta)\, d\theta + \frac{I}{\pi} = V_{dc} \left[\frac{G}{\pi} \frac{\tau}{6} + \frac{4}{\pi^2 R} \frac{\sin(\tau/2)}{\tau} \right]$$

3.100

The DC input power is then

$$P_{dc} = V_{dc} I_{dc} = V_{dc}^2 \left[\frac{G}{\pi} \frac{\tau}{6} + \frac{4}{\pi^2 R} \frac{\sin(\tau/2)}{\tau} \right]$$

3.101

resulting in the collector efficiency

$$\eta = \frac{P_0}{P_{dc}} = \frac{\dfrac{8}{\pi R} \dfrac{\sin^2(\tau/2)}{\tau^2}}{G\dfrac{\tau}{6} + \dfrac{4}{\pi R} \dfrac{\sin(\tau/2)}{\tau}} = \frac{\dfrac{8}{\pi} \dfrac{R_{ON}}{R} \dfrac{\sin^2(\tau/2)}{\tau^2}}{\dfrac{\tau}{6} + \dfrac{4}{\pi} \dfrac{R_{ON}}{R} \dfrac{\sin(\tau/2)}{\tau}}$$

3.102

The variation of the collector efficiency with the angular switching time, with R_{ON}/R as a parameter, is shown in Figure 3-22. The curves and the equations show that

1. The output power and the collector efficiency decrease as transition time increases.

2. The collector efficiency decreases as R_{ON} decreases (it is obvious that if R_{ON} decreases, the conduction losses also decrease, but this effect is not considered here).

3. The peak value of the current pulses through the active devices increas-

es as R_{ON} decreases and is independent of the transition time.

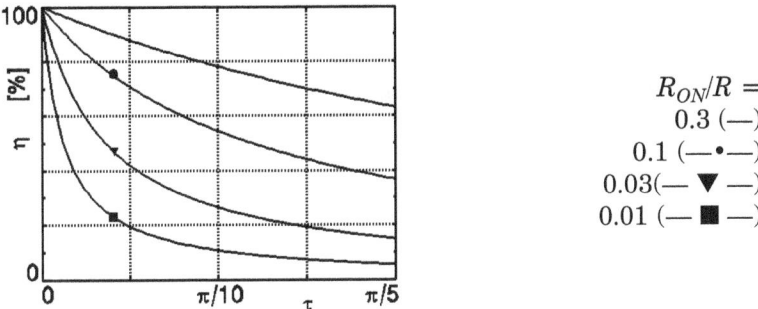

$R_{ON}/R =$
0.3 (—)
0.1 (— • —)
0.03(— ▼ —)
0.01 (— ■ —)

Figure 3-22 *Collector efficiency versus transition time for*

In a practical design, there is a major risk if the transistors' ratings are exceeded: the transistors may be destroyed. Moreover, in circuits using BJTs (which switch more slowly from ON to OFF) both transistors can be fully saturated simultaneously, shortening the supply source. The transistor failure probability is higher in this case, because a high amplitude current pulse may occur.

These problems can be eliminated by modifying the drive signals. Dead time between the signals driving the transistors in the ON state is introduced as shown in Figure 3-23. The required drive signals can be obtained using high-speed logic circuits (as suggested in [17]) or (as will be shown in Section 3.4) using a sinusoidal drive. A suitable adjustment of dead time prevents the transistors from operating in the ON state simultaneously. Moreover, a further increase of the dead time allows a higher practical efficiency, eliminating the power losses due to the parasitic capacitances (this matter is discussed in Section 3.4).

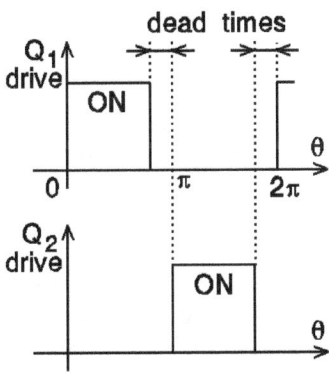

Figure 3-23 *Drive signals using dead times.*

Assuming that the active devices are not simultaneously in the ON state, and the angular transition time for each device is τ, the waveform of $v_2(\theta)$ can be approximated by a trapezoidal waveform (similar to that depicted in Figure 3-21) with rise and fall times equal to 2τ (see [2]). Decomposing $v_2(\theta)$ into a Fourier series yields the amplitude of the output current

$$I = \frac{2}{\pi} \frac{V_{dc}}{R} \frac{\sin \tau}{\tau} \qquad 3.103$$

resulting in the output power

$$P_0 = \frac{I^2}{2} R = \frac{2}{\pi^2} \frac{V_{dc}^2}{R} \left(\frac{\sin \tau}{\tau} \right)^2 \qquad 3.104$$

Because the output current waveform is only quantitatively affected, the DC input power is given by

$$P_{dc} = V_{dc} I_{dc} = V_{dc} \frac{I}{\pi} = \frac{2}{\pi^2} \frac{V_{dc}^2}{R} \frac{\sin \tau}{\tau} \qquad 3.105$$

and the collector efficiency is

$$\eta = \frac{P_0}{P_{dc}} = \frac{\sin \tau}{\tau} \approx 1 - \frac{\tau^2}{6} \qquad 3.106$$

EXAMPLE 3.13
If the switching time is 10 percent from the RF period, then Equation 3.106 yields $\eta = 93.5\%$.

Duty Ratio

The idealized operation of the Class D amplifier presented in Section 3.1 assumes a 50 percent duty ratio. In the following analysis, a CVS Class D circuit (see Figure 3-1) is considered and the assumptions made in section *Complementary Voltage Switching (CVS)* are accepted, except for that concerning the duty ratio. It is assumed below that the duty ratio, D (defined as shown in Figure 3-24), has an arbitrary value $0 < D < 1$. Note that the waveforms of Figure 3-24 depict the case $0 < D < 0.5$. If $0.5 < D < 1$, then $i_1(\theta)$ has both positive and negative values while $i_2(\theta)$ remains permanently positive.

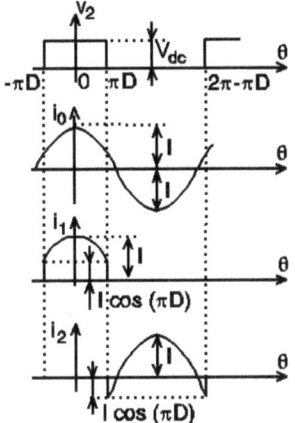

Figure 3-24 *Waveforms in CVS Class D circuit with arbitrary D (0 < D < 1).*

If $-\pi D \le \theta \le \pi D$, then Q_1 is ON and Q_2 is OFF; if $\pi D \le \theta \le 2\pi - \pi D$, then Q_1 is OFF and Q_2 is ON.

$$v_2(\theta) = \begin{cases} V_{dc} & -\pi D \le \theta \le \pi D \\ 0 & \pi D \le \theta \le 2\pi - \pi D \end{cases} \qquad 3.107$$

Decomposing $v_2(\theta)$ into a Fourier series yields

$$v_2(\theta) = DV_{dc} + \frac{2}{\pi}V_{dc}\sum_{n=1}^{\infty}\frac{\sin n\pi D}{n}\cos n\theta \qquad 3.108$$

resulting in the output current

$$i_0(\theta) = \frac{2}{\pi}\frac{V_{dc}}{R}\sin\pi D\cos\theta = I\cos\theta \qquad 3.109$$

and the output power

$$P_0 = \frac{2}{\pi^2}\frac{V_{dc}^2}{R}\sin^2\pi D = P_{0,nom}\sin^2\pi D \qquad 3.110$$

where $P_{0,nom}$ is the "nominal" output power for $D = 0.5$.

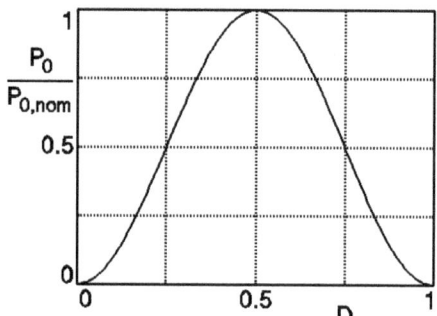

Figure 3-25 *Variation of the output power with the duty ratio, D.*

The variation of the output power (normalized to $P_{0,nom}$) with the duty ratio, D, is presented in Figure 3-25. As expected, if D approaches near 0 or 1, the output power rapidly decreases to zero. However, for small variations of D around $D = 0.5$, the output power is only slightly affected (for example, if $0.45 \leq D \leq 0.55$, the output power does not decrease under 97.55 percent from the nominal output power, obtained for $D = 0.5$).

The DC input current is given by

$$I_{dc} = \overline{i_1(\theta)} = \frac{1}{2\pi} \int_0^{2\pi} i_1(\theta)\, d\theta = \frac{2}{\pi^2} \frac{V_{dc}^2}{R} \sin^2 \pi D \qquad 3.111$$

The DC input power is

$$P_{dc} = V_{dc} I_{dc} = \frac{2}{\pi^2} \frac{V_{dc}^2}{R} \sin^2 \pi D = P_0 \qquad 3.112$$

and the collector efficiency remains (ideally) 100 percent, for any value of D. However, if $D > 0.5$, the currents through the active devices jumps when switching. This causes longer transition times and higher power dissipation during the transition times due to the lead inductances.

Power output capability is unaffected by the duty ratio $C_P = 1/(2\pi) = 0.1592$ (if P_0 decreases, in accordance to Equation 3.109, the peak collector current also decreases). Note that, if $D < 0.5$, Q_2 must be able to pass a negative current; if $D > 0.5$, then Q_1 must be able to pass a negative current. Therefore, if Q_1 and Q_2 are bipolar transistors, anti-parallel diodes are required (see Figure 3-13).

The discussion above is valid for TCVS and TCCS circuits, with just one difference. In a current switching Class D circuit operating at $D \neq 0.5$, the active devices must withstand negative collector voltages. Consequently, protection series diodes may be required (see section **Reactive Loads**).

Equation 3.109 shows that the amplitude of the load current is proportional to sin(πD). This suggests it is possible to obtain an *amplitude modulated* (AM) signal by varying duty ratio D using a standard *pulse-width modulation* (PWM) [8, 9, 22]. The desired AM signal can be obtained applying an RF PWM signal at the input of a band-pass filter (for example, a tuned circuit). Either monopolar or bipolar pulse trains can be used (see Figure 3-26). The bipolar pulse trains have certain advantages which are considered below. A linear modulation characteristic can be obtained if the pulse width varies as the inverse sine of the modulating signal. A possible solution using this principle is suggested in the circuit of Figure 3-27. The corresponding waveforms are presented in Figure 3-28.

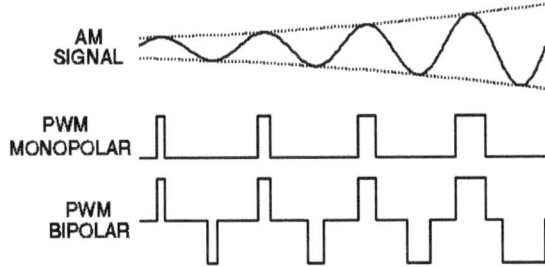

Figure 3-26 *RF PWM signals.*

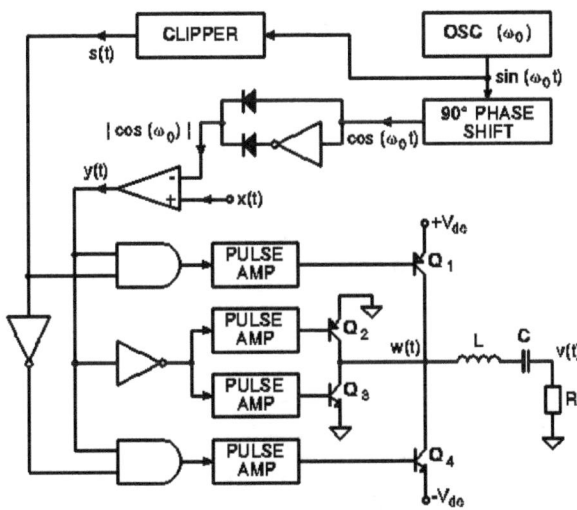

Figure 3-27 *Generating AM signals using RF PWM.*

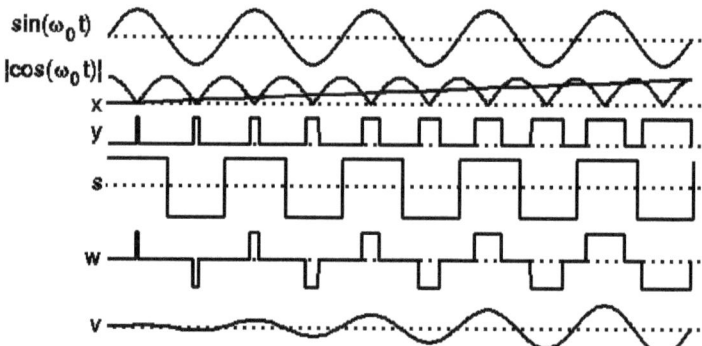

Figure 3-28 *Waveforms in the circuit of Figure 3-27.*

A sinusoidal oscillator supplies the carrier $\sin(\omega_0 t)$, where ω_0 represents the carrier frequency. This signal is phase shifted, then full-wave rectified and applied to the inverting input of a comparator. (For simplicity's sake, the modulating signal $x(t)$ is assumed to be normalized such that $0 \le x$ $(t) \le 1$, and the generated carrier is assumed to have an amplitude equal to 1). Pulses of the same polarity are generated by comparing the modulating signal and $|cos(\omega_0 t)|$. The pulse width of $y(t)$ is proportional to the inverse sine of the modulating signal $x(t)$. A ± 1 square wave $s(t)$ is generated by clipping the RF carrier $\sin(\omega_0 t)$. Pulses of proper polarity at the output are produced using the appropriate logic circuit and four transistors $Q_1 - Q_4$ acting as a three-position switch (see Figure 3-29). Figure 3-30 presents the waveforms in the three-position switch obtained with $Q_1 - Q_4$. Note that Q_2 and Q_3 ensure the closing path to ground for both polarities of the sinusoidal load current during the time interval when $\omega(t) = 0$ and Q_1 and Q_4 are OFF. In practical applications, diodes in series with the collectors of Q_2 and Q_3 are required. They prevent the collector-to-base junctions of Q_2 and Q_3 from conducting when Q_1 and Q_4 are OFF.

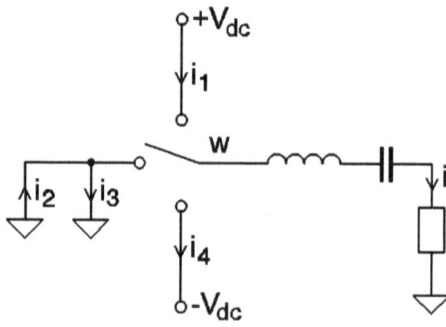

Figure 3-29 *Three-position switch obtained with $Q_1 - Q_4$.*

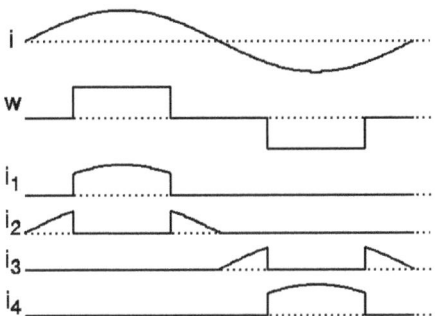

Figure 3-30 *Waveforms in the three-position switch.*

For monopolar operation, significant simplifications appear in the circuit of Figure 3-27. The DC supply voltage $-V_{dc}$ is not required, and the transistors Q_2 can be replaced by diodes connected to $+V_{dc}$, and Q_3 can be replaced by diodes connected to ground. Moreover, in a circuit operating with monopolar pulses, the number of transitions is halved; the switching losses are reduced and the efficiency at high frequencies increases. However, the bipolar operation has important advantages regarding the harmonic content in comparison with the monopolar operation.

1. The bipolar pulse train contains only odd harmonics, thus simplifying the harmonic-suppression filters.

2. If an AM single-tone signal (DSB, full carrier or suppressed carrier) is generated using bipolar pulses, the spurious products are band limited around the odd harmonics [22]. For example, the spurious products around the third harmonic occupy a three times larger bandwidth than the original modulating signal.

If the modulating frequency is low enough (below one-half of the carrier frequency), the spurious products do not cross into the passband and can be filtered out easily. This is not possible with monopolar RF PWM where the spurious products have infinite bandwidth. Many of these products fall into the passband and cannot be removed [22]. The waveforms of Figure 3-30 show that the currents through the active devices usually make jumps when switching.

Drive
In a Class D amplifier, the driving signal must be sufficient to assure that the transistors are alternately cut off or fully saturated, assuming worst-case conditions of operating temperature, gain, and such. A current switching Class D amplifier requires either a rectangular drive current

waveform (if BJTs are used) or a rectangular drive voltage waveform (if MOSFETs are used). A voltage switching Class D amplifier can be driven by either a rectangular or sinusoidal driving signal. The latter is usually preferable in practical applications because

a.   it reduces the storage time and saturation voltage of BJTs [7],
b.   it reduces the driving power requirements,
c.   it allows a simple and precise control of the dead time.

Driving BJTs

The input portion of a TCVS Class D circuit using BJTs is shown in Figure 3-31. If a sinusoidal current is driven into the primary winding of the input transformer, the currents through the secondary windings (i_{B1} and i_{B2}) are half-sinusoid, as shown in Figure 3-32.

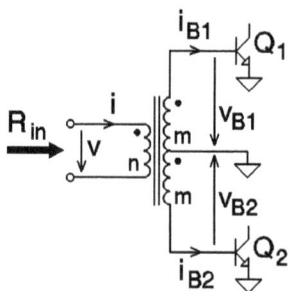

Figure 3-31 *Input portion of a TCVS Class D circuit using BJTs.*

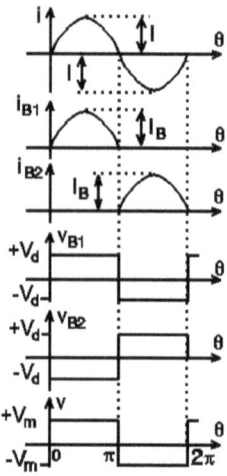

Figure 3-32 *Waveforms in the circuit of Figure 3-31.*

The current injected into a base-emitter junction causes the base-to-emitter voltage to rise to $V_d = 0.7$ V (the transistor is ON). The input transformer causes the reflection of this voltage at the input of the other transistor, where the base-to-emitter voltage is equal to $-V_d$ (the transistor is in the OFF state). Consequently, a square-wave voltage with levels $\pm V_m$, where $V_m = (n/m)V_d$, appears across the primary winding. At the switching frequency, the load impedance presented to the driver (R_{in}) is the ratio of the amplitude of the fundamental frequency component of $v(\theta)$ to the amplitude of the driving current $i(\theta)$.

$$R_{in} = \frac{4}{\pi}\left(\frac{n}{m}\right)^2 \frac{V_d}{I_B} \qquad \text{3.113}$$

The driving power is given by

$$P_{in} = \frac{2}{\pi} V_d I_B \qquad \text{3.114}$$

Some observations and practical considerations

1. The amplitude of the base current must be large enough ($> I_{Cmax}/\beta$) to sustain the peak collector current.

2. R_{in} depends only on the amplitude of the base current I_B and is not related to transistor parameters other than V_d. The driver output current must be limited to prevent transistor failure [2, 8].

3. In a TCCS Class D circuit using BJTs, the base current and voltage are rectangular waveforms. In this case

$$R_{in} = \left(\frac{n}{m}\right)^2 \frac{V_d}{I_B} \qquad P_{in} = V_d I_B \qquad \text{3.115}$$

The driver output current must be limited to prevent damage to the amplifier transistors.

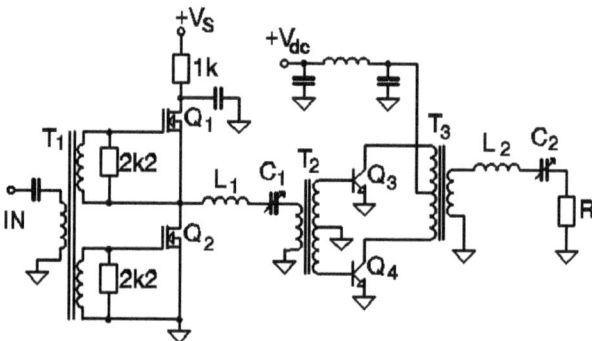

Figure 3-33 *Class D amplifier for 160, 80, and 40 meters [8].*

EXAMPLE 3.14

The Class D circuit from Figure 3-33 delivers an output power of $P_0 = 50$ W in an $R = 50$ Ω load [8]. The output stage is a TCVS circuit using 2N5262 BJTs (Q_3 and Q_4). If the DC supply voltage is $V_{dc} = 28.5$ V and the saturation voltage is $V_{CEsat} = 1$ V, the collector voltages swing between 1 and 56 volts. The output transformer, T_3 (a 4:1 matching transformer), reduces the load resistance to an equivalent 12.5-ohm load resistance presented to each transistor. With these values, the output power is $P_0 = 50$ W and the peak collector current is $I_C = 2.83$ A. Q_3 and Q_4 are driven using a sinusoidal current applied through input transformer T_2. Assuming for both BJTs, $V_d = 1$ V and $\beta = 20$,

- the peak base current is $I_B = I_C/\beta = 142$ mA,

- the driving power is $P_{in} = 90$ mW (hence, the power gain of $G_p = 27$ dB),

- the impedance presented to the driver $R_{in} = 145$ Ω (T_2 is a 16:1 matching transformer, having $n/m = 4$; if T_2 were a 1:1 transformer, the impedance presented to the driver would be too small — about 9 ohms).

A CVS Class D circuit containing SD-200 MOSFETs (Q_1 and Q_2) is used to drive the bases of Q_3 and Q_4. Because harmonic suppression is not important in this stage, the series-tuned circuit has a loaded Q of 2.5. The voltage across the primary winding of T_2 varies between ± 4 volts, and the current input to T_2 has an amplitude of about 142 mA/4 = 35 mA. Because the load resistance of this amplifier is 145 ohms and the MOSFETs have R_{ON} = 45 Ω, the DC supply voltage required for the driver (see Equation 3.59) is V_{DD} = 10.5 V. This value provides drain efficiency of about 77 percent for the driver. However, the driver is not operated from a 10.5-volt power supply because, should the base-to-emitter voltage of Q_3 and Q_4 (V_D) decrease for any reason (first, $V_D \approx$ 1 V is an approximate value; second, V_D varies with temperature, collector current, and other parameters) the output current of the driver would increase and the current would only be limited by the saturation resistances. It is better to use a higher voltage supply (V_S = 21.6 V) connected in series with a 1 kiloohm resistor (with the previously calculated values, I_{dc} = 11.1 mA and an 11.1 volt drop across the resistor). This results in a high DC input power (about 240 mW) and a very low overall drive efficiency (about 37 percent). However, the overall amplifier efficiency is not significantly affected by these supplementary losses. If V_d decreases and the output current of the driver increases, the DC input current I_{dc} also increases. As a result, the DC supply voltage at the drain of Q_1 decreases, limiting the output current of the driver. Field-effect transistors Q_1 and Q_2 are driven with a sinusoidal voltage (applied across the primary winding) having a 3-volt amplitude. The input transformer T_1 multiplies this voltage by 5, applying about 15 volts peak sinusoidal gate-to-source voltage to the MOSFETs (this voltage ensures their saturation). The 2.2 kiloohm resistors are used to produce an approximate 50-ohm input impedance.

Driving MOSFETs

To simplify the analysis, the MOSFET gate circuit is modeled by a series-RC circuit [5, 23]. The resistance and capacitance are assumed to be independent of frequency, current or voltage.[9] If the driving voltage is a rectangular waveform (see Figure 3-34), the gate-to-source voltage, v_{GS}, and the gate current, i_G, vary, as shown in Figure 3-35.

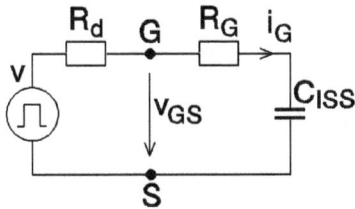

Figure 3-34 *Simplified equivalent gate circuit (square-wave drive).*

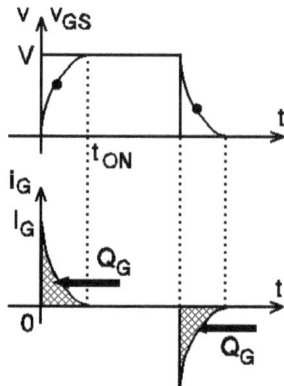

Figure 3-35 *Waveforms in the gate circuit: — v; —•—v_{GS}.*

R_G and C_{ISS} model the MOSFET input. C_{ISS} is a standard notation input capacitance of a MOS transistor and its value is presented in the data sheets. C_{ISS} is the input capacitance measured at null drain-to-source voltage, so $C_{ISS} = C_{GS} + C_{GD}$.

The voltage source, $v(t)$, in series with R_d, models the driver output. The total charge required by the gate to modify the gate-to-source voltage between 0 and V (this charge is given by the hatched area under the gate current in Figure 3-35) and the peak gate-to-source voltage V determine the energy stored in the input capacitance C_{ISS}

$$E_G = \frac{1}{2} V Q_G \qquad 3.116$$

(RF power MOSFETs or switching power MOSFETs are enhancement-mode, n-channel devices. Thus, for $v_{GS} = 0$, the transistor is OFF; voltage V is assumed high enough to ensure that the device is ON.)

When C_{ISS} is charged, the same amount of energy is dissipated into the series resistance ($R_d + R_G$). When C_{ISS} is discharged, the stored energy is also dissipated into the series resistance. Therefore, the power losses in the gate circuit are given by

$$P_G = V Q_G f \qquad 3.117$$

where f is the switching frequency.

The peak current, I_G, which is required to switch the MOSFET depends on the desired transition time, t_{ON}. It determines the output resistance of the driver from[10]

$$t_{ON} = 4\tau = 4(R_d + R_G)C_{ISS} \qquad\qquad 3.118$$

The peak current, I_{Gm}, is defined as

$$I_G = \frac{V}{R_d + R_G} \qquad\qquad 3.119$$

Note that the power losses in the gate circuit (for square-wave drive) are independent of the switching speed. The transition time, t_{ON}, is limited only by the series-gate resistance, R_G. There is an intrinsic limit regarding a MOSFET's switching speed and the maximum operating frequency with square-wave drive.

If the driving voltage is sinusoidal (see Figure 3-36) a series inductance, L_d, is introduced in the circuit to resonate with C_{ISS} at the operating frequency. L_d includes all lead and leakage inductances in the gate drive circuit; this is an important advantage when compared with the square-wave drive case, where the series inductance must be minimized. Because the nonlinearity of the input capacitance is ignored, gate current i_G is also sinusoidal (see Figure 3-37).

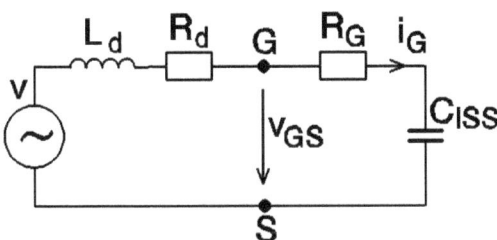

Figure 3-36 *Simplified equivalent gate circuit (sinusoidal drive).*

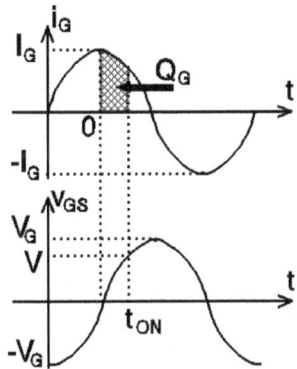

Figure 3-37 *Waveforms in the gate circuit for sinusoidal drive.*

Ignoring R_G in comparison with the reactance of C_{ISS} (a reasonable assumption at moderate frequencies), the gate-to-source voltage, v_{GS}, is shifted in phase (a 90 degree lag) relative to the i_G current waveform. To charge the gate in time, t_{ON}, with a charge, Q_G (from zero to voltage V, sufficient to assure the transistor is ON if $v_{GS} > $ V), the sinusoidal gate current (with frequency f, $\omega = 2\pi f$) must have an amplitude

$$I_G = \frac{Q_G}{\int_0^{t_{ON}} \cos \omega t \, dt} = \frac{Q_G \omega}{\sin \omega t_{ON}} \qquad \text{3.120}$$

The peak value of the gate-to-source voltage is given by

$$V_G = \frac{V}{\sin \omega t_{ON}} \qquad \text{3.121}$$

and increases as the transition time decreases.

At a given operating frequency, the transition time can be reduced by increasing peak current I_G and amplitude V_G of the gate-to-source voltage. However, the transition time cannot be reduced under an intrinsic limit of the MOSFET, which is imposed by the maximum gate-to-source voltage. Given specified operating frequency and transition time, t_{ON}, Equation 3.120 yields the peak gate current. Thus, the power losses in the gate circuit are given by

$$P_G = \frac{1}{2} I_G^2 (R_d + R_G) \qquad \text{3.122}$$

and are proportional to the total series resistance in the gate circuit. The power losses also depend on the switching speed (because of the gate current I_G) and can be reduced by decreasing the drive resistance, R_d.

Note that, for the square-wave gate drive, the energy stored in C_{ISS} is fully dissipated in its series resistance. Because the gate circuit resonates at the operating frequency for sinusoidal resonant gate drive, the energy is alternately stored in C_{ISS} and L_d. If the circuit series resistance decreases, the losses in the gate circuit also decrease. In general, the square-wave drive has more rigid requirements than the sinusoidal-wave drive (faster devices, able to pass higher currents).

EXAMPLE 3.15

To charge a 2000 pF capacitor in 10 nanoseconds to 10 volts (square-wave voltage), the total series resistance of the gate circuit must be 1.25 ohms. The driver must be able to source and sink a peak current of 8 amps. To compare these results with those obtained using sinusoidal gate drive, an operating frequency need be assumed. The waveforms presented in Figure 3-37 show that the maximum operating frequency for a given switching time, t_{ON}, occurs when switching time t_{ON} is equal to a quarter of the full cycle. In this case, for t_{ON} = 10 ns, the maximum operating frequency with sinusoidal gate drive is $1/(4t_{ON})$ = 25 MHz. In fact, this is the worst-case condition because the peak gate current, I_G, increases as the frequency increases, in accordance with Equation 3.120. The driver must be able to source and sink a peak current of about 3.14 amps, approximately 2.5 times less than that of the square-wave drive.

EXAMPLE 3.16

For the circuit in Example 3.5, the power losses in the gate circuit can be calculated using Equation 3.117. Considering a 0 to 10 volt square-wave drive, the gate charge required to turn ON an IRF540 is 39 nC [20]. Consequently, the power losses in the gate circuits of both MOSFETs are 10.6 watts (about eight times less than the power losses due to the parasitic capacitances).

EXAMPLE 3.17

Figure 3-38 depicts the input portion of a TCVS Class D circuit using MOSFETs, but the same configuration can be used for Class B operation [17]. In Class D operation, this circuit (using MRF148 MOSFETs) delivers about 100 watts at frequencies up to 60 MHz. Input transformer T_1 (a balun) ensures 180 degree phase shift between the two gate voltages. The 12.5-ohm resistors swamp the gates, diminishing the effect of the voltage-variable input capacitance and presenting an approximate constant resistive load to the driver. The bias circuits allow for separate adjustment of the quiescent drain current of each transistor (for Class B operation). When the circuit is operated in Class D, the bias circuits allow the adjustment of the gate-to-source voltage (without signal) to be slightly lower than the threshold voltage (so the transistors remain OFF when there is no signal). This reduces the drive power requirements and increases the power gain.

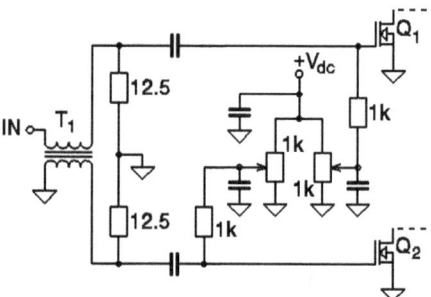

Figure 3-38 *Input portion of a TCVS circuit using MOSFETs.*

EXAMPLE 3.18

High-power MOSFET gates can be driven by a complementary pair of BJTs or MOSFETs. This solution (presented in Figure 3-39) is applied in a TCVS Class D circuit using ARF440/441 MOSFETs. It delivers 250 watts at frequencies from 1.8 to 13.56 MHz [25, 26]. The driver uses a complimentary pair of Siliconix 2N7016 (p-channel) and 2N7012 (n-channel) MOSFETs. The out-of-phase driving signals for the two halves of the amplifier are provided by inverting/noninverting Schmitt triggers. A similar solution using approximate complimentary pairs of BJTs driven by ECL circuits is presented in [16].

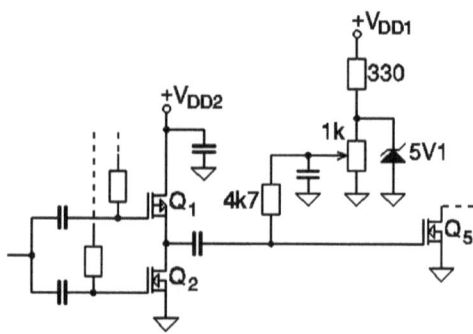

Figure 3-39 *Driving a high-power MOSFET from a complementary pair of MOSFETs [25, 26].*

To conclude, the discussion above emphasizes the following advantages of the sinusoidal gate drive:

1. The requirements for the driver active devices are less rigid in their speed and current capability.

2. All parasitic inductances (lead inductances and leakage inductances) in the gate circuit can be absorbed into resonant inductor L_d (for square-wave drive, the parasitic inductances must be minimized).

3. The power losses in the gate circuit are significantly reduced.

4. The requirements for the input transformer are less rigid (regarding its bandwidth) because the signal is sinusoidal.

Amplitude Modulation of a Class D Amplifier

A Class D amplifier can produce AM signals using a method based on RF pulse width modulation. However, the circuit is complicated (even for AM DSB full carrier signals), and the transistors must operate in hard-switching conditions that result in major losses, especially at high frequencies.

According to Equations 3.4, 3.12 and 3.22, the amplitude of the output voltage is proportional to the DC supply voltage, V_{dc}. This suggests that it is possible to obtain an AM signal using collector amplitude modulation [12]. The circuit appears in Figure 3-40.

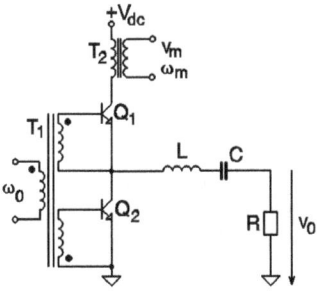

Figure 3-40 *Class D amplifier using collector amplitude modulation.*

Because wave shapes and conduction angles are not changed by the DC supply voltage, the amplitude modulation characteristic, $V_0 = V_0 (V_{dc})$, is extremely linear. The most important deviations from perfect linearity are caused by input-to-output feedthrough of the driving signal via the parasitic capacitance, which does not allow the output voltage to drop to zero when V_{dc} is reduced to zero.

Assuming a modulating signal

$$v_m(t) = V_m \cos \omega_m t$$

3.123

the supply voltage of the amplifier is

$$v_{CQ1}(t) = V_{dc} + V_m \cos \omega_m t \qquad 3.124$$

An AM DSB full carrier signal is then obtained

$$v_0(t) = \frac{2}{\pi} V_{dc}(1 + m\cos\omega_m t)\cos\omega_0 t = V_0(1 + m\cos\omega_m t)\cos\omega_0 t \qquad 3.125$$

where

$$m = \frac{V_m}{V_{dc}} \le 1 \qquad 3.126$$

is the modulation depth.

This method provides good practical results. A high overall efficiency (ideally, 100 percent) may be obtained using a Class S modulator. The same circuit can be used as a component part of an *envelope elimination and restoration* (EER) system, allowing both linear amplification and high efficiency. Another important result is provided in [27]. In a Class D amplifier using MOSFETs, the instantaneous efficiency and the average efficiency do not depend on the characteristics of the signal being amplified, because they are equal to the efficiency at peak envelope power. If BJTs are used, the average efficiency of the Class D amplifier depends on the statistical characteristics of the signal amplitude [27].

Other Practical Considerations

1. The Class D amplifier must be tuned only at very low DC supply voltages (several volts), because when the load is reactive (mistuned), large voltage spikes are generated (if BJTs without anti-parallel protection diodes are used), which usually damages the transistors.

2. If the output power increases when the drive level is increased slightly, the drive does not have sufficient power to switch the transistors.

3. The output transformer (in TCVS or TCCS circuits) is one of the critical components in practical designs because
    a. The output power can be very high, from hundreds of watts to kilowatts.
    b. Output transformer efficiency must be as high as possible because it directly affects overall circuit efficiency.
    c. The voltages across (or the currents through) the windings have rectangular waveforms and require a large bandwidth.

d. In a TCVS circuit, the parasitic capacitances associated with the output transformer must be minimized; in a TCCS circuit, the leakage inductances of the output transformer must be minimized.

For these reasons, conventional (wire-wound) transformers cannot be used at high frequencies; only transmission-line transformers perform well in these applications [2, 8, 15, 17, 25, 26, 28]. The use of transmission-line transformers resolves the problems presented above (parasitic capacitances and inductances and a large bandwidth) and allows for a high efficiency rating (98...99 percent). The main possible disadvantage concerns the transformation ratio, which must be a ratio of two small integers (restricting, to a certain extent, the matching possibilities).

4. The input transformer is a less critical component, if a sinusoidal gate drive is used. Transmission-line transformers are needed to obtain efficient performances. The input transformer from a TCVS amplifier, [16, 25, 26] or a TCCS amplifier can be replaced by high-speed logic circuits and distinct drivers, with comparable results.

The CVS circuit, without an output transformer, has the following advantages:

1. greater bandwidth (it is not affected by the output transformer),

2. reduced size and weight,

3. elimination of intermodulation distortion due to core saturation,

4. elimination of power losses in the output transformer, thus a higher overall efficiency,

5. If a transformer is required for impedance matching at the output, it can be a single-ended configuration that is smaller, less complex, and has a larger bandwidth than the push-pull transformer normally required in a TCVS or TCCS circuit.

On the other hand, the TCVS and TCCS circuits have the following advantages:

1. standard packages are intended for grounded emitter (source) applications,

2. MOSFET gates can be easily biased and their sources connected to ground,

3. input transformer can be replaced by high-speed logic circuits to obtain the out-of-phase signals and distinct drivers.

The TCCS circuit seems to have advantages at high frequencies because the parasitic capacitances do not cause power losses. However, this advantage is outweighed by the following disadvantages:

1. It requires a large inductor (an RF choke), which introduces additional power losses.

2. It requires a square-wave drive that yields higher power losses in the gate circuit.

3. The active devices switch rectangular currents, which lead to higher switching and saturation losses.

5. In practical applications, the output power, load resistance, and DC supply voltage are specified. The configurations that use an output transformer allow latitude in the choice of m/n ratio, with certain limitations. Therefore, the load resistance can be transformed into a more convenient equivalent load resistance to be presented to each transistor. This advantage is often insignificant; in many applications, harmonic-suppression filters are required that can perform the desired impedance matching.

6. The CVS circuit in Figure 3-1 is sometimes called a half-bridge Class D circuit. When the output-tuned circuit is connected between two half bridges, a full-bridge Class D circuit is obtained.

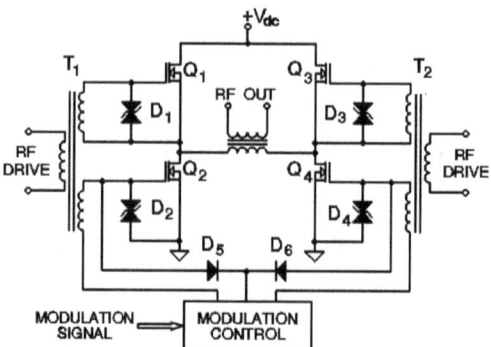

Figure 3-41 *Full-bridge Class D amplifier used in the Harris DX-10 AM transmitter.*

A full-bridge Class D amplifier is shown in Figure 3-41 [29]. This circuit is used in the Harris DX-10 AM broadcasting transmitter, which is a medium-wave transmitter with a carrier power of 10 kilowatts. Field-effect transistors Q_1 and Q_4 are driven 180 degrees out-of-phase with transistors Q_2 and Q_3. As a result, Q_1 and Q_4 are ON during half the cycle, while Q_2 and Q_3 are OFF. This produces a square-wave voltage (with an amplitude of 2 V_{dc}) across the output transformer winding, and an output power four times higher than that of a half-bridge circuit (see Equation 3.8). Back-to-back Zener diodes $D_1...D_4$ provide protection for the MOSFET inputs from transients and excessive drive. Diodes D_5 and D_6 are used to control the digital modulation.

Class B versus Class D Amplifier

1. The push-pull Class B amplifier does not require tuned circuits or harmonic-suppression filters; therefore, its bandwidth is larger than that of a Class D amplifier.

2. Class B operation demands less of the RF transistors and allows operation at frequencies higher than those at which Class D operation is possible.

3. A Class D amplifier offers higher efficiency (ideally, 100 percent) than a Class B amplifier (ideally, 78.5 percent for constant-envelope signals). In addition, the average efficiency for variable-amplitude signals is significantly lower in a Class B amplifier than in a system that includes a Class D amplifier and a Class S modulator (for example, an EER system).

4. The power output capability of a Class D amplifier is higher than that of a Class B amplifier.

5. The efficiency of a Class D amplifier is degraded minimally by reactive loads. In contrast, the efficiency of a Class B amplifier is degraded significantly by a mismatched load.

6. A Class D amplifier has much lower performance sensitivity. It also has fewer tendencies to oscillate because the transistors spend little time in the active region and the transistors' gain is low when they are either ON or OFF.

3.3 Class BD Amplifier

There are important differences (regarding the circuits and the principles of operation) between push-pull Class B and Class D amplifiers. An RF amplifier cannot normally operate in both Class B and Class D due to the constraints imposed by the tuned circuit and DC power input circuit. Consequently, if the active devices of a Class B amplifier are driven to saturation, there is no increase in output power or collector efficiency. It is also not possible to operate the active devices in a Class D amplifier as controlled-current sources.

The push-pull Class B amplifier has a linear transfer characteristic (drive envelope to output envelope) and relatively high efficiency (ideally, 78.5 percent) at peak output. However, the average efficiency for different regularly occurring signals (single-tone AM with full carrier or suppressed carrier, two-tone single sideband, quadrature-amplitude modulation, Gaussian, Gaussian AM, Rayleigh, Laplacian, Laplacian AM) is much lower [27]. The Class D amplifier has a very high collector efficiency (ideally, 100 percent), but a relatively complex system (including a Class S modulator) is required to obtain a linear transfer characteristic.

The Class BD[11] amplifier circuit [30] is similar to a current switching Class D amplifier. The circuit (see Figure 3-42) differs with the addition of a resistor, R, and DC blocking capacitor, C. The presence of a resistive AC current path from the center tap of the primary winding to ground allows the active devices to operate either as current sources or as switches.

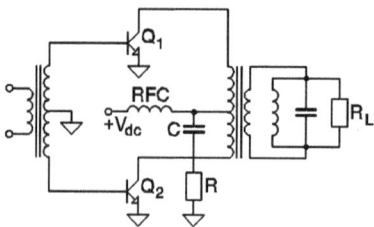

Figure 3-42 *Class BD amplifier.*

At lower output levels, Q_1 and Q_2 are alternately cut off and in the active region (acting as controlled-current sources). The circuit operates linearly, as a Class B amplifier. At higher output levels, Q_1 and Q_2 are alternately cut off (during one half of the cycle) and in the active and saturated region (during the other half of the cycle); this is called the amplifier's partially saturated region. At peak output, Q_1 and Q_2 are either cut off or saturated, as in a Class D amplifier; this is called the amplifier's fully saturated region.

The addition of resistor R does not affect the collector efficiency during linear mode operation (Class B). In fact, it decreases operational efficiency in the partially or fully saturated region. However, this resistor allows the circuit to change gradually from Class B to Class D operation, providing an increase in peak output power to $\pi^2/4$ times that obtained in Class B operation. The average efficiency for signals with large peak-to-average ratios (such as SSB speech) increases significantly [30].

3.4   Class DE Amplifier

There are many power loss mechanisms in Class D amplifiers but the power losses due to the parasitic capacitances are usually the most significant. In fact, these power losses limit the usability of a Class D amplifier at high frequencies. The only convenient way to reduce these losses is to decrease the voltage across the switches at the moment they turn ON. This is possible if dead time occurs during the period when one device has turned OFF and the other has turned ON. During the dead time, both transistors are OFF and a lossless charge/discharge mechanism occurs in the Class D circuit [4, 5, 11, 23, 31-36].

The following analysis refers to a CVS Class D circuit. The assumptions made in section **Complementary Voltage Switching (CVS) Circuit** are valid, except for the one concerning the null output capacitance of the active devices. Moreover, the parasitic capacitances are assumed to be independent of the collector voltage. Based on these assumptions, the equivalent circuits of Figure 3-43 are obtained. Figure 3-44 presents the circuit waveforms.

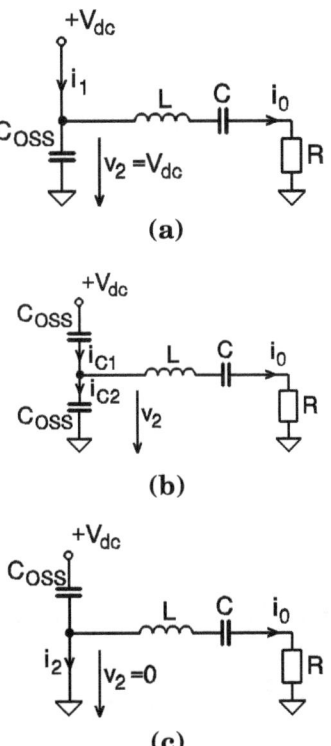

Figure 3-43 *Equivalent circuits of Class DE amplifier; (a) Q_1 = ON, Q_2 = OFF; (b) Q_1 = OFF, Q_2 = OFF; (c) Q_1 = OFF, Q_2 = ON.*

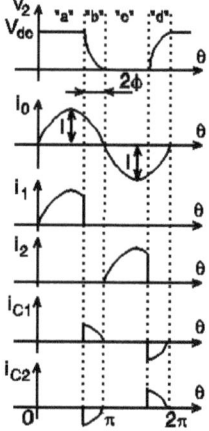

Figure 3-44 *Waveforms in Class DE amplifier; state (a):* Q_1 = *ON,* Q_2 = *OFF; state (b), (d):* Q_1 = *OFF,* Q_2 = *OFF; state (c):* Q_1 = *OFF,* Q_2 = *ON.*

Analysis and Design

Note that, in the Class DE amplifier, the series-tuned circuit does not resonate at switching frequency, f. At this frequency, it has a net reactance

$$X = 2\pi fL - \frac{1}{2\pi fC}$$

$$3.127$$

As shown below, the circuit requires this net reactance to work properly. The output current i_0 is shifted in phase relative to the fundamental frequency component of $v_2(\theta)$. The phase shift given by

$$\phi = \arctan\frac{X}{R}$$

$$3.128$$

To eliminate power losses due to parasitic capacitances, the voltage across the switches at turn ON must be zero. Equations 3.129 through 3.131 refer to the instant when switch Q_2 turns ON. These equations can be also applied (with appropriate modifications) to the instant when switch Q_1 turns ON. Analogous conditions for the instant when Q_1 turns ON can be verified easily.

$$v_2(\pi-) = 0$$

$$3.129$$

Moreover, it is convenient that the slope of the collector voltage is zero when the switch turns ON

$$\left.\frac{dv_2}{d\theta}\right|_{\theta=\pi-} = 0 \qquad\qquad 3.130$$

Conditions in Equations 3.129 and 3.130 are typical for Class E amplifiers and are discussed in detail in Chapter 4. For this reason, the Class D circuit designed to satisfy Equations 3.129 and 3.130 (and the analogous conditions at the instant $\theta = 2\pi-$) is called a Class DE circuit, soft-switching Class D circuit, or Class D circuit with Class E switching conditions.

Equation 3.130 yields successively

$$i_{C2}(\pi-) = \omega C_{OSS}\left.\frac{dv_2}{d\theta}\right|_{\theta=\pi-} = 0$$

$$i_{C1}(\pi-) = \omega C_{OSS}\left.\frac{d(V_{dc}-v_2)}{d\theta}\right|_{\theta=\pi-} \; :$$

$$= -\omega C_{OSS}\left.\frac{dv_2}{d\theta}\right|_{\theta=\pi-} = -i_{C2}(\pi-) = 0 \qquad 3.131$$

$$i_0(\pi-) = i_{C1}(\pi-) - i_{C2}(\pi-) = 0$$

$$i_2(\pi+) = i_0(\pi+) = i_0(\pi-) = 0$$

Note that, when a transistor turns ON, the voltage across it is already zero and the collector current at the start of the ON state is zero (there is no current jump at that instant). Consequently,
a. A moderately slow transition of the switches does not cause significant power losses during the transition time.
b. The zero start current helps minimize the transition time (because the actual devices have limited di/dt capability) and helps further minimize the power losses during transition.

To simplify the analysis, the waveform of $v_2(\theta)$ is approximated by a trapezoidal waveform.[12] The angular dead time is 2ϕ (see Figure 3-44). For $\pi-2\phi \le \theta \le \pi$, $v_2(\theta)$ is given by

$$v_2(\theta) = V_{dc} - \frac{1}{2\omega C_{OSS}}\int_{\pi-2\phi}^{\theta} I\sin\tau\, d\tau \qquad\qquad 3.132$$

and Equations 3.129 and 3.132 yield

$$V_{dc} = \frac{I}{2\omega C_{OSS}}(1-\cos 2\phi) \qquad\qquad 3.133$$

Note that Equation 3.130 has already been used, because it is assumed that $i_0(\theta) = I\sin\theta$ in Equation 3.132. The fundamental frequency component of

$v_2(\theta)$ is given by

$$V_2 = \frac{2V_{dc}}{\pi\phi}\sin\phi \qquad\qquad 3.134$$

resulting in the output power

$$P_0 = \frac{1}{2}V_2 I\cos\phi = \frac{1}{2}\frac{V_{dc}}{\pi\phi}I\sin 2\phi \qquad\qquad 3.135$$

Substituting Equation 3.133 in Equation 3.135 yields

$$I = \sqrt{\frac{4P_0\pi\phi\omega C_{OSS}}{\sin 2\phi(1-\cos 2\phi)}} \qquad\qquad 3.136$$

Equations 3.133 and 3.136 provide the required DC input voltage V_{dc} and the amplitude of the load current I for a given frequency, output power, devices output capacitance, and dead time. The required load resistance is then given by

$$R = \frac{2P_0}{I^2} = \frac{\sin 2\phi(1-\cos 2\phi)}{2\pi\phi\omega C_{OSS}} \qquad\qquad 3.137$$

For practical applications, it is very convenient to represent the variation of the operating parameters with the dead time graphically, selecting a suitable operating point.

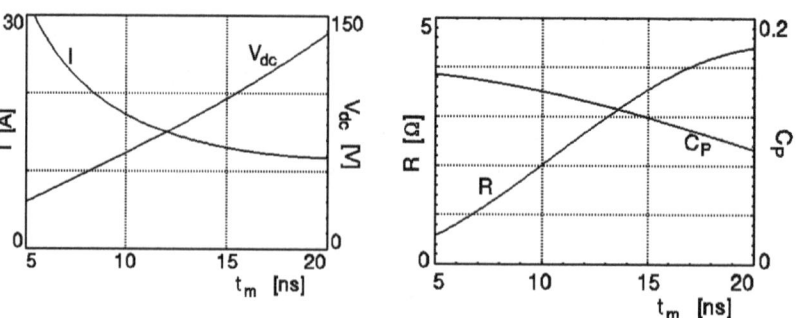

Figure 3-45 *Operating parameters versus dead time t_m (IRF540 MOSFET, $P_0 = 300\ W, f = 13.56\ MHz$).*

EXAMPLE 3.19

Using the values from Example 3.5 ($P_0 = 300$ W, $f = 13.56$ MHz, IRF540 MOSFETs with $C_{OSS} = 560$ pF at $V_{DS} = 25$ V), it is possible to plot the curves from Figure 3-45. An analysis of these curves shows that a suitable operating point is t_m = 12 ns (ϕ = $t_m f \pi$ = 0.5112 rad), V_{dc}= 75.3 V, I = 15 A, R = 2.667 Ω, and C_P = 0.1328. Assuming a loaded quality factor[13] Q = 1/(ωCR) = 5, C = 880.2 pF and L = (Q + $tan\phi$)R/ω = 174.1 nH are obtained.

Sinusoidal Drive

It is difficult to drive a Class DE circuit operating at high frequency (for example, around 10 MHz and above) using a square-wave gate drive. Precisely controlled square waves are difficult to generate, because the dead time must be controlled with high accuracy (of the order of 1 nanosecond), and an input transformer with a large bandwidth is required. However, Class DE circuits are somewhat easier to drive than Class D circuits because both transistors are OFF during the voltage transitions, and Miller's effect at both turn ON and turn OFF is reduced to almost zero.

A sinusoidal wave drive is a better choice because it reduces the driving power requirements and allows control of the dead time by the amplitude of the gate voltage [4, 5, 23]. A Sinusoidal drive can be used in Class DE circuits containing MOSFETs because MOSFETs have a threshold gate voltage level below which they are OFF; once the threshold is exceeded, the transistors quickly switch to ON. The input portion of the circuit is standard and uses an input transformer driven with sinusoidal voltage. The gate voltages of transistors are sinusoidal waves 180 degrees out of phase (see Figure 3-46).

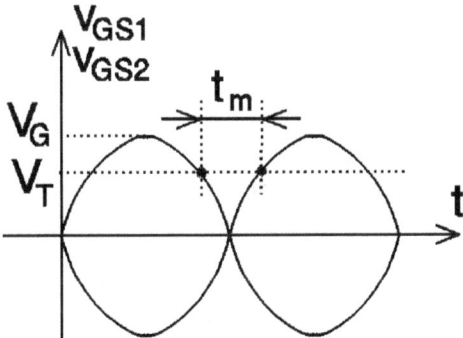

Figure 3-46 *Waveforms in the gate circuits for sinusoidal drive.*

By denoting the amplitude of the gate-to-source voltages v_{GS1} and v_{GS2} as V_G, and the threshold voltage as V_T

$$V_T = V_G \sin\left(\omega \frac{t_m}{2}\right)$$
3.138

Given V_T and t_m, Equation 3.138 yields the required V_G. By rearranging Equation 3.138 as follows:

$$t_m = \frac{2}{\omega} \arcsin\left(\frac{V_T}{V_G}\right)$$
3.139

it is possible to determine the range of dead times achievable using this method.

The first limitation for t_m is that the peak gate voltage is limited to a maximum value usually, $V_{GSmax} = \pm 20...\pm 30$ V. For example, IRF540 has $V_{GSmax} = \pm 20$ V [20] and, with $f = 13.56$ MHz and $V_T = 3.5$ V, the result is $t_{m,min} = 4.1$ ns (an angular dead time of about 20 degrees). The second limit for t_m occurs when the gate-to-source voltage is too low to turn ON the MOSFET properly. Reference [23] recommends $V_G > 6.5$ V; thus, the largest dead time is $t_{m,max} = 13.3$ ns (an angular dead time of about 65 degrees). For Example 3.19, $t_m = 12$ ns can be obtained with $V_G = 7.2$ V. For this gate voltage, the required gate charge is about 30 nC. The amplitude of the gate current $I_G = 2.56$ A is obtained from Equation 3.120. The gate resistance R_G of the IRF540 MOSFET is estimated to about 1 ohm; thus, the power losses in the gate circuit are $P_G = 3.28$ W per transistor. This power can be obtained easily and efficiently using a Class D or a Class E driver.

Tuning Procedure

Once constructed, a practical Class DE circuit will not automatically provide optimum operation (with Class E switching conditions), but will need tuning. Unfortunately, optimum operation does not occur at the point of maximum input or output power and requires a special tuning procedure [36]. This section describes a practical tuning procedure based solely upon observation of one of the active devices' voltage waveforms.

The voltage across Q_2 during the dead time is shown in Figure 3-47. It reaches a minimum value, at which time the slope of $v_2(t)$ is zero (a "trough"), and then begins to rise. When Q_2 turns ON, $v_2(t)$ drops rapidly to zero (or to V_{CEsat} if BJTs are used). Depending on the initial settings of the circuit and the presence of an anti-parallel diode during switching (for example, the parasitic diode of a power MOSFET), the trough and/or the jump at the OFF-to-ON transition may be hidden from view. In this case, the location of these features on the waveform can be estimated by extrapolating a solution from the part of the waveform that can be seen. Changes in the values of the series-tuned circuit components, or in the value of the dead time, affect the $v_2(t)$ waveform as follows (see Figure 3-47):

a. As C or L is increased (as load angle ψ is increased), the trough of the waveform moves down and to the right.

b. As the dead time is increased, the trough of the waveform moves down and to the left.

The Class DE amplifier tuning procedure is taken from Figure 3-47. The load angle and dead time are adjusted successively[14] until both $v_2(t)$ and its derivative are zero at the instant Q_2 turns ON. This automatically assures Class E switching conditions for Q_1. To avoid damage due to a large initial mistuning, tune the amplifier at low voltages, increasing V_{dc} gradually, and readjusting ψ and the dead time as needed. The increase in V_{dc} will decrease the effective value of the output capacitance of the active devices (BJTs or MOSFETs). Therefore, the dead time and ψ should be decreased slightly.

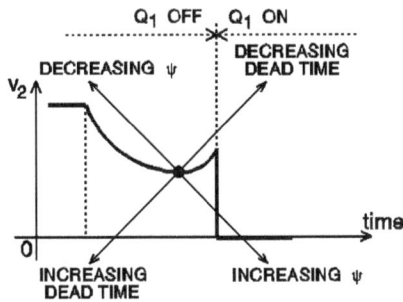

Figure 3-47 *Effects of adjusting the dead time and the load angle ψ.*

Practical Considerations

1. For angular dead times values below 50 to 60 degrees, the model presented here for the Class DE circuit (and the simplified design procedure deriving from this model) provides reasonable accuracy in practical applications. For larger dead time values, the model produces significant errors because the voltages across the output capacitances are approximated with linear ramps during the dead times.

2. Class E switching conditions may be obtained for values of the active device's output capacitance, C_{OSS}, varying between zero and a maximum value [36].

$$(\omega C_{OSS}R)_{max} = \frac{1}{2\pi} \approx 0.1592 \qquad 3.140$$

Values of C_{OSS} larger than that given by Equation 3.140 do not allow for the achievement of Class E switching conditions.

3. Assuming given values for ω and R, larger values of C_{OSS} require larg-

er dead times and larger reactive loads, providing lower output power and lower power output capability. The maximum value of C_{OSS} given by Equation 3.140 can be accommodated in a Class DE circuit operating with an angular dead time of 90 degrees (each transistor is ON for 90 degrees and OFF for 270 degrees) [32, 34].

Although large dead time values would allow an increase in the operating frequency of Class DE circuits, they are not practical. The output power and the power output capability would have very low values (for example, C_P = 0.07958 for a 90 degree angular dead time) and precise three-state switching would be difficult to achieve because the sinusoidal resonant gate drive cannot be used for large dead times.

3.5   Class D Frequency Multipliers

If the resonant circuit of a Class D amplifier is tuned to a certain harmonic (Nf, N = 2, 3, ...) of the switching frequency, f, the circuit becomes a Class D frequency multiplier. In the analysis presented below, a CVS Class D circuit (see Figure 3-1) that satisfies the simplifier assumptions given in section *Complementary Voltage Switching (CVS) Circuit* (except for the one regarding the tuning frequency of the resonant circuit) is considered. Assuming a 50 percent duty cycle, the voltage, $v_2(\theta)$, applied to the output circuit is a periodic square-wave (see Equation 3.1) and may be decomposed in a Fourier series given by Equation 3.2, containing only odd harmonics. If we assume that the series resonant circuit is tuned to an odd harmonic of the switching frequency, the output current is sinusoidal (see Figure 3-48).

$$i_0(\theta) = I\sin(2N-1)\theta = \frac{2}{\pi}\frac{V_{dc}}{R}\frac{\sin(2N-1)\theta}{2N-1} \qquad 3.141$$

and the output power is given by

$$P_0 = \frac{I^2}{2}R = \frac{2}{\pi^2}\frac{V_{dc}^2}{R}\frac{1}{(2N-1)^2} \approx \frac{0.2026}{(2N-1)^2}\frac{V_{dc}^2}{R} \qquad 3.142$$

The idealized collector efficiency is 100 percent. The power output capability is

$$C_P = \frac{P_0}{2V_{dc}I} = \frac{1}{2\pi(2N-1)} \approx \frac{0.1592}{2N-1} \qquad 3.143$$

For $N = 3$ (tripler), the equations above give

$$P_0 = \frac{2}{9\pi^2}\frac{V_{dc}^2}{R} \approx 0.02252\frac{V_{dc}^2}{R} \quad C_P = \frac{1}{6\pi} \approx 0.05305$$

3.144

and for $N = 5$

$$P_0 = \frac{2}{25\pi^2}\frac{V_{dc}^2}{R} \approx 0.008106\frac{V_{dc}^2}{R} \quad C_P = \frac{1}{10\pi} \approx 0.03183$$

3.145

Note that P_0 and C_P decrease very quickly with N (the multiplying order). This circuit cannot be used for $N > 5$.

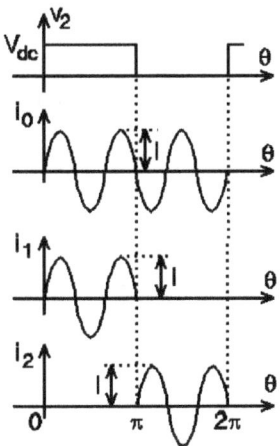

Figure 3-48 *Waveforms in CVS Class D frequency tripler (D = 0.5).*

The results given in section ***Switches Duty Ratio*** suggest another possibility for driving a Class D circuit for frequency multiplication. If an arbitrary value $(0 < D < 1)$ is assumed for the duty ratio, the voltage $v_2(\theta)$ contains both odd and even harmonics (see Equation 3.108). Suppose that the series-tuned circuit is tuned to the Nth harmonic of the switching frequency. The output current given by

$$i_0(\theta) = I\cos N\theta = \frac{2}{\pi}\frac{V_{dc}}{R}\frac{\sin N\pi D}{N}\cos N\theta$$

3.146

produces an output power

$$P_0 = \frac{2}{\pi^2}\frac{V_{dc}^2}{R}\frac{\sin^2 N\pi D}{N^2} \approx 0.2026\frac{V_{dc}^2}{R}\frac{\sin^2 N\pi D}{N^2}$$

3.147

The power output capability is

$$C_P = \frac{P_0}{2V_{dc}I} = \frac{1}{2\pi}\frac{|\sin N\pi D|}{N} \qquad\qquad 3.148$$

For any values of D and N, the collector efficiency is 100 percent. However, maximizing P_0 and C_P requires

$$|\sin N\pi D| = 1 \qquad\qquad 3.149$$

resulting in

$$D = \frac{1}{2N} \quad D = 1 - \frac{1}{2N} \qquad\qquad 3.150$$

There is no significant difference between these two values. The waveforms of $i_1(\theta)$ and $i_2(\theta)$ are interchangeable, and $i_0(\theta)$ changes its phase with 180 degrees. Some interesting cases follow:

a. Frequency doubler $(N = 2)$; $D = 1/4$ (see Figure 3-49).

$$P_0 = \frac{1}{2\pi^2}\frac{V_{dc}^2}{R} \approx 0.05066\frac{V_{dc}^2}{R} \; ; \; C_P = \frac{1}{4\pi} \approx 0.07958 \qquad\qquad 3.151$$

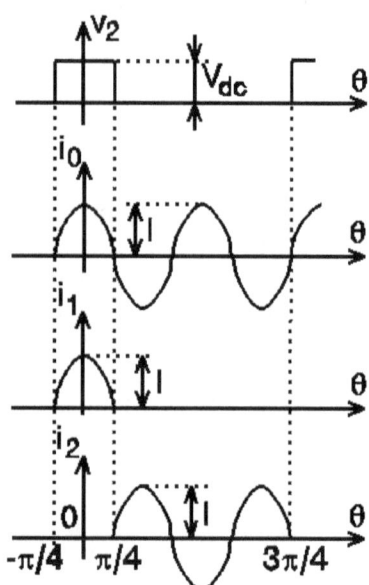

Figure 3-49 *Waveforms in CVS Class D frequency doubler (D = 0.25).*

b. Frequency tripler (N = 3); D = 1/6. P_0 and C_P are given by Equation 3.144. Their values are equal to those obtained for D = 0.5. A 50 percent duty ratio is easier to obtain and, therefore, this solution is preferable for practical designs.

In all cases, at least one of the two switches must be able to pass negative currents, so the use of MOSFETs in Class D frequency multipliers is recommended.

3.6   CAD of Class D Circuit

SPICE Simulation

Nonlinear circuit simulation can be an important tool for analysis, design, and optimization of switching-mode RF power amplifiers. The incorporation of simulation into the design phase offers several advantages over the standard prototyping method. Some of these include reduced development time and costs, increased understanding of a circuit's operation, and identification of potential problem areas. SPICE can be a versatile and useful tool for the analysis and optimization of switching-mode RF power amplifiers, as long as the transistor and circuit are modeled properly. The following precautions must be taken:

1. Significant parasitic elements must be included in the circuit description.

2. Models of active devices must be represented using subcircuits (including the device package parasitics: lead inductances and stray capacitances).

3. A selection of transient analysis options must be considered. (Although time domain techniques are somewhat inefficient, they are necessary for applications in which the active devices behave nonlinearly.)

The IRF520 SPICE model is presented in Figure 3-50 and Table 3-1. The power MOSFET is represented by a subcircuit that includes some parasitic elements (bond-wire inductances).

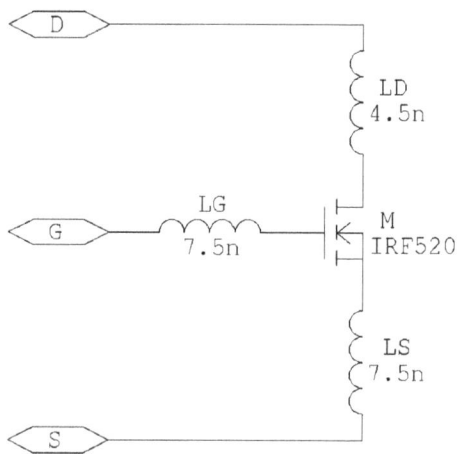

Figure 3-50 *Subcircuit schematic of IRF520.*

Table 3-1 *SPICE subcircuit net list of the IRF520 power MOSFET (courtesy of Cadence PCB Systems).*

.SUBCKT IRF520_L 1 2 3
LD 1 4 4.5n
LG 2 6 7.5n
LS 5 3 7.5n
M 4 6 5 5 IRF520
.model IRF520 NMOS (Level=3 Gamma=0 Delta=0 Eta=0 Theta=0
+ Kappa=0 Vmax=0 Xj=0 Tox=100n Uo=600 Phi=.6 Rs=.1459
+ Kp=20.79u W=.73 L=2u Vto=3.59 Rd=80.23m Rds=444.4K
+ Cbd=622.1p Pb=.8 Mj=.5 Fc=.5 Cgso=517.9p Cgdo=137.3p
+ Rg=6.675 Is=2.438p N=1 Tt=137n)
.ENDS

A Class DE amplifier using IRF520 power MOSFETs is presented in Figure 3-51 as a simulation example. This circuit was designed for a switching frequency of 5 MHz, using the procedure given in [36]; the DC supply voltage is $V_{dc} = 75$ V, the load resistance is $R_L = 10$ Ω, and the loaded Q of the series-tuned circuit is $Q = 3$.

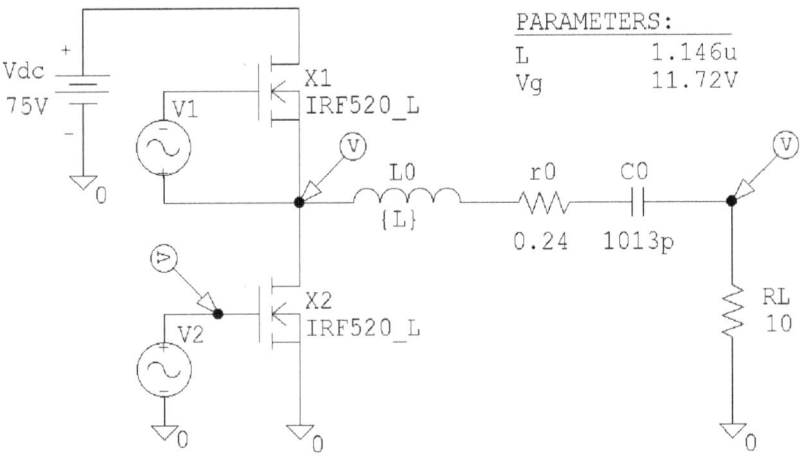

Figure 3-51 PSPICE *schematic of Class DE power amplifier.*

All simulations presented below were performed using MicroSim DesignLab Evaluation Version 7.1, 1996. The transient analysis option with sinusoidal waveform used as a stimulus was considered in the simulations. Because the circuit needed a number of cycles to stabilize, this circuit was simulated from $t = 0$ to $t = 1$ μs and only the output data from 0.8 μs to 1 μs was accumulated.

Several important circuit waveforms are depicted in Figure 3-52. From these waveforms, operating parameters, such as device stresses and output power can be obtained. Note that Class E switching conditions are not rigorously satisfied (ringing in the drain voltage waveform is caused by the parasitic inductances), and the circuit needs to be tuned by varying the drive level and the series inductance.

The benefits of using SPICE simulation to evaluate the time domain waveforms are apparent. The Class D circuits can be modeled with any degree of complexity, and all second-order effects can be taken into account simultaneously; circuits for sinusoidal or rectangular gate drive can be also modeled accurately. The simulations are slow, and it is difficult to optimize a design.

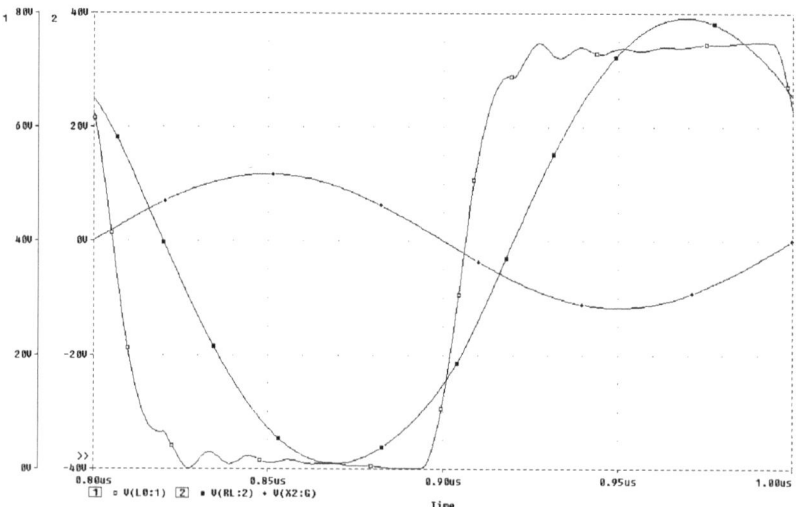

Figure 3-52 *PSPICE simulated waveforms in Class DE amplifier.*

HB-PLUS Simulation

As shown above, Class D circuits can be simulated using any general-purpose circuit simulator, such as SPICE. However, due to the characteristic features of Class D amplifiers, the DC and the AC SPICE analysis are useless in this case. The SPICE-simulation of Class D amplifiers require the use of the transient analysis option. For practical applications, the steady-state periodic response of the circuit must be simulated; the simulation is slow and it is almost impossible to optimize a design. Efficient simulation of Class D amplifier requires specialized programs such as HB-PLUS.

The HB-PLUS program (developed by Design Automation, Inc.) simulates and optimizes resonant and non-resonant half-bridge (voltage switching Class D) DC/DC converters, DC/AC inverters, and RF power generators [37]. This program can simulate the cycle-by-cycle transient response of the circuit (from any starting conditions), and the steady-state periodic response of the circuit. HB-PLUS simulates the circuit steady-state periodic time-domain response direct; a SPICE simulator requires a wait of many RF cycles (in a transient analysis) to attain an approximation of the steady-state periodic response. HB-PLUS simulates the Class D circuit very quickly. The simulation required about 0.5 second on a 33 MHz 486 computer.

HB-PLUS outputs include

1. time-domain graph plots and frequency-domain spectra of all important circuit voltage and current waveforms,

2. tabulation of DC input power, output power, efficiency, inefficiency, power dissipation in each circuit element (including the parasitic power losses in each inductor and capacitor), and transistor peak voltage and current stresses, in normal and inverse polarities.

Many of these results are not available (directly) in SPICE simulators, and some of them are very difficult to estimate after simulation.

The HB-PLUS program also allows, within specified limits, a sweep of any of the up to 28 circuit parameters and the plotting of any two output variables versus the swept parameter. Finally, one of the program's useful capabilities (which SPICE cannot provide) is the automatic adjustment of the operating parameters and the component values to achieve a defined "optimum" performance (to minimize a defined objective function).

The general model of a half-bridge amplifier used in HB-PLUS is illustrated in Figure 3-53 and includes two switches and a load network. The switch model [18, 19] is shown in Figure 3-54 and includes

1. a resistor, R_{off}, modeling the OFF resistance,

2 a resistor, R_{on}, modeling the ON resistance ($R_{DS,ON}$ for MOSFET or R_{CEsat} for BJT),

3. a voltage source, V_0, modeling V_{CEsat} of BJT,

4. a series RC circuit, $C_{out}R_{cout}$, modeling the output capacitance of the active device and its parasitic resistance,

5. an optional anti-parallel diode (the parasitic substrate diode of a silicon MOSFET, or an external diode intentionally connected from emitter to collector of a BJT), characterized by V_t and R_d.

The load network configuration is shown in Figure 3-55. It may be used to represent a nonresonant network or a resonant network containing series- or parallel-tuned circuits, narrowband matching networks (for example, T and pi), or a full T model of a transformer (including primary and secondary leakage inductances and magnetizing inductance). All inductors and capacitors a have specified loss resistance or unloaded quality factor.

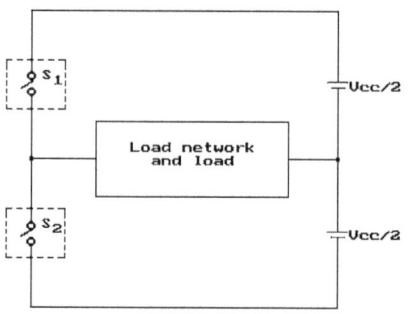

Figure 3-53 *General model of half-bridge power amplifier, courtesy of Design Automation, Inc.*

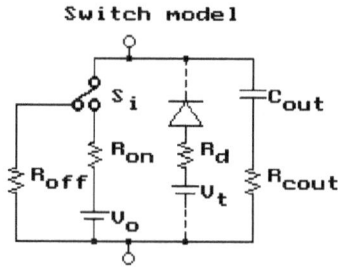

Figure 3-54 *Switch model in HB-PLUS, courtesy of Design Automation, Inc.*

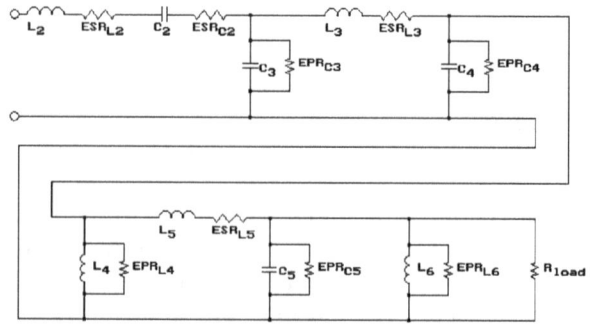

Figure 3-55 *Load network and load in HB-PLUS, courtesy of Design Automation, Inc.*

Some results obtained for the circuit presented in Example 3.11 are given below. It is assumed that $R_{off} = 1$ MΩ and, in accordance with the results given in [18] and [19], $R_{cout} = 5 R_{on}$. Figure 3-56 presents the two screens for entering circuit parameters (the load network is reduced to a series-tuned circuit). Figure 3-57 shows the powers, the efficiency, and the switch stresses. Figure 3-58 depicts the waveforms of the voltage across S_1 (drain-to-source), the active current through S_1 (the current in the intrinsic active part of the transistor), the voltage across the load, and the total current through S_1 (as would be measured by a current probe on the transistor drain lead).

```
━ Half-Bridge power amplifier/converter (HB) ━━━━━━ March 24, 1996 16:50 ━
Class D amplifier/IRF540/300W/13.56Mhz/Q=5

        ENTER CIRCUIT PARAMETERS and TITLE  SCREEN-1

   Switching frequency (f)................[Hz]:      1.356E+07
   DC supply voltage (Vcc or Vdd).......[volts]:     75
   Fraction of period, turn-on of Sw1 to Sw2...:     0.5
   Switch 1 parameters
      Duty ratio of switch 1...................      0.5
      Ron1.............................[ohms]:       0.077
      Roff1............................[ohms]:       1E+06
      Cout1...........................[farads]:      5.6E-10
      Cout1 series resistance (Rcout1)....[ohms]:    0.385
      Saturation offset voltage (Vo1)....[volts]:    0
   Switch 2 parameters
      Duty ratio of switch 2...................      0.5
      Ron2.............................[ohms]:       0.077
      Roff2............................[ohms]:       1E+06
      Cout2...........................[farads]:      5.6E-10
      Cout2 series resistance (Rcout2)....[ohms]:    0.385
      Saturation offset voltage (Vo2)....[volts]:    0
SELECTION: <,>      ENTRY: ALPHANUM.      EXECUTE: <PgDn>    ABORT: <ESC>

SELECTING PARAMETER...
━ Half-Bridge power amplifier/converter (HB) ━━━━━━ March 24, 1996 16:51 ━
Class D amplifier/IRF540/300W/13.56Mhz/Q=5
             ENTER CIRCUIT PARAMETERS SCREEN-2

   Load resistance (Rload).............[ohms]:       3.376
   L2................................[henries]:      1.9812E-07
      Qu of L2 at switching frequency........:       150
   C2.................................[farads]:      6.9533E-10
      Qu of C2 at switching frequency........:       1000








SELECTION: <,>      ENTRY: ALPHANUM.      EXECUTE: <PgDn>    ABORT: <ESC>
```

Figure 3-56 *Entering circuit parameters in HB-PLUS, courtesy of Design Automation, Inc.*

```
══ Half-Bridge power amplifier/converter (HB) ══════ March 24, 1996 16:51 ══
Class D amplifier/IRF540/300W/13.56Mhz/Q=5

                    EFFICIENCY AND POWERS

        Collector/drain efficiency (Pout/Pin)...[%]      74.258
        Collector/drain ineff'y (Pin-Pout)/Pin..[%]      25.742
        DC power input (Pin)................[watts]      403.79
        Power output (Pout)................[watts]      299.85
        Power loss in L2...................[watts]        9.9948
        Power loss in C2...................[watts]        1.4992
        Power loss in Ron1.................[watts]       15.601
        Power loss in Rcout1...............[watts]       30.625
        Power loss in Ron2.................[watts]       15.601
        Power loss in Rcout2...............[watts]       30.625




                                    "DISPLAY RESULTS" MENU: <PgUp>

DISPLAYING COMPUTED RESULTS...

══ Half-Bridge power amplifier/converter (HB) ══════ March 24, 1996 16:51 ══
Class D amplifier/IRF540/300W/13.56Mhz/Q=5


                        SWITCH STRESSES

        Switch-1 peak voltage (normal)......[volts]      74.859
        Switch-1 peak voltage (inverse).....[volts]       0.14125
        Switch-2 peak voltage (normal)......[volts]      74.859
        Switch-2 peak voltage (inverse).....[volts]       0.14125
        Switch-1 peak current (normal)....[amperes]     277.42
        Switch-1 peak current (inverse)...[amperes]       5.3638E-05
        Switch-2 peak current (normal)....[amperes]     277.42
        Switch-2 peak current (inverse)...[amperes]       5.3638E-05




                                    "DISPLAY RESULTS" MENU: <PgUp>

DISPLAYING COMPUTED RESULTS...
```

Figure 3-57 *Powers, efficiency, and switch stresses, courtesy of Design Automation, Inc.*

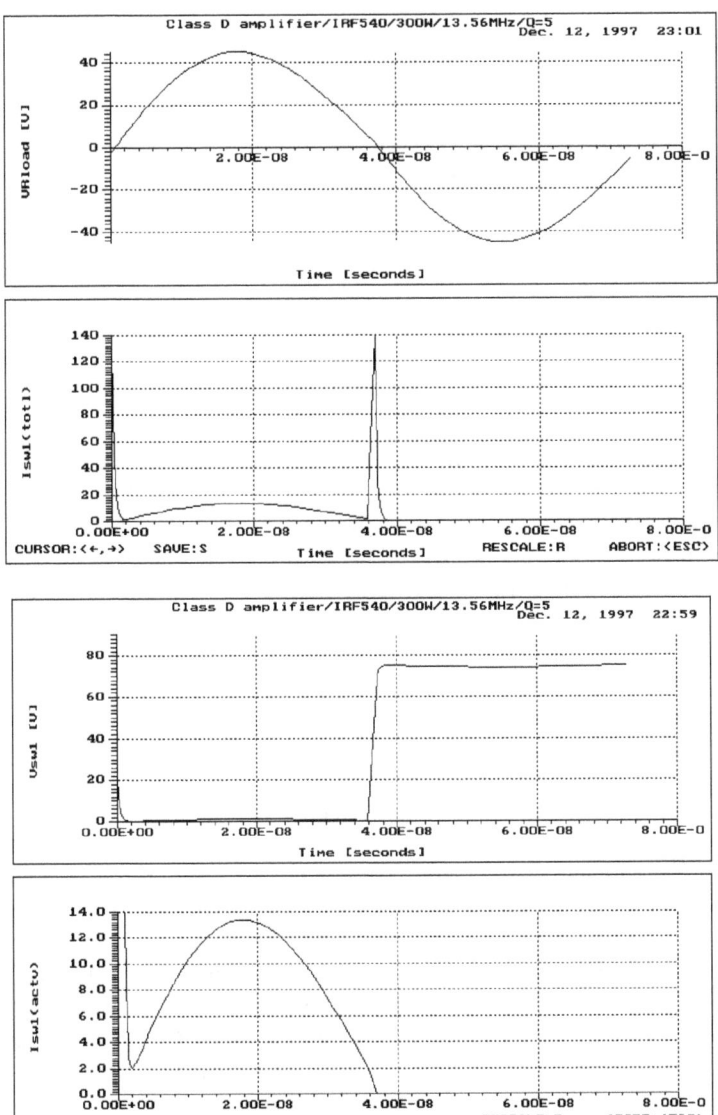

Figure 3-58 *Waveforms in HB-PLUS: voltage across S_1, active current through S_1, voltage across the load, and total current through S_1, courtesy of Design Automation, Inc.*

3.8 Notes

1. If a conventional (wire-wound) transformer is used, the m/n turns ratio can have almost any desired value. However, in practice, wideband transmission line transformers must be used rather than the wire-wound transformers. As a result, the m/n turns ratio is the ratio of two integers (2, 3, 4,..., 1/2, 1/3, 1/4, ...).

2. In practice, use of a bypass capacitor is recommended on the left end of the RF choke (see Figure 3-8). This prevents RF signal penetration in the power supply line.

3. If the active devices act as ideal switches (having zero ON resistance), the parasitic capacitances C_{p1} and C_{p2} are charged and discharged instantaneously.

4. If the collector voltage is below the base OFF voltage (provided by the driver), the base-collector junction is forward-biased and the bipolar transistor is placed in the active inverted mode. Consequently, an appreciable inverted collector current may flow through the transistor, reducing efficiency and possibly damaging the transistor. Another potentially destructive situation may occur if the load pulls the base to be negative enough so that the BV_{EBX} rating is exceeded.

5. Note that when a switch turns ON, the voltage across it is zero (see Figure 3-9). On the other hand, the substantial currents flowing through the parasitic capacitances also flow through their series resistances causing substantial power dissipation.

6. The main causes of power losses in a voltage switching Class D circuit are the conduction losses (due to R_{ON}) and the switching losses (due to C_p). The other power losses usually can be neglected because they are insignificantly affected by R and V_{dc}.

7. The switching times specified in data sheets are usually for switching rectangular current waveforms into resistive loads. In the voltage switching Class D circuits, the switched current is a portion of a sinusoid without jumps when switching (if the circuits operate correctly). If a sinusoidal drive is used, the switching times may be only about one-fourth of the data sheets values. In the current switching Class D amplifiers, the switched current is rectangular, resulting in a slower transition. Note that these data are available only for switching transistors and not for RF transistors.

8. In practical applications, a high efficiency requires $R \gg R_{ON}$; for example, if $R > 10\,R_{ON}$, then $\eta > 90\%$. During the transition, $i(\theta)$, shown in Equation 3.96, dominates the load current and $v_2(\theta)$ varies approximately linearly.

9. The input capacitance of a MOSFET transistor varies significantly during the transition. This is due to the Miller feedback effect of the gate-drain capacitance and the fast increase of the gate-drain capacitance when the drain-to-gate voltage becomes negative (at the instant of OFF-to-ON switching) [24]. The best way to predict the gate drive requirements is by using the data sheet values of the gate charge required to turn the transistor ON or OFF.

10. It is easy to calculate that after $\tau = 4\tau$, the voltage across C_{ISS} is about 98 percent from its final value.

11. The term Class BD has also been occasionally used for Class S amplifiers or modulators [16, 30].

12. Using the procedure suggested in [33] and [35], it is possible to calculate exact expressions for $v_2(\theta)$ and its fundamental component. However, for practical component values and operating parameters, the results obtained closely agree with those presented here, and the small differences can be easily compensated for when the circuit is tuned.

13. In this case, the series-tuned circuit does not resonate on the switching frequency and the quality factor can be defined as the ratio of either capacitive or inductive reactance to series resistance. These definitions give different values; if $Q = \omega L/R = 5$, then $L \approx 156.5$ nH and $C = 1/[\omega R(Q - tan\phi)] \approx 991.4$ pF.

14. A sinusoidal resonant gate drive is usually preferred in practice. In this case, increasing the drive level will decrease the dead time.

3.7 References

1. Clarke, K. K. and D. T. Hess. *Communication Circuits: Analysis and Design*. Reading, MA: Addison-Wesley, 1971.

2. Krauss, H. L., C. V. Bostian, and F. H. Raab. *Solid-State Radio Engineering*. New York: John Wiley & Sons, 1980.

3. Smith, J. *Modern Communication Circuits*. New York: McGraw-Hill, 1986.

4. Kazimierczuk, M. K. and D. Czarkowski. *Resonant Power Converters*. New York: John Wiley & Sons, 1995.

5. Albulet, M. *Amplificatoare de Radiofrecventa de Putere*. Bucuresti: Matrix Rom, 1996.

6. Baxandall, P. J. "Transistor Sine-Wave LC Oscillators, Some General Considerations and New Developments." *Proc. IEE*, vol. 106, part B, suppl. 16 (May 1959): 748–58.

7. Chudobiak, W. J. and D. F. Page. "Frequency and Power Limitations of Class D Transistor Amplifiers." *IEEE J Solid-State Circuits*, vol. SC-4 (February 1969): 25–37.

8. Raab, F. H. "High Efficiency RF Power Amplifiers." *Ham Radio*, vol. 7, no. 10 (October. 1974): 8–29.

9. Raab, F. H. "High Efficiency Amplification Techniques." *IEEE Circuits and Systems Newsletter*, vol. 7, no. 10 (December 1975): 3–11.

10. Raab, F. H. and D. J. Rupp. "A Quasi-Complementary Class-D HF Power Amplifier." *RF Design*, vol. 15, no. 9 (September 1992): 103–110.

11. Sokal, N. O., I. Novak, and J. Donohue. "Classes of RF Power Amplifiers A through S, How They Operate, and When to Use Them." *Proceedings of the RF Expo West 1995*, (San Diego, CA) (1995): 131–138.

12. Senak, Jr, P. "Amplitude Modulation of the Switched-Mode Tuned Power Amplifier." *Proc. IEEE*, vol. 53 (October 1965): 1658–1659.

13. Kazimierczuk, M. and J. S. Modzelewski. "Drive-Transformerless Class-D Voltage Switching Tuned Power Amplifier." *Proceedings of the IEEE*, vol. 68, no. 6 (June 1980): 740–741.

14. Clemente, S. "An Introduction to International Rectifier P-Channel HEXFETs." *Application Note 940B* International Rectifier, Inc., El Segundo, CA, 15.

15. Raab, F. H. "MOSFET Power Amplifier for Operation from 160 Through 6 Meters." *Ham Radio*. vol. 11, no. 11 (November 1978): 12–17.

16. Granberg, H. O. "Applying Power MOSFETs in Class D/E RF Power Amplifier Design." *RF Design*, vol. 8, no. 6 (June 1985): 42–47.

17. Raab, F. H. and D. J. Rupp. "HF Power Amplifier Operates in Both Class B and Class D." *Proceedings of the RF Expo West 1993*, (San Jose, CA) (March 1993): 14–124.

18. Sokal, N. O. and R. Redl. "Model for Switching Power Transistor Output Port, for Analyzing High-Frequency Resonant Power

Converters." *Power Electronics Conference.* (Boxborough, MA) April 1987.

19. Sokal, N. O. and R. Redl. "Power Transistor Output Port Model." *RF Design.* vol. 10, no. 6, (June 1987): 45–48, 50, 51, 53.

20. HEXFET Power MOSFET Designer's Manual, *International Rectifier, Inc.,* El Segundo, CA, 1993.

21. Raab, F. H. "Class-D Power Amplifier Load Impedance for Maximum Efficiency." *Proceedings of the RF Technology Expo 85,* (Anaheim, CA), (January 1985): 287–295.

22. Raab, F. H. "Radio Frequency Pulsewidth Modulation." *IEEE Trans. Commun.,* vol. COM-21, no. 8 (August 1973): 958–966.

23. El-Hamamsy, S. A. "Design of High-Efficiency RF Class-D Power Amplifier." *IEEE Trans. Power Electron.,* vol. 9 (May 1994): 297–308.

24. Everard, J. K. A. and A. J. King. "Broadband Power Efficient Class E Amplifiers with a Non-Linear CAD Model of the Active MOS Device," *J. of IERE,* vol. 57, no. 2 (March/April 1987): 52–58.

25. Dierberger, K., F. H. Raab, and L. Max. "Simple and Inexpensive High-Efficiency Power Amplifier Using New APT MOSFETs." *RF Expo East 1994,* Orlando, FL, (November 1994).

26. Dierberger, K. et al. "High-Efficiency Power Amplifiers for 13.56 ISM and HF Communications." *RF Design,* vol. 18, no. 5 (May 1995): 28, 30, 32, 34–36.

27. Raab, F. H. "Average Efficiency of Power Amplifiers." *Proceedings of RF Technology Expo 1986.* (Anaheim, CA) (1986): 473–486.

28. Sevick, J. *Transmission Line Transformers.* Atlanta: Noble Publishing, 1996.

29. Dick, B. "Digital Modulation: DX-10 AM Transmitter." *Broadcast Engineering,* no. 3, (March 1988): 302–318.

30. Raab, F. H. "The Class BD High-Efficiency RF Power Amplifier." *IEEE J. Solid-State Circuits.* vol. SC-12 (June 1977): 291–298.

31. Zhukov, S. A. and V. B. Kozyrev. "Poluprovodnikovye Pribory v Tekhnike Electrosvyazi." *Svyaz,* vol. 15 (1975): 95–107.

32. Koizumi, H. et al. "A Class D Type High Frequency Tuned Power Amplifier with Class E Switching Conditions." *IEEE Int. Symp. Circ. Syst.,* vol. 5 (London, UK) (1994): 105–108.

33. Hamill, D. C. "Impedance Plane Analysis of Class DE Amplifier." *Electronic Letters,* vol. 30, no. 23 (November 1994): 1905–1906.

34. Koizumi, H. et al. "Class DE High-Efficiency Tuned Power Amplifier." *IEEE Trans. Circuits Syst. - I,* vol. 43 (January 1996): 51–60.

35. Hamill, D. C. "Class DE Inverters and Rectifiers for DC-DC Conversion." *27th IEEE Power Electron. Spec. Conference,* (Baveno, Italy) (1996): 854–860.

36. Albulet, M. "An Exact Analysis of Class DE Amplifier at Any Output *Q*," *IEEE Trans. Circuits Syst.I,* vol. 46, (October 1999): 1228–1239.

37. *HB-PLUS User Manual,* Design Automation, Inc., Lexington, MA, 1993.

4

Class E Power Amplifiers

It is desirable to obtain high RF power amplifier efficiency in many practical applications. At least one, if not all, of the following requirements is important: low power consumption (especially for battery-operated equipment), low temperature rise of the components, high reliability, small size, and light weight. Although the parasitic energy losses in, for example, inductors and capacitors are sometimes important, the major power loss in power amplifiers is usually due to power dissipated in the transistor(s). This was true for the Class A, B, and C PAs (their theoretical collector efficiency was calculated in Sections 2.1 through 2.3). Minimization of the power dissipated in the transistor(s) can be achieved by [1, 2]

a. minimizing the voltage across the transistor when current flows through it, and the current through the transistor when voltage exists across it,

b. minimizing the time intervals when large currents and voltages are applied simultaneously to the transistor.

Increased Class C amplifier efficiency is obtained by applying option a. above. During the large collector pulse, the collector-emitter voltage is low; on the other hand, the transistor is cut off when the collector voltage is high. The Class D PA further increases collector efficiency over that of the Class C amplifier by applying both options a. and b. First, the transistors are alternately cut off or saturated; either the voltage during the ON state or the current during the OFF state is negligible (see Figure 4-1).[1] Second, the switching times (when appreciable voltage and current are applied simultaneously to the transistor) are minimized in respect to the RF period; this usually requires fast switching transistors.

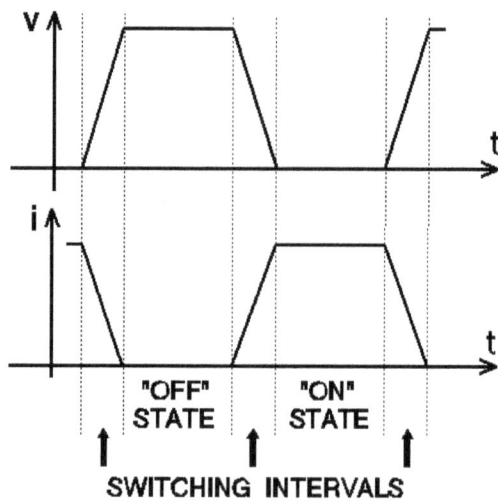

Figure 4-1 *Voltage and current waveforms in a switching-mode amplifier (driving a resistive load).*

The Class E circuit is a single-ended switching-mode amplifier (although it is possible to derive push-pull configurations easily). The following discussion will therefore refer only to single-ended circuits. The block diagram is given in Figure 4-2 [1, 2].

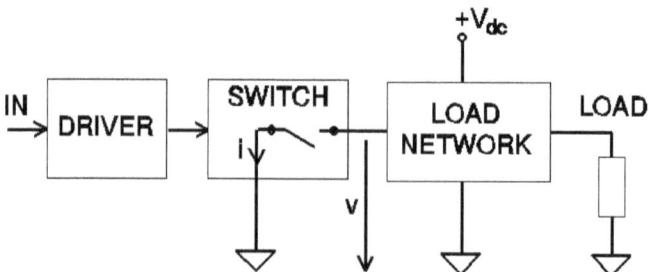

Figure 4-2 *Block diagram of a single-ended switching-mode power amplifier [1, 2].*

The circuit includes an active device driven to perform as a single-pole single-throw switch (at the desired output frequency), a load network created by reactive elements, and the load resistor. The ideal operating conditions described above may be approached if the circuit is designed so the optimum waveforms in Figure 4-3 are obtained.

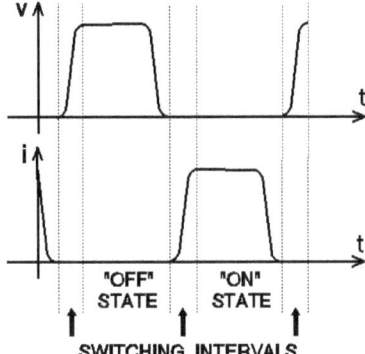

Figure 4-3 *Optimum waveforms (for maximum collector efficiency) of voltage across the switch (top) and current through the switch (bottom) [1, 2].*

Condition a. is easily fulfilled by choosing an active device with negligible leakage current in the OFF state and negligible voltage drop across it in the ON state. Condition b. is fulfilled by avoiding the simultaneous imposition of substantial voltage and substantial current on the switch. This can be achieved even if the switching times are a significant part of the RF period.

Assuming the idealized waveforms in Figure 4-3 are obtained, the following conditions are also met:

1. The voltage across the switch returns to zero at the end of the OFF state (just before the switch turns ON), which avoids the energy dissipation caused by the simultaneous imposition of substantial voltage and current on the switch, during the OFF→ON transition. Moreover, a real switch is always shunted by a parasitic capacitance (including intrinsic switch capacitance and circuit stray capacitance) that must be discharged in each RF period at the OFF→ON transition (see the discussion in section **Shunt Capacitances or Series Inductances at Switches** in Chapter 3 for a Class D amplifier). Because the voltage across the switch returns to zero before switching occurs, the discharge losses are eliminated.

2. The voltage across the switch reaches zero at the end of the OFF state with zero slope (i.e., $dv/dt = 0$). Consequently, the current through the switch at the beginning of the ON state is zero. This helps to minimize the transition time (because the actual devices have limited di/dt capabilities) and, hence, further minimize the power losses during transition. The zero slope condition also allows for a slight mistuning of the amplifier and a slow switch turn-ON, without severe collector efficiency loss.

3. The current through the switch returns to zero at the end of the ON state (just before the switch starts turning OFF). This avoids the energy dissipation caused by the simultaneous imposition of substantial voltage and current on the switch during the ON→OFF transition. Moreover, a real switch always has a series parasitic inductance that may cause power losses if the current through the switch jumps at the instant that switching occurs (see the discussion on *Series Inductances* in Chapter 3 for a Class D amplifier). Because the current through the switch returns to zero before switching, these losses are eliminated.

4. The current through the switch reaches zero at the end of the ON state with zero slope (i.e., $di/dt = 0$). The voltage across the switch at the start of the OFF state is zero. This allows slight mistuning of the amplifier and slow turn-OFF of the switch without severe loss of efficiency.

A switching-mode power amplifier (or frequency multiplier) that meets all the conditions above, achieves a very high collector efficiency. This is true even at very high frequencies, when the switching times are an important part of the RF period. However, note that the waveforms in Figure 4-3 cannot be obtained practically. The switching losses can be reduced significantly in some designs (for example, in Class E circuits), but they cannot be fully eliminated. For example, the Class E configurations can achieve waveforms somewhat similar to those depicted in Figure 4-3, but only at the switch-ON or switch-OFF time. This eliminates switching losses only at the OFF→ON or ON→OFF transition. At the other transition point, the switching losses cannot be eliminated. This means that conditions 1. through 4. above cannot be satisfied simultaneously in a real amplifier. However, it is possible (and this is actually the case in a Class E PA) to satisfy either conditions 1. and 2., or conditions 3. and 4.

There is an important distinction to remember when comparing Class A to D amplifiers with Class E types. Class A to D amplifiers are defined by a particular circuit topology or the value of a certain parameter (the conduction angle). However, the Class E circuit is based on the principle of obtaining high efficiency without relying on a certain circuit configuration or specific parameters. Sokal and Sokal emphasized this fact by defining the Class E amplifier as a switching-mode single-ended circuit that meets the following three basic criteria [1, 2, 3]:

- The switch voltage is brought back to zero when the switch turns ON (condition 1, above).

- The slope of the switch voltage is zero when the switch turns ON (condition 2, above).

- The rise of the voltage across the switch is delayed until after the transistor is OFF. (This is actually caused by the presence of a capacitor in parallel at the switch, which delays the rise of the voltage while the current is falling during the turn-OFF transition.)

This definition of Class E circuits is somewhat restrictive and can be broadened as follows: Reference [4] defines the Class E circuit as a tuned amplifier composed of a single-pole single-throw switch and a load network. Although this definition is slightly confusing,[2] it allows a Class E circuit to be described as one that is not optimized or mistuned (intentionally or not). If the circuit meets the three conditions above, it is called an *optimum Class E circuit*. Otherwise, it is called a *suboptimum Class E circuit*. Various topologies can be used to build a Class E amplifier.

The three basic criteria given above are met by the so-called *zero voltage switching* (ZVS) Class E amplifiers. In the ZVS circuits, the OFF→ON transition occurs at zero voltage across the switch. The first two conditions above are known as *the Class E switching conditions* for ZVS amplifiers. Dual circuits, meeting dual criteria (the ON→OFF transition of the switch occurs at zero current through the switch), are called ZCS (*zero current switching*) Class E amplifiers. The Class E switching conditions for these circuits are 3. and 4. above. Note that the ZCS amplifiers are usually less attractive for practical applications (especially at high frequencies) because they require a switch with negligible capacitance. This chapter is mostly dedicated to the basic ZVS Class E circuit that was first described by Sokal and Sokal [1, 2].

4.1   Idealized Operation of the Class E Amplifier

The basic topology of the Class E amplifier is shown in Figure 4-4(a) [1, 2, 4, 5]. The circuit includes a transistor, Q, operated as a switch,[3] a shunt capacitor, C_2, an RF choke, RFC, a series-tuned output circuit, $L_0 C_0$, and the load resistor, R. C_1 is the parasitic capacitance in parallel at the switch (including intrinsic transistor output capacitance and circuit stray capacitance).

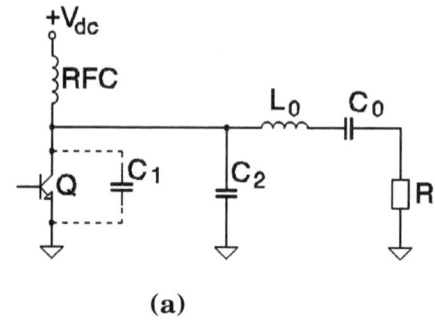

(a)

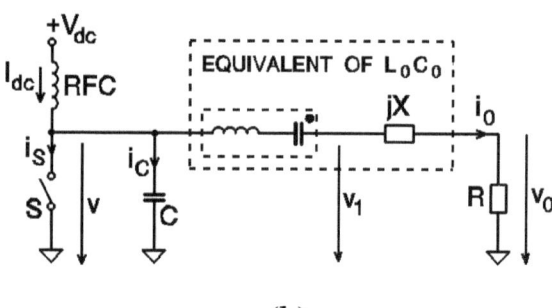

(b)

Figure 4-4 *(a) Basic circuit of a Class E amplifier and (b) equivalent idealized circuit (• = ideal series-tuned circuit).*

The simple equivalent circuit shown in Figure 4-4(a) is based on the following assumptions [4].

- The RF choke RFC is ideal: zero series DC resistance and infinite reactance at the operating frequency. The RF choke therefore allows only a constant (DC) input current. Note that in practical circuits, the use of a bypass capacitor is recommended to prevent penetration of the RF signal into the power supply line.

- The series resonant circuit, $L_0 C_0$, is usually not tuned to the operating frequency, f, having at this frequency a net series reactance jX (produced by the difference in the reactances of the inductor and capacitor of the series-tuned circuit).

$$X = 2\pi f L_0 - \frac{1}{2\pi f C_0} \qquad 4.1$$

For modeling purposes, it is convenient to consider $L_0 C_0$ as a resonant circuit tuned to the operating frequency, f, in series with a net reactance,

jX [see Figure 4-4(b)]. The series resonant circuit [marked with a "•" in Figure 4-4(b)] is assumed ideal; as a result, the load current is sinusoidal.

- The active device acts as an ideal switch: zero saturation voltage, zero saturation resistance, and infinite OFF resistance; the switching action is instantaneous and lossless. The transistor can withstand both positive and negative collector voltage (v) and can pass both positive and negative collector current (i_S).

- The total shunt capacitance $C = C_1 + C_2$ is independent of the collector voltage.

- All components are ideal. (The possible parasitic resistances of L_0 and C_0 can be included in load resistance R; the possible parasitic reactance of the load can be included in either L_0 or C_0.)

Based on the assumptions above, the equivalent circuit of Figure 4-4 (b) is obtained. Note that the voltage, $v_1(\theta)$, has no physical meaning, as it is only a convenient reference point for the analysis below. The analysis of the Class E amplifier is based on first finding its steady-state waveforms (see Figure 4-5). This is a much more difficult task than for the amplifiers analyzed previously, because there is no clear voltage or current source in the Class E circuit. Moreover, all the parameters of interest (for example, the collector voltage, the collector current, the output current, the DC input current) are interrelated via nonlinear equations.

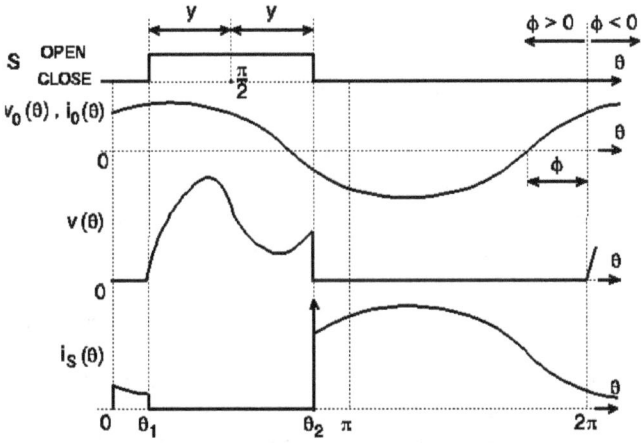

Figure 4-5 *Waveforms in the Class E amplifier.*

Circuit Analysis

The output voltage and current are sinusoidal and are given by

$$i_0(\theta) = \frac{V_0}{R} \sin(\theta + \phi) \qquad 4.2$$

$$v_0(\theta) = V_0 \sin(\omega t + \phi) = V_0 \sin(\theta + \phi) \qquad 4.3$$

where $\theta = \omega t = 2\pi f t$, ϕ is defined in Figure 4-5, and the parameters V_0 and ϕ are determined later. The fictitious voltage, $v_1(\theta)$, is also a sinusoidal waveform, but is not in phase with $v_0(\theta)$ because of reactance jX.

$$v_1(\theta) = v_0(\theta) + v_X(\theta) = V_0 \sin(\theta + \phi) + \frac{X}{R} V_0 \cos(\theta + \phi) =$$

$$= V_1 \sin(\theta + \phi_1) \qquad 4.4$$

where $v_X(\theta)$ is the voltage across jX,

$$V_1 = V_0 \sqrt{1 + \left(\frac{X}{R}\right)^2} = \rho V_0 \qquad 4.5$$

is the amplitude of $v_1(\theta)$, and

$$\phi_1 = \phi + \psi = \phi + \arctan\left(\frac{X}{R}\right) \qquad 4.6$$

is its initial phase.

The difference between the DC input current, I_{dc}, and the output current, $i_0(\theta)$, flows into capacitor C when switch S is open, and through the switch when it is closed. If the voltage across the capacitor is a value other than zero when the switch closes, the capacitor is discharged and the switch dissipates the stored energy. Because the ON resistance of the switch is assumed equal to zero in this model, the discharge of the capacitor is instantaneous (a Dirac current pulse as depicted in Figure 4-5). In a real circuit, the ON resistance of the switch is nonzero and thus the discharge requires a nonzero time interval. However, the energy to be dissipated does not depend on the switch resistance or on the particular discharge waveforms. As long as the time required to discharge the capacitor is negligible in respect to the RF period, the model given here remains valid.

When the switch is OFF, the current $i_C(\theta) = I_{dc} - i_0(\theta)$ charges capacitor C.

$$v(\theta) = \frac{1}{\omega C} \int_{\theta_1}^{\theta} i_C(\tau) d\tau \qquad\qquad 4.7$$

where θ_1 is the instant at which the switch opens.

For this analysis, it is convenient to choose the center of the OFF interval as $\pi/2$ and describe the waveforms in terms of the OFF angle $2y$ (see Figure 4-5).[4] The switching moments are $\theta_1 = \theta/2 - y$ and $\theta_2 = \pi/2 + y$ (plus/minus an integer number of 2π). Denoting with B the susceptance of C at the switching frequency f ($B = \omega C$), the collector voltage is given by

$$v(\theta) = \begin{cases} \dfrac{1}{\omega C} \displaystyle\int_{\frac{\pi}{2}-y}^{\theta} \left[I_{dc} - \dfrac{V_0}{R} \sin(\tau + \phi) \right] d\tau & \theta_1 \le \theta \le \theta_2 \\ 0 & \text{otherwise} \end{cases}$$

$$v(\theta) = \begin{cases} \dfrac{1}{B} \left[\begin{array}{l} I_{dc}\left(-\dfrac{\pi}{2} + y\right) + \dfrac{V_0}{R}\sin(\phi - y) + \\ +I_{dc}\theta + \dfrac{V_0}{R}\cos(\theta + \phi) \end{array} \right] & \theta_1 \le \theta \le \theta_2 \\ 0 & \text{otherwise} \end{cases} \qquad 4.8$$

At the switching frequency, f, there is no voltage drop across the ideal series-tuned circuit in Figure 4-4. Therefore, the fictitious voltage, $v_1(\theta)$, which is a sinewave of phase ϕ_1, is the fundamental of the collector voltage, $v(\theta)$, and its amplitude is given by

$$V_1 = \frac{1}{\pi} \int_0^{2\pi} v(\theta) \sin(\theta + \phi_1) d\theta \qquad\qquad 4.9$$

Integrating Equation 4.9 yields

$$V_1 = -2\left[\frac{I_{dc}}{\pi B}\left(\frac{\pi}{2} - y\right) + \frac{V_0}{\pi BR}\sin(y - \phi) \right]\cos\phi_1 \sin y +$$

$$+ \frac{I_{dc}}{\pi B}\left(-2\sin\phi_1 \sin y + \pi\cos\phi_1 \sin y + 2y\sin\phi_1 \cos y\right) - \qquad 4.10$$

$$- \frac{V_0}{2\pi BR}\left[\sin(2\phi + \psi)\sin 2y - 2y\sin\psi\right]$$

and using Equations 4.5 and 4.6,

$$\rho V_0 + V_0 \left[\frac{\sin(2\phi + \psi)\sin 2y - 2y\sin\psi}{2\pi BR} + \frac{2\sin(y - \phi)\cos\phi_1 \sin y}{\pi BR} \right] =$$

$$= \frac{I_{dc}}{\pi B} \left[2y\cos\phi_1 \sin y + (2y\cos y - 2\sin y)\sin\phi_1 \right]$$

4.11

and then

$$V_0 = I_{dc} R h(\phi, \psi, y, B, R, \rho)$$

4.12

where

$$h(\phi, \psi, y, B, R, \rho) =$$

$$= \frac{2y\sin y\cos\phi_1 + (2y\cos y - 2\sin y)\sin\phi_1}{\pi BR\rho + \frac{1}{2}\sin(2\phi + \psi)\sin 2y - y\sin\psi + 2\sin(y - \phi)\cos\phi_1 \sin y}$$

4.13

The fundamental component of $v(\theta)$ has no cosine (quadrature) component with respect to phase ϕ_1.

$$0 = \frac{1}{\pi} \int_0^{2\pi} v(\theta)\cos(\theta + \phi_1)d\theta$$

4.14

As a results,

$$0 = -2\sin\phi_1 \sin y \left[\frac{I_{dc}}{\pi B}\left(y - \frac{\pi}{2}\right) + \frac{V_0}{\pi BR}\sin(\phi - y) \right] +$$

$$+ \frac{I_{dc}}{\pi B}\left(-2\cos\phi_1 \sin y - \pi\sin\phi_1 \sin y + 2y\cos\phi_1 \cos y\right) -$$

$$- \frac{V_0}{2\pi BR}\sin 2y\cos(2\phi + \psi) + \frac{yV_0 \cos\psi}{\pi BR}$$

4.15

and then

$$V_0 = I_{dc} R g(\phi, \psi, y)$$

4.16

where

$$g(\phi, \psi, y) = \frac{2y\sin\phi_1 \sin y - 2y\cos\phi_1 \cos y + 2\cos\phi_1 \sin y}{-2\sin(\phi - y)\sin y\sin\phi_1 - \frac{1}{2}\sin 2y\cos(2\phi + \psi) + y\cos\psi}$$

4.17

Equations 4.12 and 4.16 yield

$$g(\phi,\psi,y) = h(\phi,\psi,y,B,R,\rho) \qquad \text{4.18}$$

Assuming that the component values and switch duty ratio are specified, Equation 4.18 allows to determine the initial phase, ϕ. Equation 4.18 may be rewritten as

$$\frac{q_1 \sin\phi + q_0 \cos\phi}{r_2 \sin^2\phi + r_1 \sin\phi\cos\phi + r_0 \cos^2\phi} =$$
$$= \frac{-q_0 \sin\phi + q_1 \cos\phi}{s_2 \sin^2\phi + s_1 \sin\phi\cos\phi + s_0 \cos^2\phi} \qquad \text{4.19}$$

where

$$q_0 = 2(\sin y - y\cos y)\cos\psi + 2y\sin y\sin\psi$$
$$q_1 = 2y\sin y\cos\psi + 2(y\cos y - \sin y)\sin\psi$$
$$r_0 = (y - \sin y\cos y)\cos\psi + 2\sin^2 y\sin\psi$$
$$r_1 = 2\sin^2 y\cos\psi$$
$$r_2 = (y - \sin y\cos y)\cos\psi \qquad \text{4.20}$$
$$s_0 = \pi BR\rho - (y - \sin y\cos y)\sin\psi + 2\sin^2 y\cos\psi$$
$$s_1 = -2\sin^2 y\sin\psi$$
$$s_2 = \pi BR\rho - (y - \sin y\cos y)\sin\psi$$

Rearranging Equation 4.19,

$$\alpha_3 \sin^3\phi + \alpha_2 \sin^2\phi\cos\phi + \alpha_1 \sin\phi\cos^2\phi + \alpha_0 \cos^3\phi = 0 \qquad \text{4.21}$$

where

$$\alpha_0 = q_0 s_0 - q_1 r_0$$
$$\alpha_1 = q_1 s_0 + q_0 s_1 + q_0 r_0 - q_1 r_1$$
$$\alpha_2 = q_1 s_1 + q_0 s_2 + q_0 r_1 - q_1 r_2 \qquad \text{4.22}$$
$$\alpha_3 = q_1 s_2 + q_0 r_2$$

With $\alpha_1 = \alpha_3$ and $\alpha_0 = \alpha_2$, Equation 4.21 becomes

$$\left(\tan\phi + \frac{\alpha_2}{\alpha_3}\right)\left(\tan^2\phi + 1\right) = 0 \qquad \text{4.23}$$

and its solution is given by

$$\phi = \arctan\left(-\frac{\alpha_2}{\alpha_3}\right) = \arctan\left(-\frac{\alpha_0}{\alpha_1}\right) \qquad 4.24$$

Equation 4.24 allows calculation of ϕ, and then Equation 4.17 gives $g(\phi, \psi, \psi)$. Note that the initial phase, ϕ, may have any value between $-\pi$ and π. However, Equation 4.24 provides values of ϕ only between $-\pi/2$ and $\pi/2$. For this reason, it may be necessary to add or subtract π to the calculated value of ϕ. The correct value of ϕ is that which provides a positive value for g in Equation 4.17.

Because there is no DC voltage across the RF choke, the average (DC) value of $v(\theta)$ must be V_{dc}. Using Equations 4.8 and 4.16. Denoting (for simplicity) $g = g(\phi, \psi, \psi)$,

$$V_{dc} = \frac{1}{2\pi} \int_{0}^{2\pi} v(\theta) d\theta =$$

$$= \frac{I_{dc}}{2\pi B} \int_{\frac{\pi}{2}-y}^{\frac{\pi}{2}+y} \left[y - \frac{\pi}{2} + g\sin(\phi - y) + \theta + g\cos(\theta + \phi) \right] d\theta = \qquad 4.25$$

$$= \frac{I_{dc}}{\pi B} \left[y^2 + yg\sin(\phi - y) - g\sin\phi\sin y \right] = I_{dc} R_{dc}$$

where R_{dc} is the equivalent resistance the amplifier shows to the DC power supply

$$R_{dc} = \frac{y^2 + g\left[y\sin(\phi - y) - \sin\phi\sin y \right]}{\pi B} \qquad 4.26$$

Using Equations 4.16 and 4.25, the output power (dissipated in the load resistor, R) is obtained

$$P_0 = \frac{1}{2}\frac{V_0^2}{R} = \frac{I_{dc}^2 g^2 R}{2} = \frac{V_{dc}^2 g^2 R}{2R_{dc}^2} \qquad 4.27$$

The DC input power is given by

$$P_{dc} = V_{dc} I_{dc} = \frac{V_{dc}^2}{R_{dc}} \qquad 4.28$$

and the collector efficiency is

$$\eta = \frac{P_0}{P_{dc}} = \frac{g^2}{2} \frac{R}{R_{dc}} \qquad \text{4.29}$$

The maximum collector voltage (during the OFF state) is calculated by first using Equations 4.16 and 4.25 in Equation 4.8. The following expression is obtained for the collector voltage:

$$v(\theta) = \begin{cases} \dfrac{V_{dc}}{BR_{dc}} \left\{ \theta + y - \dfrac{\pi}{2} + g\left[\sin(\phi - y) + \cos(\theta + \phi)\right] \right\}, & \theta_1 \leq \theta \leq \theta_2 \\ 0 \; \text{otherwise} \end{cases} \qquad \text{4.30}$$

The time when the maximum voltage occurs results from

$$0 = \left. \frac{dv(\theta)}{d\theta} \right|_{\theta = \theta_{v\max}} = \frac{V_{dc}}{BR_{dc}} \left[1 - g\sin(\theta_{v\max} + \phi) \right] \qquad \text{4.31}$$

This value given by

$$\theta_{v\max} = \arcsin\left(\frac{1}{g}\right) - \phi \qquad \text{4.32}$$

Finally, with θ_{vmax} known, the maximum collector voltage can be obtained from Equation 4.30. If θ_{vmax} given by the previous equation is outside of the interval $\theta_1 \leq \theta_{vmax} \leq \theta_2$, then the maximum value of the collector voltage occurs at the moment of switch-ON (θ_2). It is also possible to have a negative collector voltage peak during the OFF state at

$$\theta_{v\min} = \pi - \arcsin\left(\frac{1}{g}\right) - \phi \qquad \text{4.33}$$

Because $i_S(\theta) = I_{dc} - i_0(\theta)$ (during the ON state), the maximum value of the collector current is given by

$$I_{S\max} = I_{dc} + \frac{V_0}{R} = I_{dc} + \frac{I_{dc}Rg}{R} = I_{dc}(1 + g) \qquad \text{4.34}$$

assuming that this maximum value (see Figure 4-5) occurs during the ON state (when the switch is closed). Otherwise, Equation 4.34 is not valid, and I_{Smax} is obtained either at switch-ON or switch-OFF time. If the voltage across the shunt capacitance is not zero when the switch turns ON, I_{Smax} is theoretically infinite. The current pulse that discharges the shunt capacitor may have a much higher value than that given by Equation 4.34.

Finally, the power output capability is given by

$$C_P = \frac{P_0}{V_{\max} I_{S\max}} \qquad 4.35$$

If the voltage across the switch is not zero at the instant of turn-ON, C_P is zero in theory (because $I_{S\max} \to \infty$) and practical values may be very low.

EXAMPLE 4.1

The component values of a Class E amplifier operating at a DC supply voltage of $V_{dc} = 10$ V are $R = 50\ \Omega$, $C = 700$ pF, $C_0 = 325$ pF, and $L_0 = 80\ \mu$H; the switching frequency is $f = 1$ MHz and the switch duty ratio is $D = 0.5$. First, Equation 4.1 yields $X = 12.9475\ \Omega$, thus $\psi = 0.25338$. Then, Equations 4.17, 4.24, and 4.26 provide the values of $\phi = -0.20091$, $g = 1.33174$, and $R_{dc} = 49.4536\ \Omega$, respectively. The output power, the DC input power, and the collector efficiency are given by Equations 4.27 to 4.29: $P_0 = 1.8130\ \Omega$, $P_{dc} = 2.0221$ W, and $\eta = 89.659\%$. According to Equation 4.32, the maximum collector voltage occurs at $\theta = 1.05031$: $V_{\max} = 28.7289$ V. The voltage across the switch at the instant of turn-ON can be calculated from Equation 4.30, with $y = \pi/2$. Next, $v(\pi) = 24.4426$ V is obtained, and the power dissipated by discharging the shunt capacitor is given by $P_{dis} = (1/2)C(v(\pi))^2 f = 0.2091$ W. Note that $P_{dis} = P_{dc} - P_0$, because this is the only power loss in the idealized model presented here. Finally, Equation 4.34 provides the maximum collector current: $I_{S\max} = 0.47150$ A. However, because the voltage across the switch is not zero when the switch turns ON, the current pulse discharging the shunt capacitor (not shown in Figure 4-6) will have a much higher value. The calculated waveforms of this circuit are depicted in Figure 4-6.

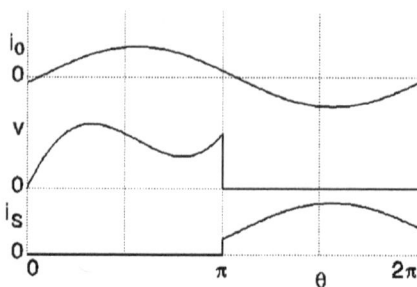

Figure 4-6 *Waveforms in the circuit of Example 4.1.*

Circuit Design

The idealized model of the Class E amplifier presented above includes only lossless elements (other than the load resistor). Thus, there may be only one loss mechanism in this model: the discharge of the shunt capacitor at the instant the switch turns ON. Thus, a collector efficiency of 100 percent is possible if the component values and operating parameters are chosen so the collector voltage drops to zero at the end of the OFF state (that is, at the instant when the switch turns ON). In this section, the design equations of a Class E circuit working with unity collector efficiency will be determined. The normalized slope of the collector voltage at turn-ON and the switch duty ratio will be considered as design parameters. Their effect on the design values and circuit performance will be also analyzed.

The requirement for 100 percent efficiency is

$$v\left(\frac{\pi}{2} + y\right) = 0 \qquad\qquad 4.36$$

Using Equation 4.30,

$$\cos\phi = \frac{y}{g\sin y} \qquad\qquad 4.37$$

The normalized slope of the collector voltage at turn-ON is defined as

$$\xi = \frac{1}{V_{dc}} \left.\frac{dv(\theta)}{d\theta}\right|_{\theta=\frac{\pi}{2}+y} \qquad\qquad 4.38$$

Using Equation 4.30,

$$\xi = \frac{1}{R_{dc}B}\left[1 - g\cos(y + \phi)\right] \qquad\qquad 4.39$$

Substituting $R_{dc}B$ from Equation 4.26 into 4.39, yields

$$(\pi\cos y - \xi y \sin y)\cos\phi + (-\pi\sin y + \xi y \cos y - \xi\sin y)\sin\phi =$$
$$= \frac{\pi - \xi y^2}{g} \qquad\qquad 4.40$$

If a value for ξ is somewhat arbitrarily chosen, Equations 4.37 and 4.40 form an equation system that provides the unknowns g and ϕ. By dividing the left and right sides of Equation 4.40 by the left and right sides of Equation 4.37, we can write

$$\tan\phi = \frac{\dfrac{\sin y}{y} - \cos y}{\dfrac{\xi y}{\pi}\cos y - \left(1 + \dfrac{\xi}{\pi}\right)\sin y}$$ 4.41

and

$$g = \frac{y}{\cos\phi\,\sin y}$$ 4.42

Because $\eta = 1$, Equation 4.29 yields

$$R_{dc} = \frac{g^2 R}{2}$$ 4.43

Substituting it in Equation 4.26, the design equation for the shunt capacitance is obtained

$$B = \frac{2\left[y^2 + yg\sin(\phi - y) - g\sin\phi\sin y\right]}{\pi g^2 R}$$ 4.44

Finally, Equation 2.18 is used to find ψ. Denoting

$$w_1 = -2g\sin(\phi - y)\sin y - 2y\sin y$$ 4.45
$$w_2 = 2y\cos y - 2\sin y$$
$$w_3 = -g\sin y\cos y$$

we obtain

$$\psi = \arctan\frac{w_1\sin\phi + w_2\cos\phi + w_3\cos 2\phi + gy}{w_2\sin\phi + w_3\sin 2\phi - w_1\cos\phi}, \quad \psi \in \left(-\frac{\pi}{2}\,\frac{\pi}{2}\right)$$ 4.46

EXAMPLE 4.2

Design a Class E amplifier operating at a DC supply voltage of $V_{dc} = 10$ V, with a load resistance $R = 50$ W. The switching frequency is $f = 1$ MHz, the switch duty ratio is $D = 0.5$, and the normalized slope of the switch voltage at the instant of turn-ON is $\xi = -2$. Equations 4.41 to 4.44 and 4.46 provide the values of $\phi = -1.05213$, $g = 3.16869$, $R_{dc} = 251.014$ Ω, $B = 0.00349$ S ($C = 555.4$ pF), and $\psi = 1.15983$, respectively. With these values known, the output power and the DC input power are given by Equations 4.27 and 4.28: $P_0 = P_{dc} = 0.39838$ W, because $\eta = 1$. Assuming that the quality factor of the series-tuned circuit is $Q = 10$, the component values are $C_0 = 1/(2\pi f Q R) = 318.3$ pF and $L_0 = (Q + tan\psi)R/(2\pi f) = 97.84$ µH. According to Equation 4.32, the maximum collector voltage occurs at $\theta = 1.3732$: $V_{max} = 32.0692$ V. Finally, Equations 4.34 and 4.35 provide the maximum collector current, $I_{Smax} = 0.16607$ A, and the power output capability, $C_P = 0.07480$. The calculated circuit waveforms are shown in Figure 4-7.

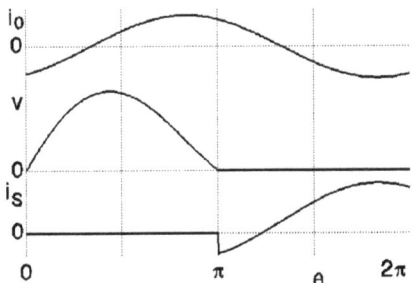

Figure 4-7 *Waveforms in the circuit of Example 4.2.*

The equations above show that the Class E circuit can provide 100-percent efficiency for a variety of values of y and ξ. The effect of the choice of y and ξ on the design values and circuit performance is presented below. For clarity, use the switch-OFF duty ratio, D, instead of y. The switch-OFF duty ratio is defined as the ratio between the duration of the OFF interval and the period of the signal. D and y are interrelated, $y = \pi D$, with D varying between 0 and 1. (Many articles on Class E circuits use the switch-ON duty ratio, defined as the ratio of the ON time interval to the period of the signal.) To simplify the calculations and the presentation of the results, use the standard normalized values of $V_{dc} = 1$ V and $R = 1$ Ω. This normalization is possible because the voltages in the circuit are proportional with V_{dc}, whereas the currents are proportional with V_{dc}/R. Two cases of practical interest will be detailed below. They are: the switch duty ratio $D = 0.5$ (that is, equal ON and OFF time intervals), and the normalized slope $\xi = 0$ (assuring high efficiency in practical circuits operating at high-frequencies).

a. $D = 0.5$, $R = 1$ Ω, $V_{dc} = 1$ V. The variation of the design values of B and ψ (for 100 percent efficiency) versus the normalized slope ξ are shown in Figures 4-8 (a) and (b). Figures 4-8 (c) and (d) depict the output power, P_0, and the power output capability, C_P, versus ξ. Both P_0 and C_P decrease rapidly toward zero as ξ approaches $-\pi$. Figure 4-9 presents the calculated waveforms of the collector voltage and current, with ξ as a parameter ($\xi = -\pi, -\pi/2, 0, \pi/2$, and π). The values of ξ are restricted between $-\pi$ and π because, for $D = 0.5$, Equation 4.44 shows that if $\xi < -\pi$, then B is be negative. Values of ξ above π are physically possible, but of little practical interest because they require large positive and negative collector voltage swings.

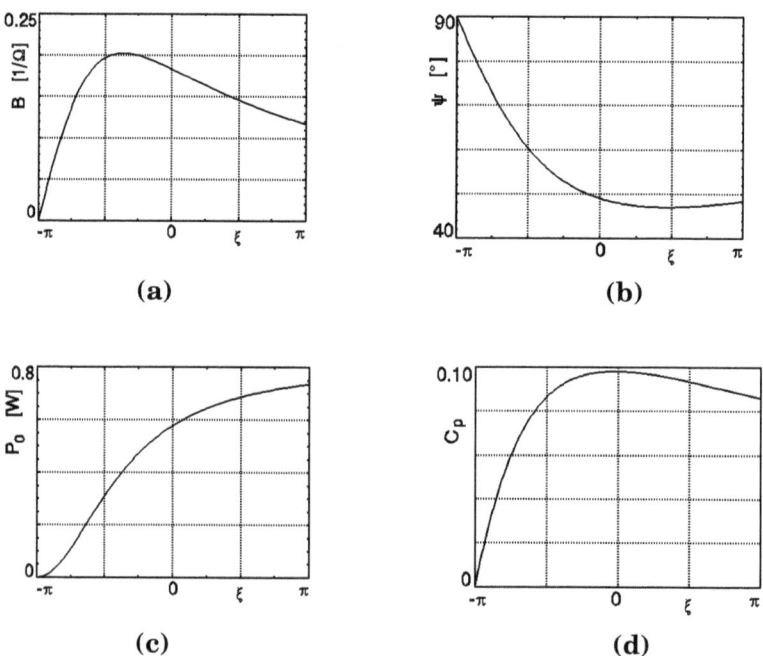

Figure 4-8 *(a), (b) Design values of B and ψ, (c) output power P_0, and (d) power output capability C_P, versus ξ (D = 0.5, V_{dc} = 1 V, R = 1 Ω, η = 1).*

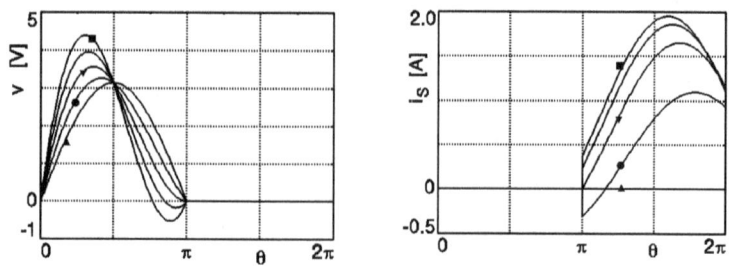

Figure 4-9 *Waveforms of collector voltage and current with ξ as a parameter (D = 0.5, V_{dc} = 1 V, R = 1 Ω, η = 1).* —▲— $\xi = -\pi$; —•— $\xi = -\pi/2$; — ▼ — $\xi = 0$; — $\xi = \pi/2$; — ■ — $\xi = \pi$.

b. $\xi = 0.5$, $R = 1\ \Omega$, $V_{dc} = 1$ V. The variation of the design values of B and ψ (for 100 percent efficiency) versus the normalized slope ξ are shown in Figures 4-10 (a) and (b). The output power, P_0, and the power output capability, C_P, are depicted in Figures 4.10 (c) and (d). Figure 4-11 presents the calculated waveforms of the collector voltage and current, with D as a parameter ($D = 0.2$, 0.5, and 0.8). Observe that the maximum value of C_P ($C_P = 0.098089$) is obtained for $D = 0.5$. (Exact calculations actually show that maximum C_P is obtained for $D = 0.511$ and this maximum is $C_p = 0.098159$. However, the difference is insignificant so $D = 0.5$ is usually considered the operating point providing maximum power output capability. On the other hand, a switch duty ratio of 0.5 is very convenient for practical applications and can be obtained easily.)

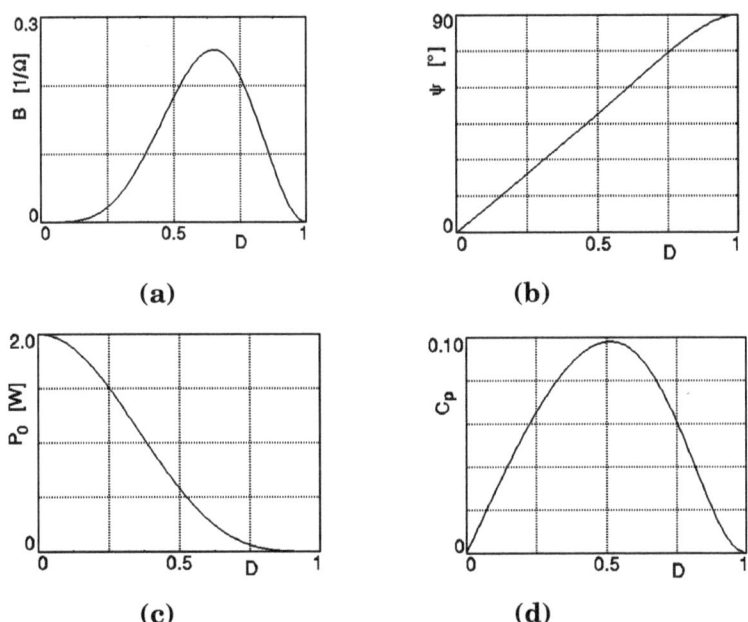

(a) (b)

(c) (d)

Figure 4-10 (a), (b) Design values of B and ψ, (c) output power P_0, and (d) power output capability C_P versus D ($\xi = 0$, $V_{dc} = 1$ V, $R = 1\ \Omega$, $\eta = 1$).

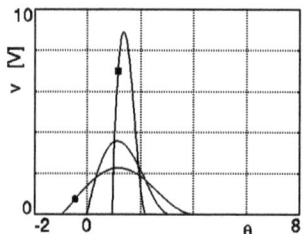 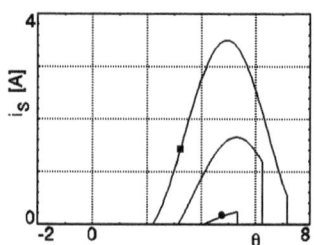

Figure 4-11 *Waveforms of collector voltage and current with D as a parameter ($\xi = 0$, $V_{dc}=1$ V, $R = 1 \Omega$, $\eta = 1$). — ■ — D = 0.2; — D = 0.5; —•— D = 0.8.*

The optimum operating point of the Class E amplifier is obtained for $D = 0.5$ and $\xi = 0$. In this case, the following design and performance values can be established.

$$\phi = -\arctan\frac{2}{\pi} \approx -32.4816° \approx -0.56691\,rad$$

$$R_{dc} = \frac{\pi^2 + 4}{8}R \approx 1.7337R$$

$$B = \frac{8}{\pi(\pi^2 + 4)R} \approx \frac{1}{5.4466R} \approx \frac{0.1836}{R}$$

$$\psi = \arctan\frac{\pi(\pi^2 - 4)}{16} \approx 49.0524° \approx 0.85613\,rad$$

$$X = R\tan\psi = \frac{\pi(\pi^2 - 4)}{16}R \approx 1.1525R$$

$$V_0 = \frac{4}{\sqrt{\pi^2 + 4}}V_{dc} \approx 1.0741V_{dc}$$

$$P_0 = P_{dc} = \frac{8}{\pi^2 + 4}\frac{V_{dc}^2}{R} \approx 0.5768\frac{V_{dc}^2}{R}$$

$$V_{\max} = 2\pi\left[\frac{\pi}{2} - \arctan\left(\frac{\pi}{2}\right)\right]V_{dc} \approx 3.5620V_{dc}$$

$$I_{S\max} = \left(1 + \frac{\sqrt{\pi^2 + 4}}{2}\right)I_{dc} \approx 2.8621I_{dc} \approx$$

$$\approx \frac{8}{\pi^2 + 4}\left(1 + \frac{\sqrt{\pi^2 + 4}}{2}\right)\frac{V_{dc}}{R} \approx 1.6509\frac{V_{dc}}{R}$$

$$C_P \approx 0.098089$$

4.47

Although the Class E circuit can theoretically operate with unity efficiency for a large variety of values for D and ξ, optimum operation with a switch duty ratio of $D = 0.5$ and a normalized slope of $\xi = 0$ is usually aimed for in practice. (For the idealized model discussed here, unity efficiency requires only that $v(\theta_2) = 0$.) A switch duty ratio of 0.5, together with the Class E switching conditions,

$$v(\theta_2) = 0$$
$$\left.\frac{dv(\theta)}{d\theta}\right|_{\theta=\theta_2} = 0$$

4.48

assures the following benefits:

1. The voltage across the switch is zero at the end of the OFF state, $v(\theta_{2-})$ = 0. (Note that, because the switching action is assumed to be instantaneous, the instant θ_2 represents both the end of the OFF state and the beginning of the ON state. If more clarity is required in a certain context, use θ_{2-} to stand for the end of the OFF state and $\theta_{2}+$ to represent the start of the ON state. According to this definition, θ_2 in Equation 4.48 should be replaced by θ_{2-}.) There is no power loss due to the discharge of the shunt capacitance at the instant of turn-ON. High current pulses and the possibility of transistor destruction by secondary breakdown are also avoided. In practice, any real switch has a non-zero shunt capacitance (the intrinsic transistor output capacitance and the stray circuit capacitance), so these problems cannot be avoided if the zero voltage switching condition is not satisfied. As discussed in Chapter 3, the discharge losses limit the usability of the Class D amplifier at high frequencies. The Class E circuit performs much better at high frequencies because the parasitic capacitance can be included in the load network.

2. The slope of the collector voltage at the instant of turn-ON (at the end of the OFF state) is zero. As a result, the current through the shunt capacitance immediately before the switch closes is also zero $i_C(\theta_{2-})=0$. Thus, $i_S(\theta_{2}+) = 0$, i.e., the collector current in the ON state starts from zero (see Figures 4-9 and 4-11). This provides the following benefits in practical circuits:

   • Even if the OFF→ON switching time is an important fraction of the RF period, the power dissipated in the transistor during OFF→ON switching is low because there are no voltage or current jumps. In this context, note that during the ON→OFF transition only the collector current has a jump, while the collector voltage remains low and increases significantly only after the collector current has dropped to zero. As a result, the power loss during the ON→OFF transition is also reduced. The shunt capacitance plays an important role because it does not allow a fast rise of the collector voltage when the switch turns OFF.

- A slight mistuning of the amplifier will not cause a significant decrease of the collector efficiency. Assume that the OFF→ON switching begins before the collector voltage reaches zero. The power dissipated in the transistor during the switching interval will be reasonably low, because the collector current is low at the beginning of the ON interval.

- The lack of a current jump at the instant of turn-ON helps to minimize the transition and further reduces the power loss during the transition, because the actual transistors have limited di/dt capabilities.

- If ξ is negative, the collector current (in the ON state) has both positive and negative values, and a MOSFET transistor may be suitable for implementing the switch. If ξ is positive, the collector voltage (in the OFF state) has both positive and negative values and a BJT may be suitable for implementing the switch (although it has somewhat limited capabilities to withstand negative collector voltages). Transistor-diode combinations may be also used to implement switches able to pass negative currents or withstand negative voltages. If $\xi = 0$, then both the collector voltage and current are positive and the switch can be implemented using just one transistor (either BJT or MOSFET).

3. The power output capability has the maximum value.

Harmonic Content of the Output Current

The idealized operation of the Class E amplifier presented above is based on the assumption that the quality factor of the series-tuned circuit is high enough so the load current is sinusoidal. However, in most practical applications, series-tuned circuit $L_0 C_0$ is unable to provide the required attenuation for the output current harmonics and additional filtering circuits may be required. A similar analysis to the one presented in Chapter 3 for the Class D amplifier shows that the collector efficiency of a circuit employing real LC components decreases as the loaded Q of the series-tuned circuit increases. Practical values of Q are typically below 10 and are insufficient to provide the required degree of harmonic suppression for many applications (for example, −20 to −60 dBc).

Note that the series-tuned circuit in a Class E amplifier does not usually resonate at the operating frequency. A net reactance is required to obtain the optimum operation. Therefore, it is possible to define the loaded quality factor either as [4]

$$Q = \frac{1}{\omega C_0 R} \qquad 4.49$$

or as [1]

$$Q = \frac{\omega L_0}{R} \qquad 4.50$$

The values of Q given by these two equations are interrelated by means of net reactance, X, given by Equation 4.1, but the numerical values are different. The definition in Equation 4.49 is preferred here; the results obtained can be modified easily to allow the use of Q given by Equation 4.50. The harmonics of the collector voltage are given by the Fourier series [4].

$$c_n \sin(n\theta + \phi_n) = a_n \cos n\theta + b_n \sin n\theta, \text{ where}$$

$$a_n = \frac{1}{\pi} \int_0^{2\pi} v(\theta) \cos n\theta \, d\theta \qquad 4.51$$

$$b_n = \frac{1}{\pi} \int_0^{2\pi} v(\theta) \sin n\theta \, d\theta$$

Using Equation 4.30, the result is [4]

$$a_n = \frac{V_{dc}}{R_{dc}B} \left\{ \left[y - \frac{\pi}{2} + g \sin(\phi - y) \right] q_2 + q_4 + g q_6 \right\} \qquad 4.52$$

$$b_n = \frac{V_{dc}}{R_{dc}B} \left\{ \left[y - \frac{\pi}{2} + g \sin(\phi - y) \right] q_1 + q_3 + g q_5 \right\}$$

where

$$q_1 = \begin{cases} \dfrac{2}{n\pi}(-1)^{\frac{n-1}{2}} \sin ny, & n \text{ odd} \\[2mm] 0, & n \text{ even} \end{cases}$$

$$q_2 = \begin{cases} 0, & n \text{ odd} \\[2mm] \dfrac{2}{n\pi}(-1)^{\frac{n}{2}} \sin ny, & n \text{ even} \end{cases} \qquad 4.53$$

$$q_3 = \begin{cases} (-1)^{\frac{n-1}{2}} \dfrac{\sin ny}{n}, & n \text{ odd} \\[3mm] \dfrac{2}{n\pi}(-1)^{\frac{n}{2}}\left[\dfrac{\sin ny}{n} - y\cos ny\right], & n \text{ even} \end{cases}$$

$$q_4 = \begin{cases} \dfrac{2}{n\pi}(-1)^{\frac{n+1}{2}}\left(\dfrac{\sin ny}{n} - y\cos ny\right), & n \text{ odd} \\[3mm] (-1)^{\frac{n}{2}}\dfrac{\sin ny}{n}, & n \text{ even} \end{cases}$$

4.54

$$q_5 = \begin{cases} -\dfrac{\sin\phi}{\pi}\left(\dfrac{\sin 2y}{2} + y\right), & n = 1 \\[3mm] \dfrac{(-1)^{\frac{n+1}{2}}\sin\phi}{\pi}\left[\dfrac{\sin(n+1)y}{n+1} + \dfrac{\sin(n-1)y}{n-1}\right], & n > 1, \ n \text{ odd} \\[3mm] \dfrac{(-1)^{\frac{n}{2}}\cos\phi}{\pi}\left[\dfrac{\sin(n+1)y}{n+1} - \dfrac{\sin(n-1)y}{n-1}\right], & n \text{ even} \end{cases}$$

4.55

$$q_6 = \begin{cases} -\dfrac{\cos\phi}{\pi}\left(\dfrac{\sin 2y}{2} - y\right), & n = 1 \\[3mm] \dfrac{(-1)^{\frac{n+1}{2}}\cos\phi}{\pi}\left[\dfrac{\sin(n+1)y}{n+1} - \dfrac{\sin(n-1)y}{n-1}\right], & n > 1, \ n \text{ odd} \\[3mm] \dfrac{(-1)^{\frac{n+2}{2}}\sin\phi}{\pi}\left[\dfrac{\sin(n+1)y}{n+1} + \dfrac{\sin(n-1)y}{n-1}\right], & n \text{ even} \end{cases}$$

As an example, the harmonic content (expressed in dBc) of an optimally operated Class E amplifier is depicted in Figure 4-12.

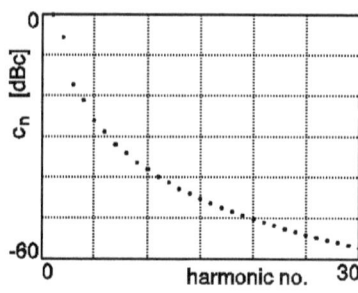

Figure 4-12 *Harmonic spectrum of the collector voltage of an optimum Class E amplifier.*

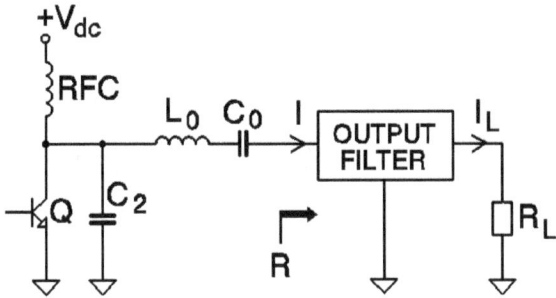

Figure 4-13 *Class E PA with an additional filter.*

The requirements to design the harmonic suppression filters (see Figure 4-13) can be determined as follows. Assuming that the series-tuned circuit has the loaded quality factor Q, given by Equation 4.49, the net reactance of $L_0 C_0$ at the frequency nf (that is, for the nth harmonic of the operating frequency) is given by

$$X_n = n\omega L_0 - \frac{1}{n\omega C_0} = R\left[Q\left(n - \frac{1}{n}\right) + 1.1525n\right] \qquad 4.56$$

where R is the equivalent load resistance of the Class E amplifier (seen at the input port of the output filter, Figure 4-13, at the switching frequency f). For $n = 1$, this equation yields $X_1 = X$ given by Equation 4.47. X_n increases roughly in proportion with n, while the input impedance of the filter decreases quickly with n. Typically, for any $n > 1$, the input impedance is negligible in comparison with X_n. Denoting the amplitude of the harmonic currents at the input of the filter with I_n ($n = 1, 2,...$), an approximation can be made [6].

$$\frac{I_n}{I_1} \approx \frac{c_n}{c_1} \frac{\sqrt{X_1^2 + R^2}}{X_n} = \frac{c_n}{c_1} \frac{1.5259}{Q\left(n - \frac{1}{n}\right) + 1.1525n} \qquad 4.57$$

With I_n/I_1 known, calculate the required transfer factor for the filter

$$\frac{(I_L / I)_n}{(I_L / I)_1} \leq \frac{S_n}{I_n / I_1} \qquad 4.58$$

where $(I_L/I)_n$ represents the transfer factor for the nth harmonic, and S_n represents the required suppression into the load (for instance, –60 dBc means $S_n = 1/1000$).

If a certain filter topology is chosen, Equation 4.58 may be used as a design requirement. Other design requirements may be related to, for example, impedance matching and bandwidth. After a first-cut design is developed, more exact values for the input impedance of the filter at different frequencies can be found. The approximation in Equation 4.57 may be corrected and the design refined; it is also possible to check whether the required suppression is obtained.

4.2   Practical Considerations

The analysis presented in Section 4.1 is based on several simplifying assumptions, which are not always acceptable. For example,

- real transistors have nonzero switching times, parasitic reactances, and nonzero ON resistance (and/or nonzero saturation voltage),

- the output current is not a pure sine wave, because the series-tuned circuit has a finite loaded Q,

- the reactive components have finite unloaded quality factor,

- accidental (or maybe intentional) mistuning or frequency variations are possible.

The effect of all these factors on design values and circuit performance may be impossible to evaluate. A quasi-complete model of the Class E circuit would provide extremely complicated equations that would be difficult to solve, even when using numerical techniques. (Keep in mind that the analysis requires the determination of the steady-state waveforms in the circuit. As shown in the previous section, even a simplified model leads to complicated equations.)

It is important to determine the effect each circuit imperfection has on circuit operation. A qualitative approach to this problem and insight into the physical understanding of the circuit that ignores the exact equations to be used for analysis and design follows. A convenient combination of the individual effects can be used to estimate the total effects of more than one of these factors. A more exact design solution for practical applications can be obtained using numerical techniques and dedicated CAD software (see Section 4.7).

Nonideal LC Components

An optimally operating Class E amplifier is considered in the analysis below. Assume that the series-tuned circuit, L_0C_0, has a loaded quality factor Q defined by Equation 4.49 and high enough to determine a sinusoidal output current. If Q_{L0} is the unloaded quality factor of L_0 and Q_{C0} is the unloaded quality factor of C_0, the equivalent series resistance of the L_0C_0 circuit is given by

$$r = \frac{\omega L}{Q_{L0}} + \frac{1}{\omega C Q_{C0}} = \frac{RQ + X}{Q_{L0}} + \frac{RQ}{Q_{C0}} \qquad 4.59$$

This resistance is placed in series with load resistance R. The equivalent load of the Class E amplifier is $R + r$, and the output power dissipated in load resistance R is given by (see Equation 4.47)

$$P_0 = \frac{8}{\pi^2 + 4} \frac{V_{dc}^2 R}{(R+r)^2} \approx 0.5768 \frac{V_{dc}^2 R}{(R+r)^2} \qquad 4.60$$

The collector efficiency can be calculated from

$$\eta = \frac{R}{R+r} = \frac{1}{1 + \dfrac{r}{R}} \qquad 4.61$$

The observations made in Chapter 3 for the Class D amplifier remain valid for the Class E amplifier. The power loss due to the inductor usually dominates the power loss due to the capacitor, so for an optimally operated Class E amplifier, it can be approximated that

$$\eta \approx \frac{1}{1 + \dfrac{Q + 1.1525}{Q_{L0}}} \qquad 4.62$$

The power loss due to the nonzero DC resistance of the RF choke is calculated using the equivalent circuit shown in Figure 4-14. Assume an optimally operated Class E circuit, with component values and circuit performance given by Equation 4.47. Considering only the effect of R_{RFC}, the collector efficiency is given by

$$\eta = \frac{R_{dc}}{R_{dc} + R_{RFC}} \approx \frac{1}{1 + \dfrac{R_{RFC}}{1.7337R}} \qquad 4.63$$

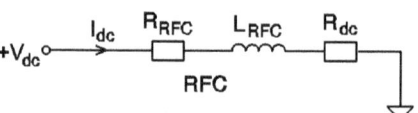

Figure 4-14 *Equivalent circuit used to calculate the power loss due to* R_{RFC}.

The increase of the load resistance causes an increase in the collector efficiency. However, if R increases, L_{RFC} must also be increased. This determines the increase of R_{RFC}; the ratio R/R_{RFC} remains almost constant. In a practical design, the power losses caused by R_{RFC} are determined solely by the volume available for the RF choke.

The power loss in shunt capacitor C can be calculated as follows. During the OFF state, the current $i_C(\theta) = I_{dc} - i_0(\theta)$ flows through the capacitor and through its series loss resistance r_C. The power dissipated in r_C is given by

$$P_{dC} = \frac{r_C}{2\pi} \int_{\theta_1}^{\theta_2} i_C^2(\theta) d\theta \qquad 4.64$$

Assuming r_C is small enough so it does not affect the circuit waveforms of an optimally operated Class E PA,

$$P_{dC} = \frac{r_C}{2\pi} \int_0^{\pi} \left(I_{dc} - \frac{V_0}{R} \sin(\theta + \phi) \right)^2 d\theta =$$

$$= \frac{\pi^2 + 28}{2(\pi^2 + 4)} \frac{r_C}{R} P_0 \approx 1.3652 \frac{r_C}{R} P_0 \qquad 4.65$$

However, in a practical circuit, things are a bit more complicated because C includes the output capacitance of the transistor (which at high frequencies may be an important part of C). Therefore, a more exact model of the circuit should include two RC series groups connected in parallel.

Nonideal RF Choke

The ideal RF choke used in the equivalent circuit of Figure 4-4 has zero DC resistance and infinite reactance at the operating frequency. The effect of the nonzero DC resistance was discussed in the previous section. The purpose of this section is to discuss the effect of the finite reactance of a real RF choke on the design values and circuit performance of the Class E amplifier. More exactly, what will be discussed here is the effect of a finite DC feed inductance still large enough to behave like an RF choke. Section 4.4 shows that Class E circuits do not require an RF choke in the DC supply path and that Class E operation with a finite (and small) DC feed inductance is possible and sometimes desirable.

In practical circuits, it is inconvenient to use RF chokes with very large inductances (larger than normally required in that circuit) because

- The size, weight, and cost of the RF choke increases as its inductance increases.

- The DC resistance of the RF choke is also large, creating major power losses.

- A large number of turns may cause a high parasitic capacitance of the RF choke.

Therefore, when designing an RF choke, it is important to find a good tradeoff. Make sure the inductance is as low as possible, but still large enough to achieve the desired behavior of the circuit and the calculated performance. A detailed mathematical analysis of the problem is beyond the scope of this book, but the most important results are included here.

A practical recommendation is to choose the RF choke reactance large in comparison to the reactance of the shunting capacitor, i.e. [3],

$$L_{RFC} > \frac{10}{\omega^2 C} \qquad 4.66$$

For a Class E PA operating optimally, this condition is equivalent to

$$\omega L_{RFC} > \frac{10}{0.1836} R \approx 55R \qquad 4.67$$

In addition, C may be increased slightly to compensate for the finite value of the DC feed inductance

$$\omega C \approx \frac{0.1836}{R} + \frac{0.7}{\omega L_{RFC}} \qquad 4.68$$

Another recommendation is to choose the RF choke inductance so [7, 8]

$$\omega L_{RFC} > 30R \qquad 4.69$$

If this condition is satisfied, the variation of the design values and circuit performance of the Class E amplifier is below 10 percent with respect to the "ideal" values given in Section 4.1 for a circuit employing an RF choke.

Transition Time

The analysis of the idealized operation of the Class E amplifier presented in Section 4.1 is based on the assumption of an instantaneous and lossless switching of the active device. However, real transistors have nonzero transition times and, especially at high frequencies, the switching time may be a significant part of the RF period. If the Class E switching conditions are met, the power loss during the OFF→ON transition is negligible, even for slow transitions because the collector voltage has dropped back to zero at the end of the OFF state and the collector current in the ON state starts from zero.

The situation is completely different at the ON→OFF transition. The collector current decreases from $i_S(0) = I_{OFF}$ to zero during the ON→OFF transition (see Figure 4-15).

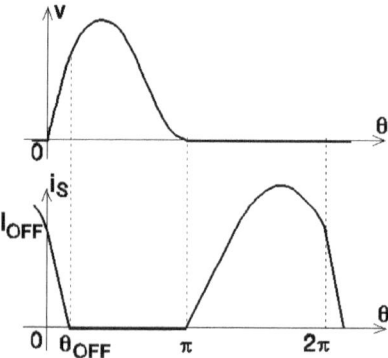

Figure 4-15 *Waveforms in a Class E circuit with a nonzero ON→OFF switching time.*

A simple estimation of the power loss during the ON→OFF transition can be made by assuming that the collector current varies linearly from I_{OFF} to zero. Moreover, the switching time is assumed negligible in respect to the RF period, so the circuit performance and the waveforms obtained for the idealized model presented in Section 4.1 are not affected.

Labeling the angular switching time as θ_{OFF} (in radians),[5] for $0 \leq \theta \leq \theta_{OFF}$, the collector current is given by

$$i_S(\theta) = I_{OFF}\left(1 - \frac{\theta}{\theta_{OFF}}\right) \qquad 4.70$$

The capacitor charging current is zero during the ON state and varies linearly between zero and I_{OFF} during the ON→OFF transition.

$$i_C(\theta) = I_{OFF} \frac{\theta}{\theta_{OFF}}$$ 4.71

The collector voltage during the transition is given by

$$v(\theta) = \frac{1}{B} \int_0^\theta i_C(\tau) d\tau = \frac{I_{OFF}}{2B\theta_{OFF}} \theta^2$$ 4.72

The power dissipated during transition is

$$P_d = \frac{1}{2\pi} \int_0^{\theta_{OFF}} v(\theta) i_S(\theta) d\theta = \frac{I_{OFF}^2 \theta_{OFF}^2}{48\pi B}$$ 4.73

For an optimum amplifier operating with $D = 0.5$, the equations in Section 4.1 show that $I_{OFF} = 2I_{dc}$, hence

$$P_d = \frac{\theta_{OFF}^2}{12} P_0$$ 4.74

The collector efficiency may be estimated as

$$\eta \approx 1 - \frac{\theta_{OFF}^2}{12}$$ 4.75

A slightly different equation is presented in References [1, 2, 3, and 10] (ignoring the saturation voltage)

$$\eta = \frac{1 - \dfrac{\theta_{OFF}^2}{6}}{1 - \dfrac{\theta_{OFF}^2}{12}}$$ 4.76

A more exact analysis is carried out in Reference [11] (assuming a linear variation of the collector current during the ON→OFF transition) and in Reference [12] (assuming, more realistically, an exponential decay of the current). The results are close to the ones given by the equations above. For example, for $\theta_{OFF} = 30°$, Equations 4.75 and 4.76 along with the model described in Reference [11] yield $\eta = 97.7\%$. The model in Reference [12] yields $\eta = 96.8\%$. For $\theta_{OFF} = 60°$, Equations 4.75 and 4.76 along with the models in References [11] and [12] yield $\eta = 90.9\%$, 89.9%, 90.8%, and 86.6%, respectively. The differences among these values are quite small for reasonably small transition times. The collector efficiency remains reasonably high, even if the transition time is an important fraction of the RF period.

Series Inductance at the Switch

At the instant of the ON→OFF transition, the collector current jumps between I_{OFF} and 0, while the current through the shunt capacitance jumps between 0 and I_{OFF}. The parasitic series inductances L_C, L_E, L_{c1}, and L_{c2} (Figure 4-16) cause power losses in the Class E circuit. With $L_s = L_C + L_E + L_{c1} + L_{c2}$ standing for the total wiring inductance of the loop, the power losses are given by [9]

$$P_d = \frac{1}{2} L_s I_{OFF}^2 f \qquad\qquad 4.77$$

where $I_{OFF} = 2I_{dc}$ for an optimally operated Class E amplifier with $D = 0.5$.

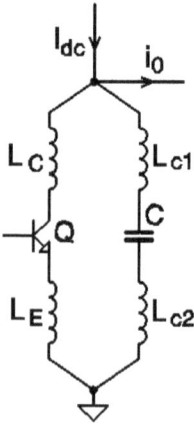

Figure 4-16 *Parasitic series inductances of transistor and shunt capacitances.*

This power is dissipated in the transistor and may be reduced by decreasing the parasitic inductances.

Effects of Circuit Variations

Variations in component values, operating frequency, and duty cycle can have an effect on the performance of the Class E amplifier [13]. The results of these variations can be used to estimate the behavior of the circuit in different situations, determine its tolerance to the dispersion of component values, and provide information for the correct tuning of the circuit.

The mathematical equations used in this analysis are those presented in Section 4.1. In all cases, the simplifying assumptions given in Section 4.1 are valid and a Class E circuit operating optimally, with $D = 0.5$ is considered. The design values and circuit performance are given by Equation 4.47. Standard normalized values of $V_{dc} = 1$ V and $R = 1 \Omega$ are used for convenience. Only one of the circuit parameters is varied at a time, within some reasonable limits; the other values remain fixed.

Series Reactance of the Output Circuit

Any possible change of the load reactance or mistuning of the series-tuned circuit can be modeled by varying load angle ψ between –90 and +90 degrees. The results are presented in Figures 4.17 (the output power and the DC input power), 4.18 (the collector efficiency), and 4.19 (the collector voltage and current waveforms). Note that the current pulse, which discharges the shunt capacitance at turn-ON (if the voltage across it is not zero at that instant), is not represented in Figure 4-19.

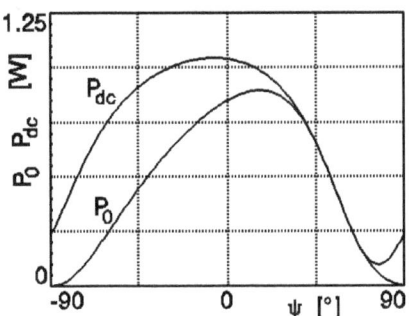

Figure 4-17 *Output power P_0 and DC input power P_{dc}, versus ψ ($V_{dc} = 1$ V, $R = 1 \Omega$, $D = 0.5$).*

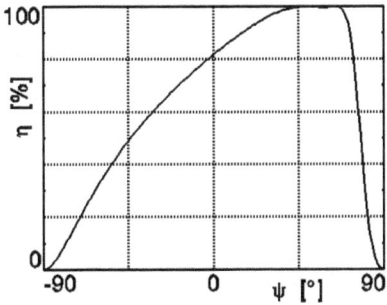

Figure 4-18 *Collector efficiency versus ψ ($V_{dc} = 1$ V, $R = 1 \Omega$, $D = 0.5$).*

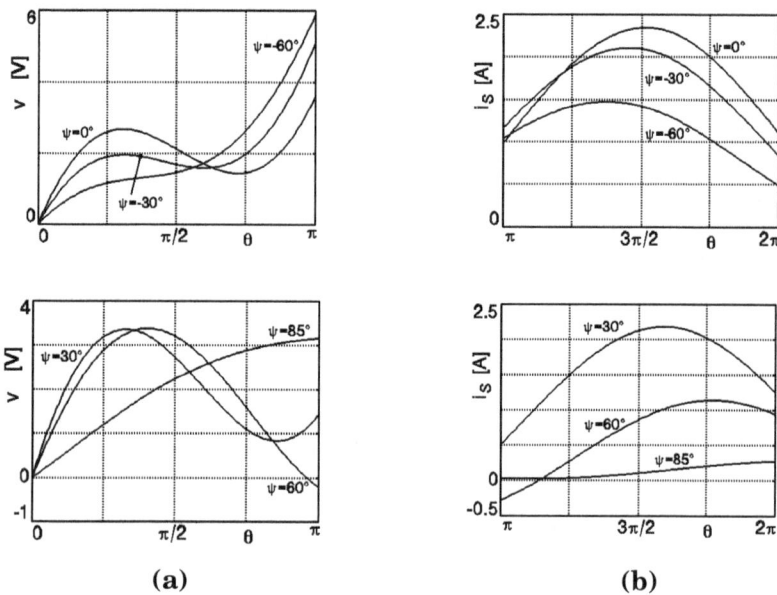

Figure 4-19 *Waveforms of (a) collector voltage, and (b) collector current, with ψ as a parameter ($V_{dc} = 1\ V$, $R = 1\ \Omega$, $D = 0.5$).*

The graphs in Figure 4-17 show that the peak output power occurs for $\psi \approx 10°$, and the peak of the DC input power occurs for $\psi \approx -5°$. According to Equation 4.47, the optimum operation of the circuit occurs for $\psi \approx 49°$. Consequently, the tuning procedure of the Class E amplifier cannot be based on measurements of the output power and DC supply current (as is usual for other classes of amplification). The practical tuning procedure of the Class E circuit is provided in the **Tuning Procedure of the Class E amplifier** section. Figure 4-18 suggests that the collector efficiency remains close to 100 percent for values of ψ varying between 40 and 70 degrees. The efficiency reaches 100 percent for $\psi \approx 65°$, but in this case the circuit does not operate optimally because $\xi \neq 0$.

The waveforms depicted in Figure 4-19 show that

- If ψ exceeds the optimal value (49.05 degrees) even slightly, a small negative voltage swing appears as well as a negative collector current. The use of an anti-parallel diode (connected between collector and emitter) in Class E circuits employing BJTs is recommended. Although this diode is not required in a normal operation, it may help to protect the transistor when the circuit is tuned or in case of accidental mistuning.

- If $\psi < 0°$ or $y > 70°$, the collector voltage at turn-ON is high, causing low collector efficiency.

- If the load is extremely reactive (that is, ψ is very close to ±90 degrees), the shunt capacitor is charged (during the OFF state) at an almost constant current. The collector voltage tends toward a ramp shape. The output power approaches zero. The DC input power tends to be determined by the shunt capacitance.

Shunt Capacitance

Changes in the shunt capacitance ($0.01 \le B \le 1$) caused by mistuning (very often, C includes an adjustable capacitor) or technological dispersion of the output capacitance of the transistor are modeled in this section. The results are presented in Figures 4-20 (the output power and the DC input power), 4-21 (the collector efficiency), and 4-22 (the collector voltage and current waveforms).

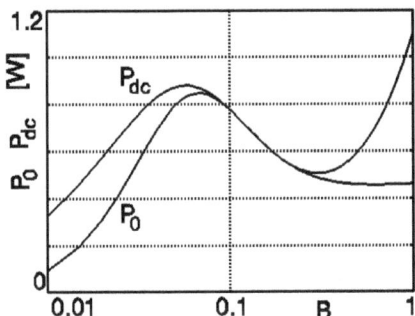

Figure 4-20 *Output power P_0 and DC input power P_{dc} versus B ($V_{dc} = 1$ V, R = 1 Ω, D = 0.5).*

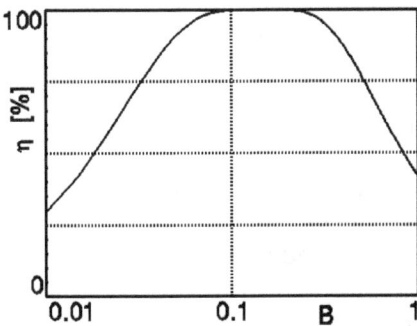

Figure 4-21 *Collector efficiency versus B ($V_{dc} = 1$ V, R = 1 Ω, D = 0.5).*

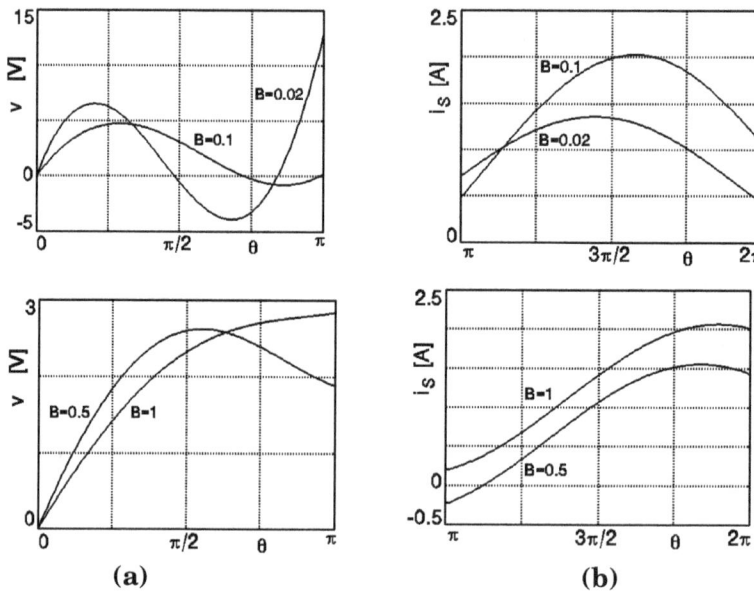

Figure 4-22 *Waveforms of (a) collector voltage, and (b) collector current, with B as a parameter (V_{dc} = 1 V, R = 1 Ω, D = 0.5).*

As is the case with ψ, P_0 and P_{dc} cannot be used as a tuning indicator of the shunt capacitance. Circuit efficiency remains very close to unity for values of B between 0.06 and 0.3. For $B \approx 0.1144$, the efficiency reaches 100 percent, but operation is not optimal ($\xi \neq 0$).

The waveforms of Figure 4-22 show that

- For low values of B, a negative collector voltage swing appears and the transistor must be able to withstand it.

- Negative values are possible for the collector current if B exceeds the optimum value. An anti-parallel diode is recommended in circuits using BJTs.

- If B is large, the shunt capacitor dominates the series-tuned circuit and the collector voltage tends toward a ramp shape. The output power remains almost constant, and the efficiency decreases quickly.

Switching Frequency

By varying the switching frequency, it is possible to evaluate the broadband capability of the Class E amplifier and the usable bandwidth for a certain carrier frequency, f_0. A frequency variation is modeled by varying the shunt susceptance, B (which is proportional to frequency), and the phase angle, ψ, simultaneously. With the loaded quality factor of the series-tuned circuit designated Q (defined by Equation 4.49), with f_0 as the center frequency, and with f as the operating frequency, the variation of ψ with f is given by

$$\psi = \arctan\left(\frac{X}{R}\right) = \arctan\left[(Q + 1.1525)\frac{f}{f_0} - Q\frac{f_0}{f}\right] \qquad 4.78$$

This analysis is based on the assumption that the load current is sinusoidal, and usually requires a large loaded quality factor value. However, a sinusoidal output current may be obtained for low values of Q (even $Q = 0$) if a filter is inserted between the series-tuned circuit and the load (see Figure 4-13). The input impedance of the filter must be very high for the harmonics of the output current.

The results are presented in Figures 4-23 (the output power and the DC input power for $Q = 0$), 4-24 (the collector efficiency), and 4-25 (the waveforms of collector voltage and current for $Q = 0$).

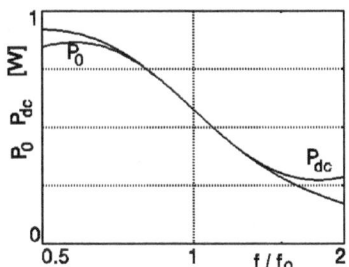

Figure 4-23 Output power P_0 and DC- input power P_{dc}, versus f/f_0 ($V_{dc} = 1\ V$, $R = 1\ \Omega$, $D = 0.5$, $Q = 0$).

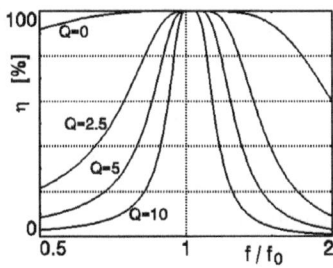

Figure 4-24 Collector efficiency versus f/f_0 with Q as a parameter ($V_{dc} = 1\ V$, $R = 1\ \Omega$, $D = 0.5$).

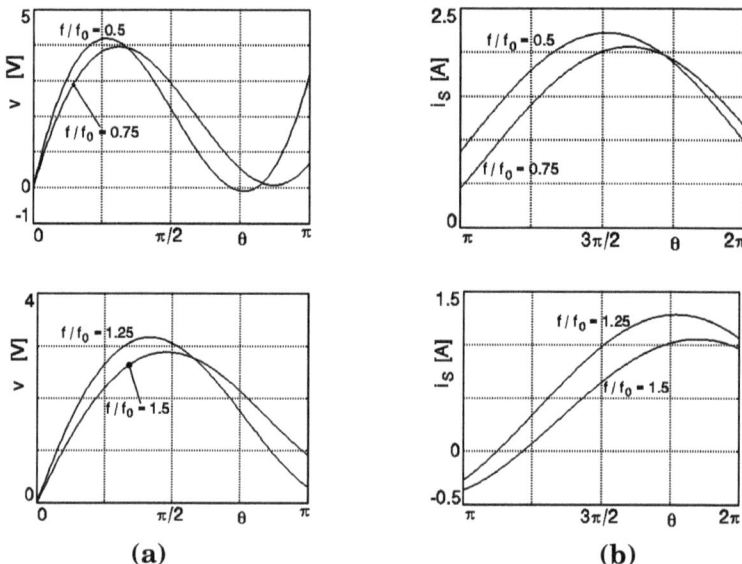

(a)   (b)

Figure 4-25 *Waveforms of (a) collector voltage, and (b) collector current, with f/f₀ as a parameter (V_{dc}= 1 V, R = 1 Ω, D = 0.5, Q = 0).*

These results show that broadband operation of the Class E amplifier is not possible. However, if $Q = 0$, the circuit can operate (without retuning) over an octave bandwidth with a theoretical collector efficiency of 95 percent or better. The obvious disadvantage is the relatively large variation of the output power versus frequency. It is difficult to filter out the output current harmonics in practical circuits.

Load Resistance

The load resistance variation (at a fixed operating frequency) is of particular interest in power electronics applications; for example, Class E inverters working on variable loads. In radio communications applications, the load resistance is usually constant, but a change in R may still occur due to technological dispersion or component failures. A simultaneous and related change of R and X may simulate the VSWR on the antenna feeder.

The results are presented in Figures 4-26 (the output power and the DC input power), 4-27 (the collector efficiency), and 4-28 (the waveforms of collector voltage and current).

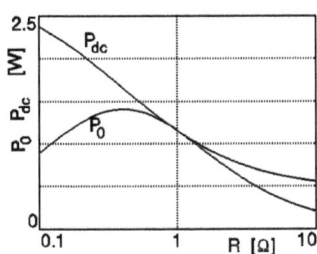

Figure 4-26 *Output power P_0 and DC input power P_{dc} versus R (V_{dc} = 1 V, D = 0.5).*

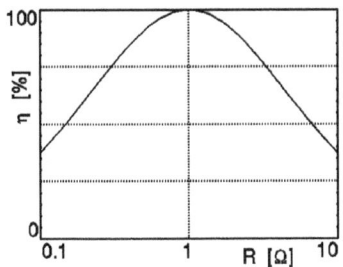

Figure 4-27 *Collector efficiency versus R (V_{dc} = 1 V, D = 0.5).*

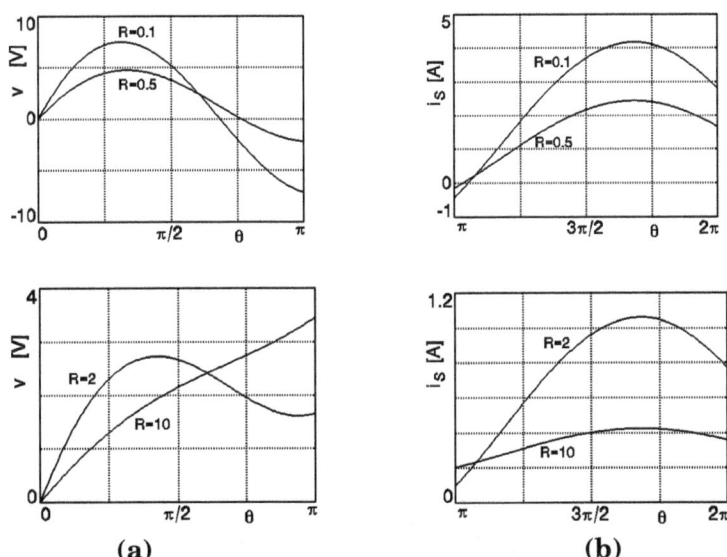

Figure 4-28 *Waveforms of (a) collector voltage and (b) collector current, with R as a parameter (V_{dc} = 1 V, D = 0.5).*

The graphs in Figures 4-26 and 4-27 show strong variations of P_0, P_{dc}, and η with load resistance. The waveforms in Figure 4-28 show that low values of R ($R < 1$) determine negative collector current and voltages; large values of R cause the shunt capacitance to dominate the series-tuned circuit. The collector voltage tends toward a ramp shape, and the efficiency decreases very quickly.

Duty Cycle

This analysis allows the investigation of

*   the possibilities of stabilizing the output voltage in a DC/DC converter using a Class E inverter by controlling the switch duty ratio,

*   the possibilities of obtaining an amplitude modulated signal by controlling the switch duty ratio (as shown in Chapter 3, in section **Duty Ratio** for a Class D amplifier),

*   the sensitivity of the Class E circuit with accidental variations of the switch duty ratio.

The results are shown in Figures 4-29 (the output power and the DC input power), 4-30 (the collector efficiency), and 4-31 (the waveforms of collector voltage and current). In all these graphs, D represents the switch-OFF duty ratio.

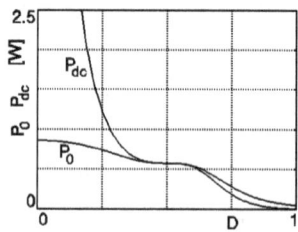

Figure 4-29 *Output power P_0 and DC input power P_{dc} versus D (V_{dc} = 1 V, R = 1 Ω).*

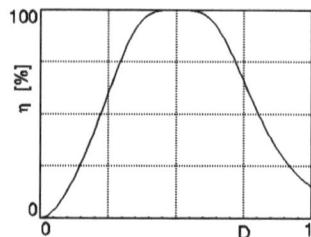

Figure 4-30 *Collector efficiency versus D (V_{dc} = 1 V, R = 1 Ω).*

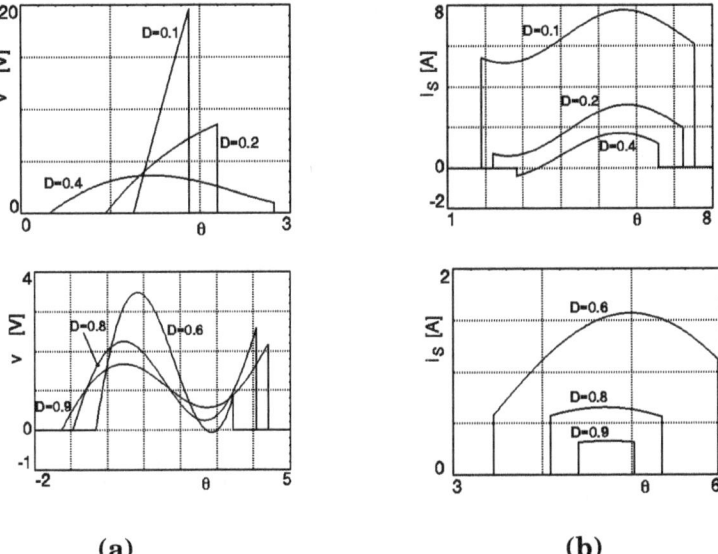

(a)   **(b)**

Figure 4-31 *Waveforms of (a) collector voltage, and (b) collector current, with D as a parameter (V_{dc} = 1 V, R = 1 Ω).*

The curves in Figures 4-29 and 4-30 show that the output power remains almost constant and the efficiency exceeds 95 percent for D, varying between 0.4 and 0.6. Therefore, the circuit tolerates reasonable variations of the switch duty cycle around its nominal value of 0.5 quite well. If D approaches zero, then P_{dc} tends to infinity because the switch shunts the DC supply to ground. It is interesting to observe (without an intuitive explanation) that if $D\rightarrow0$, then P_0 tends toward a nonzero value, and if $D\rightarrow1$, then P_0, P_{dc}, and η tend toward nonzero values. The waveforms in Figure 4-31 show that some values of D cause negative collector voltages or currents.

When looking at the possibility of amplitude modulation of the output signal by varying D, it is important to note that output voltage V_0 varies with D as the square of P_0. The variation of V_0 with D is nonlinear, and a zero output amplitude is not possible. Moreover, the collector efficiency decreases quickly for extreme values of D, and the transistor must be able to withstand negative voltages or pass negative currents. Amplitude nonlinearity might be corrected by envelope feedback [5], allowing the use of the circuit for low-index full-carrier AM signals. However, the phase of the output signal (ϕ) is not constant with D, which requires phase-locking of the carrier generator and output signal. This method raises excessive complications for practical use. Amplitude modulation may be obtained easily using the collector amplitude modulation (see Section 4.3), with simple circuits and excellent performance.

Saturation Voltage of the Transistor

If the active device of a Class E amplifier is a bipolar transistor with a saturation voltage of V_{sat}, then the results obtained in Section 4.1 must be modified as follows. $V_{eff} = V_{dc} - V_{sat}$ replaces V_{dc} in all equations, except in those giving the peak collector voltage and the DC input power. Therefore, the amplitude of the output voltage and the output power are given by [9]

$$V_0 = \frac{4}{\sqrt{\pi^2 + 4}}(V_{dc} - V_{sat})$$

(4.79)

$$P_0 = \frac{8}{\pi^2 + 4}\frac{(V_{dc} - V_{sat})^2}{R}$$

The DC input power is

$$P_{dc} = V_{dc}I_{dc} = V_{dc}\frac{V_{dc} - V_{sat}}{R_{dc}} = \frac{8}{\pi^2 + 4}\frac{V_{dc}(V_{dc} - V_{sat})}{R}$$

(4.80)

yielding the collector efficiency

$$\eta = 1 - \frac{V_{sat}}{V_{dc}} = \frac{V_{eff}}{V_{dc}}$$

(4.81)

The waveforms of the voltages and currents in the circuit are only quantitatively affected (their amplitudes are reduced because $V_{dc} - V_{sat}$ replaces V_{dc}), except for the collector voltage, whose waveform is depicted in Figure 4-32. The peak collector voltage is

$$V_{max} \approx 3.5620(V_{dc} - V_{sat}) + V_{sat}$$

(4.82)

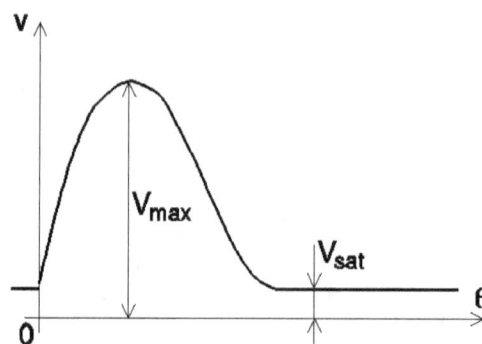

Figure 4-32 *Waveform of collector voltage for nonzero saturation voltage.*

Finally, note that the Class E switching conditions must be changed to take into account that for V_{sat}, the condition $v(\theta_2) = V_{sat}$ must replace $v(\theta_2) = 0$.

ON Resistance of the Transistor

The active device ON resistance is given below as R_{ON}. It is assumed constant during the ON state and independent of the current through (or the voltage across) it. The other assumptions from Section 4.1 remain valid. Assume that R_{ON} is small enough in comparison to R, so the waveforms calculated in Section 4.1 are not affected by R_{ON}. The simple analysis in this section is based on the idealized model presented earlier. The power dissipated in the active device can be calculated from [9]

$$P_d = \frac{R_{ON}}{2\pi} \int_{\theta_2}^{2\pi+\theta_1} i_S^2(\theta)\,d\theta = \frac{R_{ON}}{2\pi} \int_{\theta_2}^{2\pi+\theta_1} \left[I_{dc} - \frac{V_0}{R}\sin(\theta+\phi) \right]^2 d\theta \qquad 4.83$$

Using Equations 4.16 and 4.27,

$$P_d = P_0 \frac{R_{ON}}{R_{dc}} \frac{\left(g^2+2\right)(\pi-y) + 4g\cos\phi\sin y - g^2\cos 2\phi \sin y \cos y}{\pi g^2} \qquad 4.84$$

For an optimally operated Class E amplifier with $D = 0.5$ (that is, $y = \pi/2$), the dissipated power is given by

$$P_d = \frac{\pi^2 + 28}{2\left(\pi^2 + 4\right)} \frac{R_{ON}}{R} P_0 \approx 1.3652 \frac{R_{ON}}{R} P_0 \qquad 4.85$$

and the collector efficiency is

$$\eta \approx \frac{P_0}{P_0 + P_d} \approx \frac{1}{1 + 1.3652\dfrac{R_{ON}}{R}} \qquad 4.86$$

For a given transistor (R_{ON} fixed), an increase in the load resistance yields an increase in efficiency. On the other hand, if the load resistance increases

- the output power decreases

- the RF choke reactance must be increased

- the required shunt susceptance decreases (limiting the maximum operating frequency of the circuit). Note that MOSFETs with low R_{ON} usually have a large output capacitance and cannot be used at frequencies that are too high.

Loaded Q of the Tuned Circuit

Taking into account the finite loaded Q of the series-tuned circuit (and possibly, the switch ON resistance and/or other circuit parameters), the analysis and design of the Class E amplifier follows a more complicated procedure than that presented in Section 4.1 for an infinite Q. The analysis technique is based on Liou's work [14] and is described briefly below in relation to a model that includes the effects of finite output Q and nonzero ON resistance, R_{ON}. Curves depicting the variation of the design values and circuit performance with both Q and R_{ON} are also presented. The intention is to present an approach to the problem and some qualitative results that will provide additional physical insight, not to provide accurate algorithms or design equations for practical applications. For practical purposes, the use of a dedicated computer program such as HEPA-PLUS (presented in Section 4.7) is recommended. Such a program can provide quick and accurate results using a more complex circuit model. However, a good physical understanding of the operation and knowledge of the expected behavior of the circuit when the parameters are varied is critical for a successful design using numerical simulation tools.

With the assumptions of an infinite output Q and zero ON resistance removed (but keeping the other assumptions from Section 4.1), the equivalent circuits in Figure 4-33 are obtained.

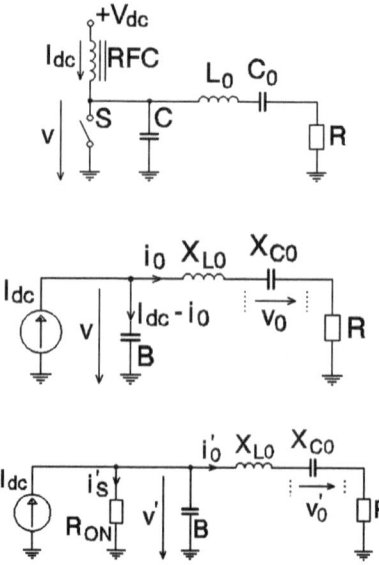

Figure 4-33 *Basic Class E amplifier (top) and equivalent circuits for the OFF (middle) and ON states (bottom).*

The first step is to write and solve (in the time- or s-domain) the differential equations describing the two equivalent circuits for the OFF and ON states. The solutions for these differential equations are functions of the circuit element values (and switch duty ratio), and the initial circuit conditions for steady-state operation (note that the initial conditions for the state variables are not known at this time). The next step is to eliminate the initial conditions using the boundary conditions that hold when the switch changes state (the voltages across capacitors and the currents through inductors must be continuous waveforms). A linear system of easily solved equations is obtained and, given the initial conditions, the circuit steady-state waveforms are determined as functions of the circuit element values. The output and DC input power, collector efficiency, harmonic content, and other parameters, can then be found without difficulty.

Design of a circuit for optimum Class E operation is more complicated (from a mathematical point of view). The two design conditions in Equation 4.48 are added to the equation system involving the initial conditions. The result is a system of nonlinear equations that can be solved for two circuit element values; all the others are handled as parameters. Circuit steady-state waveforms and circuit performance are then derived as described above.

Although this analysis and design procedure seems simple and straightforward, it is time-consuming and yields complicated equations that can be solved only by numerical algorithms. More details and results, as well as mathematical equations can be found in many papers that use this method to analyze the Class E configuration [15–23].

The figures below present the results for a Class E amplifier operating with a switch duty ratio of $D = 0.5$. Standard normalized values of $V_{dc} = 1$ V and $R = 1\ \Omega$ are considered. The output Q is defined according to Equation 4.49 and the value of R_{ON} is normalized to R.

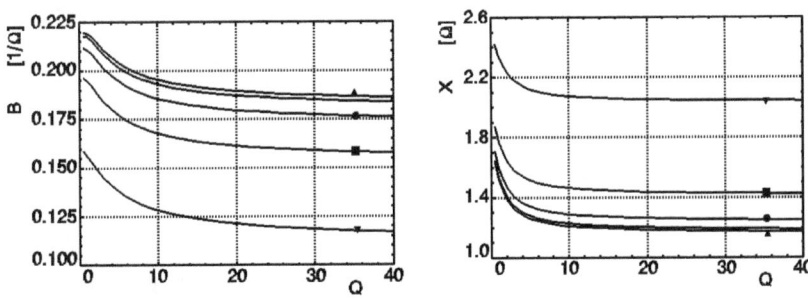

Figure 4-34 Design values of B and X versus Q, with R_{ON} as a parameter ($V_{dc} = 1$ V; $R = 1\ \Omega$); —▲— $R_{ON} = 0.01$; —— $R_{ON} = 0.03$; —•— $R_{ON} = 0.1$; —■— $R_{ON} = 0.3$; —▼— $R_{ON} = 1$.

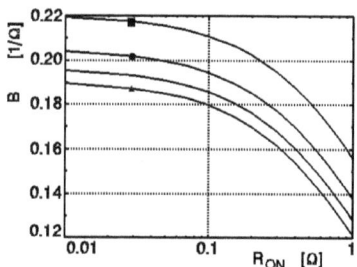 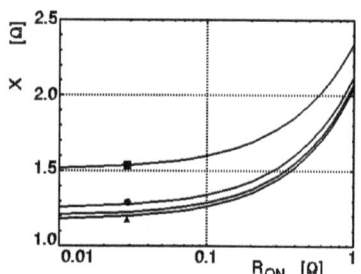

Figure 4-35 *Design values of B and X versus R_{ON}, with Q as a parameter (V_{dc} = 1 V; R = 1 Ω); —▲— Q = 20; ——— Q = 10; —•— Q = 5; —■— Q = 1.*

Figures 4-34 and 4-35 present the variation of the design values of B and X (the net series reactance) with Q and R_{ON}. These design values are strongly affected by Q (for low values of $Q < 10...15$) and R_{ON} (for high values of $R_{ON} > 0.1$).

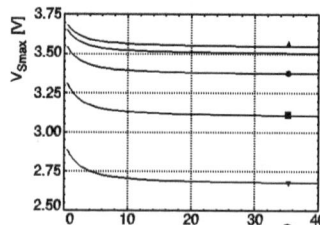 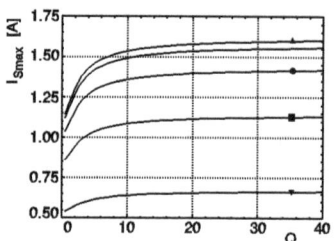

Figure 4-36 *Device stress versus Q, with R_{ON} as a parameter (V_{dc} = 1 V; R = 1 Ω); —▲— R_{ON} = 0.01; ——— R_{ON} = 0.03; —•— R_{ON} = 0.1; —■— R_{ON} = 0.3; —▼— R_{ON} = 1.*

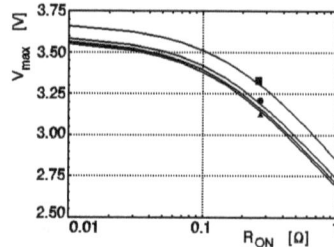 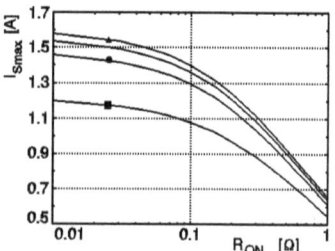

Figure 4-37 *Device stress versus R_{ON}, with Q as a parameter (V_{dc} = 1 V; R = 1 Ω); —▲— Q = 20; ——— Q = 10; —•— Q = 5; —■— Q = 1.*

Figures 4-36 and 4-37 present the variation of the active device stresses (V_{max} and I_{Smax}) with Q and R_{ON}. Both V_{max} and I_{Smax} are affected significantly by Q and R_{ON} for low values of Q ($Q < 5$) or high values of R_{ON} ($R_{ON} > 0.1$).

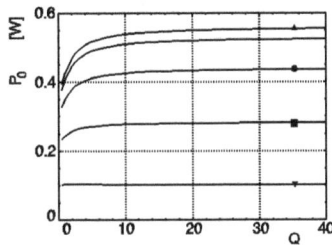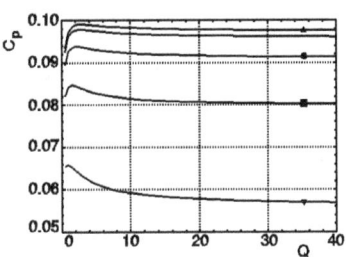

Figure 4-38 *Output power P_0 and power output capability C_P versus Q, with R_{ON} as a parameter (V_{dc} = 1 V; R = 1 Ω); —▲— R_{ON} = 0.01; ——— R_{ON} = 0.03; —•— R_{ON} = 0.1; —■— R_{ON} = 0.3; —▼— R_{ON} = 1.*

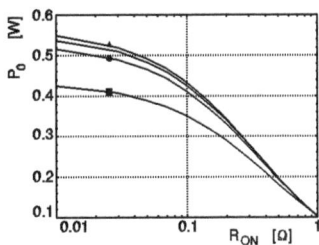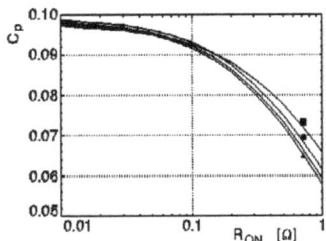

Figure 4-39 *Output power P_0 and power output capability C_P versus R_{ON}, with Q as a parameter (V_{dc} = 1 V; R = 1 Ω); —▲— Q = 20; ——— Q = 10; —•— Q = 5; —■— Q = 1.*

Figures 4-38 and 4-39 present the variation of output power P_0 and power output capability C_P with Q and R_{ON}. As expected, P_0 and C_P decrease with increasing R_{ON}. The power output capability reaches a maximum value for $Q \approx 2...3$, depending on the value of R_{ON}.

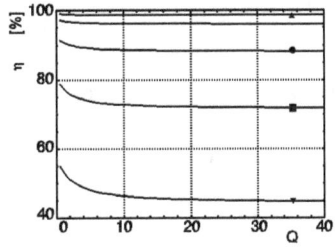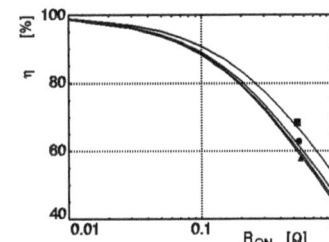

Figure 4-40 *(a) Collector efficiency versus Q, with R_{ON} as a parameter (V_{dc} = 1 V; R = 1 Ω); —▲— R_{ON} = 0.01; ——— R_{ON} = 0.03; —•— R_{ON}=0.1; —■— R_{ON}=0.3; —▼— R_{ON} = 1; (b) collector efficiency versus R_{ON}, with Q as a parameter (V_{dc} = 1 V; R = 1Ω); —▲— Q = 20; ——— Q = 10; —•— Q = 5; —■— Q = 1.*

Figure 4-40 presents the variation of the theoretical collector efficiency with Q and R_{ON}. As expected, η decreases quickly with increasing R_{ON}. Also, the collector efficiency decreases slightly as Q increases; the dependence is stronger for higher values of R_{ON}.

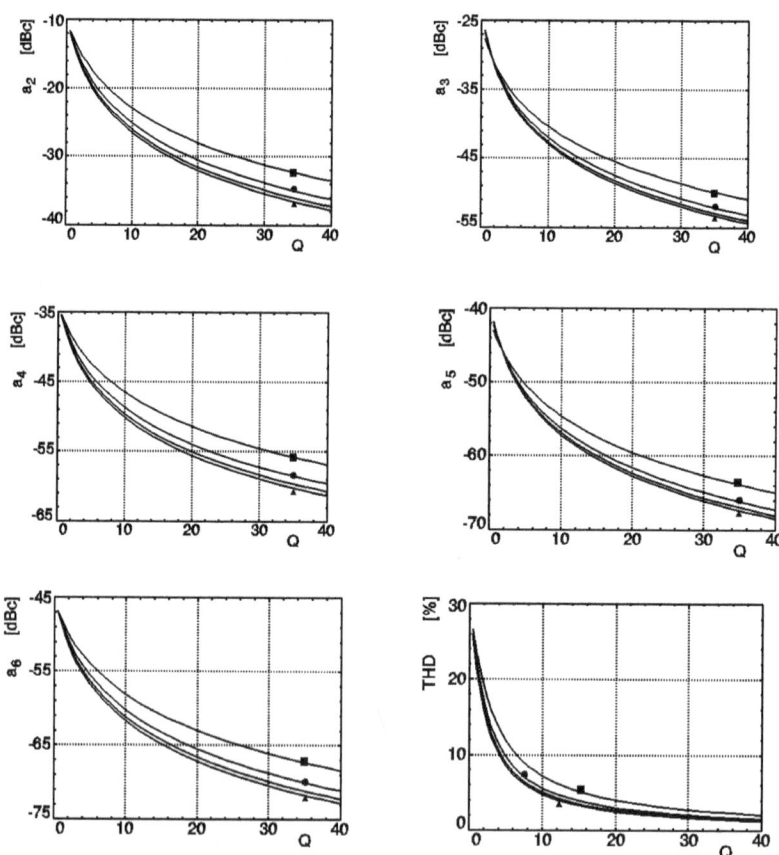

Figure 4-41 *Harmonics level and total harmonic distortion of the output current versus Q, with R_{ON} as a parameter (V_{dc} = 1 V; R = 1 Ω; —▲— R_{ON} = 0.01; ——— R_{ON} = 0.1; —•— R_{ON} = 0.3; —■— R_{ON} = 1.*

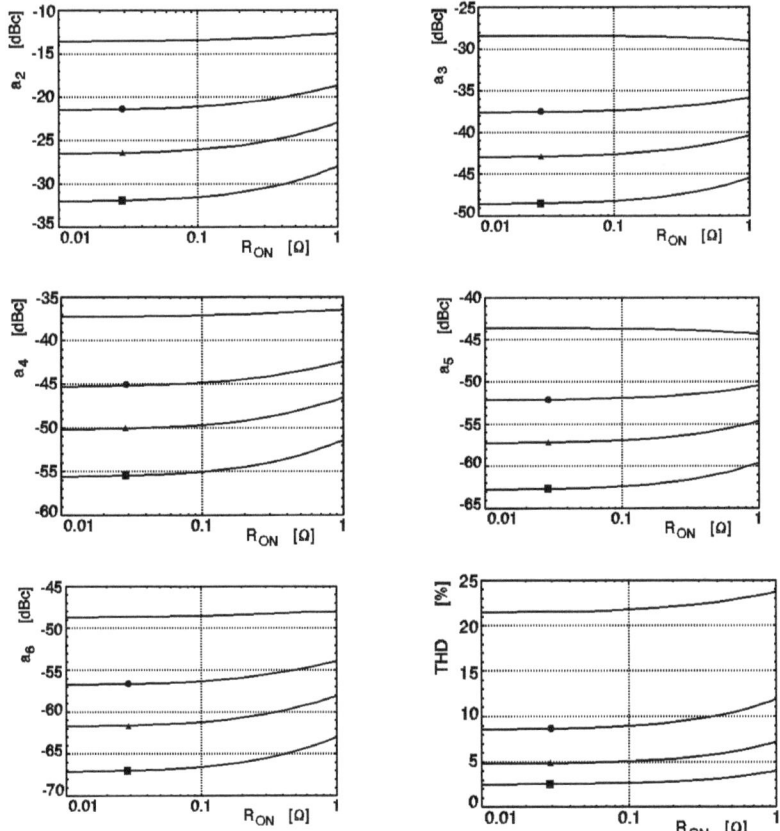

Figure 4-42 *Harmonics level and total harmonic distortion of the output current versus R_{ON}, with Q as a parameter ($V_{dc} = 1$ V; $R = 1$ Ω); —■— Q = 20; —▲— Q = 10; —•— Q = 5; ——— Q = 1.*

Figures 4-41 and 4-42 present the variation of the harmonic content (a_2 to a_6 expressed in dBc) and the *total harmonic distortion* (THD) with Q and R_{ON}. THD is calculated as

$$THD = 100 \frac{\sqrt{I_{02}^2 + I_{03}^2 + \ldots}}{I_{01}} \quad [\%]$$

4.87

where I_{0n} is the amplitude of the nth harmonic of the output current (I_{01} is the fundamental). As expected, the harmonic content decreases as Q increases, and as R_{ON} increases (an increase in R_{ON} lowers the *overall* Q when the switch is ON).

Tuning Procedure of the Class E Amplifier

The tuning procedure for a Class E amplifier is very different from the one used for other amplifiers (Class A, B, C, or D). The section on *Effects of Circuit Variations* emphasized the impossibility of using the voltage across the load, the output power, or the DC power as tuning indicators. A possible solution for this tuning problem involves measuring the collector efficiency; however, this is a laborious procedure.

The standard tuning procedure requires only an oscilloscope to visualize the collector voltage waveform [24]. Figure 4-43 presents a typical waveform of the collector voltage in a mistuned Class E amplifier (note that $V_{sat} = 0$ if a MOS transistor is used).

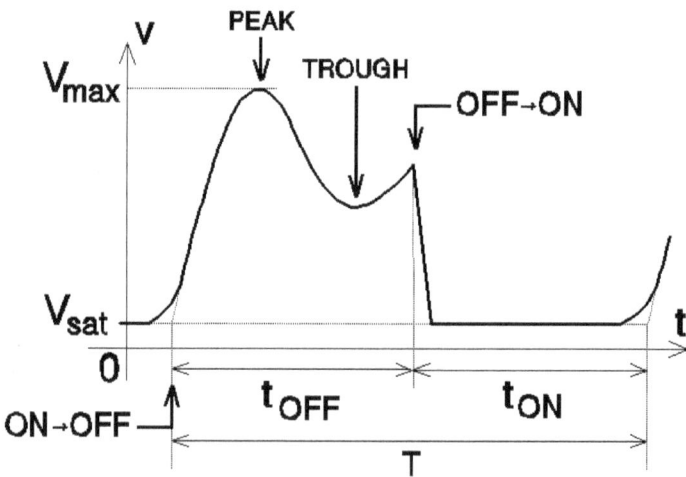

Figure 4-43 *Typical waveform of collector voltage in a mistuned Class E amplifier [24].*

During the OFF state, the collector voltage has a peak and often a trough (a minimum value, at which time the slope is zero). If the trough is missing, its position may be estimated by extrapolation, as suggested by the dotted lines in Table 4.1. The trough may be missing for two reasons: a) the transistor turns ON before the collector voltage reaches zero slope, and b) the circuit may include an anti-parallel diode at switch (BJT + diode, or MOS transistor) that clips the negative voltages, which makes the trough invisible.

The Class E circuit is tuned correctly for optimum operation if

- the collector voltage at turn-ON is zero (or V_{sat}, if a BJT is used),
- the slope of the collector voltage at turn-ON is zero,
- the switch duty ratio is 50 percent.

Table 4-1 *Tuning procedure of the Class E amplifier.*

Collector voltage v at time of zero slope	dv/dt at transistor turn-ON (indicated by the vertical arrow)		
	dv/dt>0 increase (C series C_0)	dv/dt=0 keep same (C series C_0)	dv/dt<0 decrease (C series C_0)
v>0 (or Vsat); decrease C/C_0	increase C_0	decrease C, increase C_0	decrease C
v=0 (or Vsat); keep same C/C_0	increase C, C_0 in same proportion	correct values for C, C_0	decrease C, C_0 in same proportion
v<0 (or Vsat); increase C/C_0	increase C	increase C, decrease C_0	decrease C_0

When the circuit is tuned, the procedure usually involves adjusting the shunt capacitance, C (which includes an adjustable capacitor), the net reactance of the series-tuned circuit (L_0 or C_0 are adjustable components), and the switch duty ratio D (especially at high frequencies, to achieve exactly 50-percent). The effect of tuning a load network component on the collector voltage waveform (more exactly, on the position of the trough) is shown in Figure 4-44. Although its effect is shown here, the load resistance, R, cannot usually be adjusted.

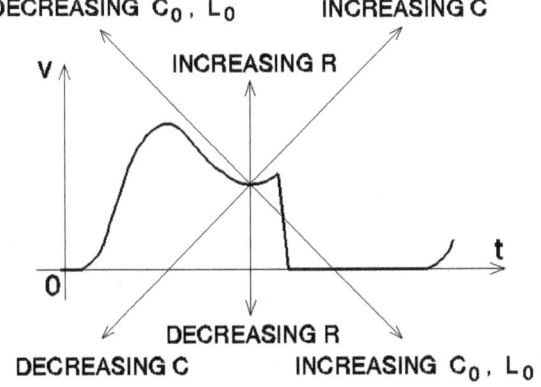

Figure 4-44 *Effect of tuning on the collector voltage waveform [24].*

The waveforms in Figures 4-19, 4-22, 4-28, and 4-45 provide explanations for the effects shown in Figure 4-44. Note that Figure 4-45 presents the same waveforms as Figures 4-19 and 4-22, except that the values of B and ψ are closer to optimum.

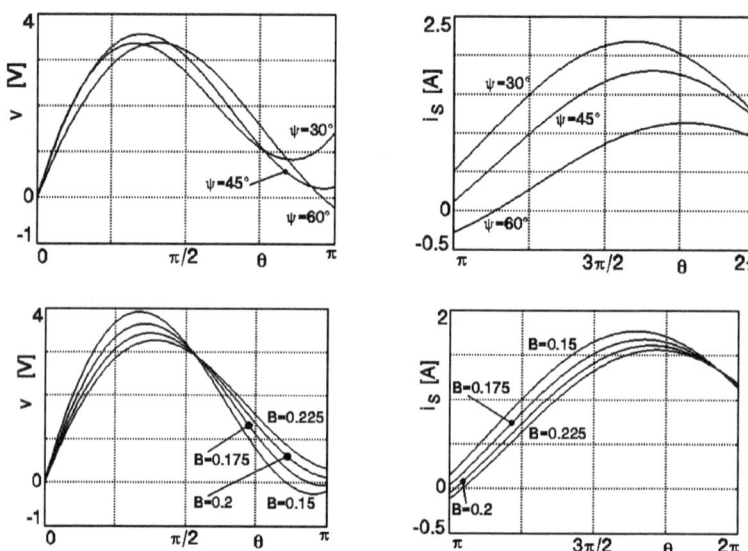

Figure 4-45 *Waveforms of (a) collector voltage, and (b) collector current, for values of B and ψ around their optimal values.*

With knowledge of the effects of possible tuning adjustments on the voltage waveform, circuit tuning is achieved, as discussed below and shown in Table 4-1. (Assume that the load resistance, the operating frequency, and the series inductance, L_0, have been selected previously or adjusted for their nominal values.)

1. A low DC supply voltage is applied to the circuit (4 to 5 volts or less, as necessary), to avoid transistor destruction. (If the initial mistuning is large enough, an excessive collector voltage or power dissipation may occur). The switch duty ratio is first adjusted for 50 percent (see Figure 4-43). The ON→OFF transition is always visible on the collector waveform. If the OFF→ON transition is not clearly visible, try a preliminary adjustment using the base waveform (this adjustment is not very accurate, especially for BJTs, which are affected by the storage time).

2. Look for the trough of the collector voltage, or if the trough is missing, extrapolate its position. Compare the value of collector voltage v at the minimum point (when $dv/dt = 0$) with V_{sat} (less, equal, greater). Also,

determine the sign of the slope of the collector voltage (dv/dt) at turn-ON (negative, zero, or positive). If it is difficult to estimate these values, consider them equal to V_{sat} or zero, respectively.

3. Tune C and C_0 according to the indications in Figure 4-44 and Table 4-1, until the desired waveform is obtained.

4. Increase the DC supply voltage (in several steps) toward the nominal value, retuning the circuit after each step. This is necessary because the effective output capacitance of the transistor decreases as the collector voltage increases.

5. Make a final check by increasing the value of C slightly to generate an easily visible marker at turn-ON. This allows for verification that the duty ratio is at the desired value. Finally, return C to the value that eliminates the marker.

It may be useful to monitor the collector current waveform during the tuning process. If the circuit is tuned correctly, the collector current starts from zero at the OFF→ON transition (condition $dv/dt = 0$ is satisfied), and there are no current spikes at turn-ON (condition $v = 0$ is met). Several precautions are necessary when a current probe is used.

- A correction of the waveforms may be needed if the delays introduced by the current and voltage probes are different.

- The current probe must be connected in the collector circuit of the transistor (not in the emitter circuit). The series impedance introduced by the current probe in the collector circuit must be kept as low as possible.

- The zero level of the collector current is unknown because the current probe is AC coupled. However, the zero level of the current can be determined knowing that the collector current is zero when the slope of the collector voltage is zero ($dv/dt = 0$); for example, when the collector voltage attains its peak positive value.

- The current probe monitors the sum of the active current through the switch, i_S, and the current through the parasitic capacitance of the transistor. Consequently, the waveform observed on the oscilloscope appears slightly different from current i_S through the switch shown in Figure 4-5. The difference is usually significant during the OFF state, when the current is not zero and varies as dv/dt. A method of visualizing only the active current of the transistor can be found in Reference [24].

4.3   Amplitude Modulation of the Class E Amplifier

As was discussed earlier, with Class B, C, and D amplifiers (see Chapters 2 and 3), collector modulation is the main form of amplitude modulation used in solid-state RF amplifiers. This technique can be used directly to obtain an AM *double sideband* (DSB) signal with a modulation depth of $m < 1$. Collector modulation can be also used to generate any type of AM signal in an *envelope elimination and restoration* (EER) system. The basic circuit of a collector-modulated Class E RF amplifier and the corresponding waveforms are shown in Figures 4-46 and 4-47.

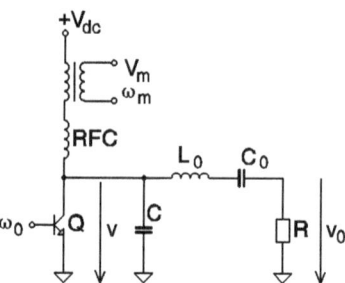

Figure 4-46 *Circuit of collector-modulated Class E amplifier.*

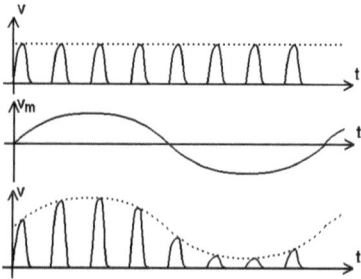

Figure 4-47 *Waveforms in collector AM of the Class E amplifier: CW collector voltage (top), modulating signal (middle), and AM collector voltage (bottom).*

As with the Class D amplifier, collector efficiency is very high (ideal is 100 percent) and constant with the modulation depth. The amplitude-modulation characteristic is very linear, because the conduction angle and wave shapes are not affected by the collector DC supply voltage. Deviations from perfect linearity are the result of nonlinear variations in V_{sat} or R_{ON}, base-to-collector feedthrough of the driving signal (see Figure 2-101), and voltage-variable output capacitance. An additional distortion mechanism occurs in amplitude-modulated Class E amplifiers [25] and is caused by the series-tuned circuit that does not resonate on the carrier frequency. Because the circuit resonant frequency (f_r) is lower than the carrier frequency, the two sideband components of the modulated carrier are transmitted from collector to load with different magnitudes and delays. This causes nonlinear distortion of the envelope of the AM output voltage and a decrease in the fundamental component of the envelope.

Calculation of the distortion is straightforward and uses numerical techniques for the Fourier-series expansions. Examples of the results are shown in Figure 4-48 [25]. V_{01n} is the fundamental component of the AM output voltage envelope normalized to its low-frequency value $(f_m = 0)$, and $h_2...h_4$ is the distortion caused by harmonics 2...4. Q_r is the quality factor of the series-tuned circuit at its resonant frequency.

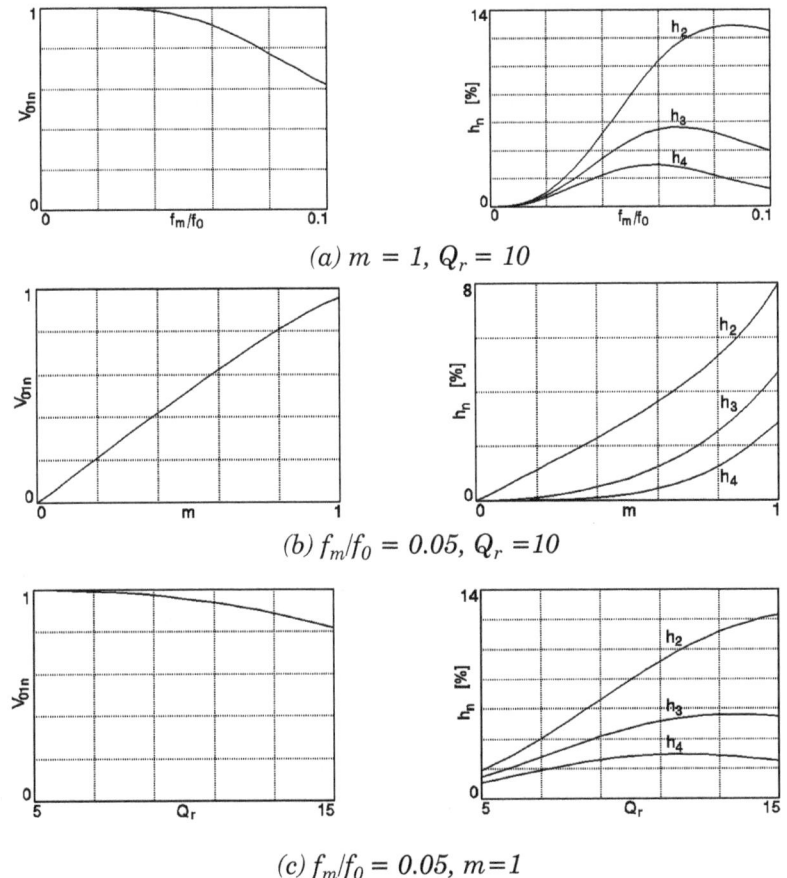

(a) $m = 1$, $Q_r = 10$

(b) $f_m/f_0 = 0.05$, $Q_r = 10$

(c) $f_m/f_0 = 0.05$, $m=1$

Figure 4-48 *Normalized fundamental component of the envelope of $v_0(t)$ and the harmonics of the envelope of $v_0(t)$ versus (a) f_m/f_0, (b) m, and (c) Q_r.*

The fundamental component of the output voltage envelope is not proportional to m and decreases as the modulating frequency and the quality factor increase. The harmonic distortion increases with the modulating frequency (for low values of f_m), m, and Q_r; however, the distortion is negligible ($h_n < 1\%$) if $f_m < 0.02f_0$ and $Q_r < 15$. Hence, this mechanism often yields negligible distortion for practical values of the signal bandwidth, carrier frequency, and quality factor. If the distortion is unacceptable, it may be reduced via a Class E circuit with the output circuit tuned on the carrier frequency, as suggested in Reference [26].

4.4 Amplifiers with Finite DC Feed Inductance

In the previous sections, it was assumed that the Class E circuit included an ideal RF choke in the DC path. However, optimum Class E operation can be also obtained using a finite (small) DC feed inductance. Amplifiers with finite DC feed inductance have a number of characteristics that make them attractive for some practical applications.

The analysis and design procedure of the Class E circuit with finite DC feed inductance is similar to that of the circuit employing an RF choke. However, this procedure is laborious (the resulting equations are much more complicated), and the presentation of results is somewhat more difficult because the number of design parameters is larger. The first analysis of the circuit is presented in Reference [27] and is based on a model assuming infinite output Q and ideal switching. Later works allowed for finite Q [19], and finite Q and nonzero ON resistance of the switch [17].

Some of the results presented in [27] emphasize those characteristics that make the finite DC feed inductance design attractive for practical applications. The model used is similar to the one given in Section 4.1 for a Class E circuit with an ideal RF choke. The RF choke is replaced by a finite inductance and the assumption regarding the ideal character of the RF choke is removed, but the other assumptions remain valid.

The results below are presented with the shunt susceptance as a parameter. Its normalized value, i.e., the ratio of the actual value of the shunt susceptance to the "conventional" value of B in a circuit with RF choke (operating in the same conditions) is used for convenience.

$$B_r \approx \frac{B}{B_{conv}} \qquad\qquad 4.88$$

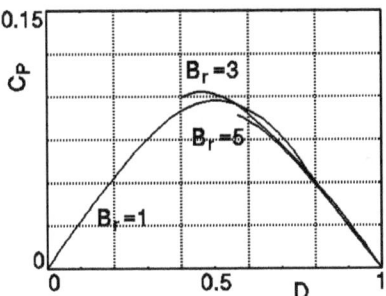

Figure 4-49 *Power output capability versus switch-OFF duty ratio, with B_r as a parameter [27].*

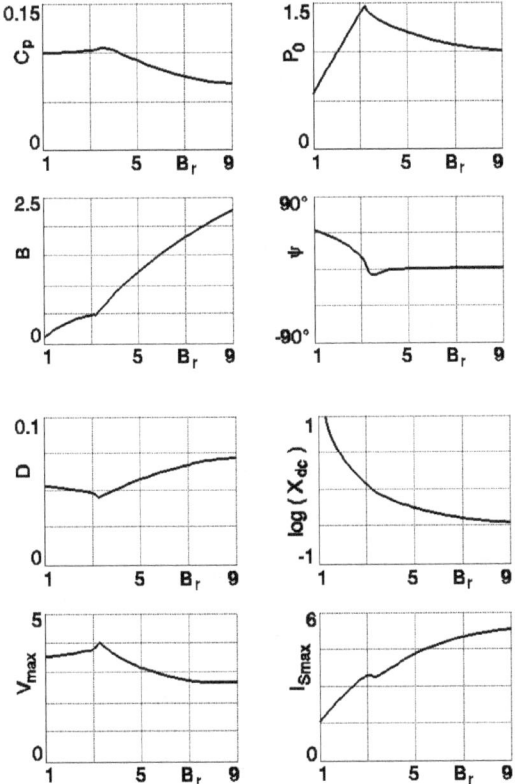

Figure 4-50 *Design values and circuit performance (for maximum C_P) versus B_r ($V_{dc} = 1$ V, $R = 1$ Ω) [27].*

Figure 4-50 shows the variation of the power output capability with the switch-OFF duty ratio, D (with B_r as a parameter), for an optimally operated Class E amplifier (which satisfies the Class E switching conditions given by Equation 4.48). Figure 4-50 presents the variation of the design values (B, ψ, D, X_{dc}) and circuit performance (V_{max}, I_{smax}, P_0, C_P) with B_r, for Class E circuits providing maximum power output capability. Maximum C_P is obtained for $B_r \approx 3.64$. If the main goal of the design is to maximize C_P, this may be considered the best operating point of a Class E amplifier. The design and performance values for this circuit are presented in Table 4-2, which also shows (for comparison) the corresponding values in a circuit with the RF choke providing maximum C_P and in the circuit with the RF choke and $D = 0.5$.

Table 4-2 *Design and performance values for three Class E circuits ($V_{dc} = 1\,V$, $R = 1\,\Omega$).*

Parameter	Circuit with finite dc-feed inductance	Circuit with RF choke	Circuit with RF choke and D=0.5
X_{dc} [Ω]	0.7218	RFC	RFC
B [$1/\Omega$]	0.6258	0.1918	0.1836
D	0.485	0.511	0.5
ψ [$°$]	-6.53	50.15	49.05
C_P	0.1049	0.0982	0.0981
P_0 [W]	1.378	0.5418	0.5768
V_{max} [V]	3.762	3.487	3.562
I_{smax} [A]	3.492	1.583	1.651

The curves in Figures 4-49 and 4-50 and Table 4-2 show that

- Optimum operation of Class E amplifier with $B_r > 1$ is not possible for some values of D.

- Optimum operation of a Class E amplifier with $B_r < 1$ is not possible. A finite DC feed inductance will tune out some part of the shunt capacitance. In a circuit with finite DC feed inductance, the shunt susceptance is always larger than in a circuit with RF choke.

- For some values of D, the power output capability may be higher in circuits with $B_r > 1$. In other words, a circuit with finite DC feed inductance may provide a higher C_P than a circuit with RF choke.

- It may be desirable to use a large value of B_r in practical designs, because this allows use of a given device (with a certain parasitic shunt capacitance) at higher frequencies, or use of a device with a larger output capacitance at the same frequency.

- Circuits with finite DC feed inductance may provide a larger RF output power than a circuit with RF choke, for the same values of V_{dc} and R.

4.5   Other Class E Configurations

The Class E circuit discussed in the previous sections is probably the most attractive configuration for practical applications. However, Class E operation can be obtained with many circuit configurations that present some advantages for specific applications.

Push-Pull Circuit

Push-pull configurations [4, 28] may be used to combine two Class E amplifiers to obtain a larger power output. A push-pull Class E configuration and its corresponding waveforms are shown in Figure 4-51.

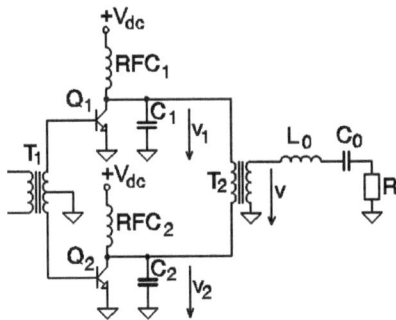

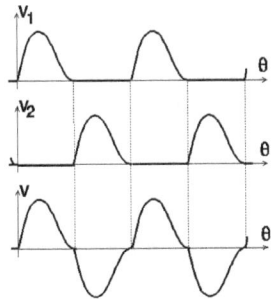

Figure 4-51 Push-pull Class E amplifier and waveforms for optimum operation.

Figure 4-52 presents two other possible push-pull circuit configurations. The two transistors are out-of-phase driven by the input transformer, T_1, but each transistor operates as if it were a single transistor Class E amplifier. When one of the two transistors is ON, the primary of the output transformer, T_2, is connected to ground. As a result, the difference between the DC input current and transformed sinusoidal output current charges the shunt capacitance of the other transistor. The voltage across the secondary winding of T_2 contains both positive and negative Class E waveforms, providing a fundamental frequency component with twice the amplitude of the fundamental frequency component of either collector waveform. Note that, otherwise, the equations presented in the previous sections remain valid.

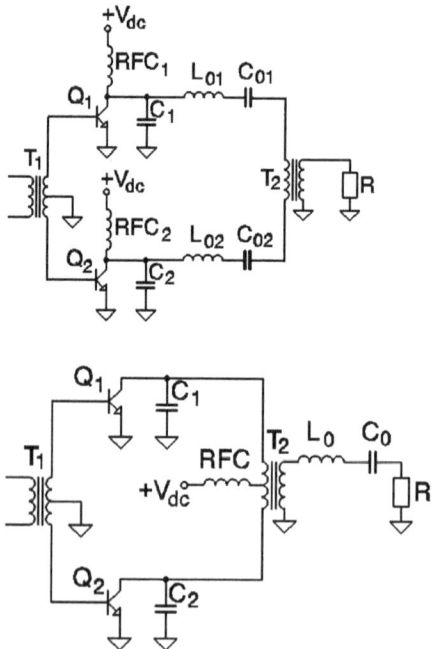

Figure 4-52 *Other push-pull Class E configurations.*

Parallel Tuned Amplifier with a Series Inductor

The circuit shown in Figure 4-53 [4], is the dual of the standard circuit presented in Section 4.1. The series inductor takes the role of the shunt capacitor, and the roles (and waveforms) of voltage and current are reversed. The analogy is straightforward and is presented in Table 4-3 [4]. Several comments regarding this duality are given below.

- The Class E switching conditions (for optimum operation) are

$$i_S(\pi) = 0$$

$$\left.\frac{di_S(\theta)}{d\theta}\right|_{\theta=\pi} = 0 \qquad\qquad 4.89$$

- To achieve optimum operation, the parallel circuit is not tuned on the switching frequency; its net susceptance is capacitive.

- If the collector current is not zero at the moment of switch-OFF, a short, high-amplitude pulse occurs on the collector voltage waveform. The current jumps cause power losses due to series inductance L_s.

- The parasitic series inductance of the switch may be included in L_s.

- If the Class E switching conditions are satisfied, the power losses at switch-OFF are eliminated. However, the power losses at switch-ON (caused by the parasitic capacitance in parallel at switch, which is discharged at switch-ON) cannot be eliminated. This is a major disadvantage, especially at high frequencies.

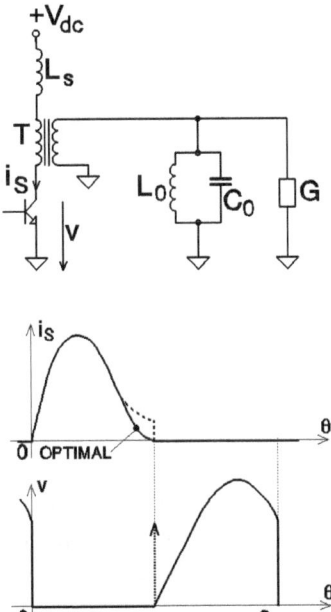

Figure 4-53 *Parallel tuned Class E amplifier with series inductor and typical waveforms.*

Another possible configuration (equivalent to that shown in Figure 4-53) is presented in Figure 4-54.

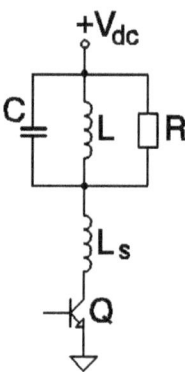

Figure 4-54 *Another parallel-tuned Class E configuration.*

Table 4-3 *Dual parameters for two Class E configurations.*

Series-tuned, shunt capacitor	Parallel-tuned, series inductor
$B=\omega C$	$X=\omega L_s$
$X=\omega L_0 - 1/(\omega C_0)$	$B=\omega C_0 - 1/(\omega L_0)$
R	$G=1/R$
ϕ	ϕ
$\psi = \arctan(X/R)$	$\psi = \arctan(B/G)$
$v(\theta)$	$i_S(\theta)$
$i_S(\theta)$	$v(\theta)$
V_{dc}	I_{dc}
I_{dc}	V_{dc}

Parallel-Tuned Amplifier With Series Capacitor

This circuit is presented in Figure 4-55 [4]. The topology is quite similar to that of Class A, B, and C amplifiers. However, here, the transistor operates as a switch, and the series capacitor is not a short circuit at the operating frequency. The waveforms for optimum operation are depicted in Figure 4-56. Note that not all waveforms in this circuit are similar to those in a series-tuned amplifier with shunt capacitor, but an analogy is still possible (see Table 4-4) [4].

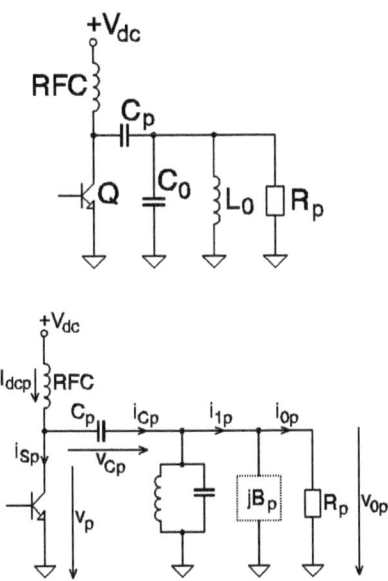

Figure 4-55 *Basic circuit of a parallel-tuned amplifier with series capacitor, and its equivalent circuit.*

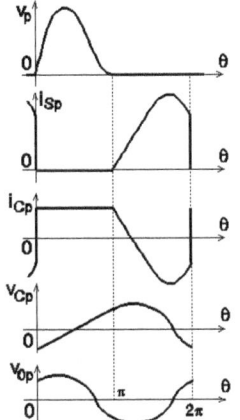

Figure 4-56 *Waveforms in a parallel-tuned amplifier with series capacitor (optimum operation).*

Table 4-4 *Dual parameters for two Class E configurations [4].*

Series-tuned, shunt capacitor		Parallel-tuned, series capacitor	
general	optimum	general	optimum
$D=1-D_p$	0.5	$D_p=1-D$	0.5
$B=\omega C=1/X_p$	0.1836	$X_p=1/(\omega C_p)=1/B$	5.4466
$R=1/(B^2R_p)$	1	$R_p=X_p^2/R$	29.66
$X=B_pRR_p$	1.1525	$B_p=X/(RR_p)$	0.0388
ψ	49.052°	ψ	49.052°
ϕ	-32.482°	ϕ	-32.482°
P_0	0.5768	P_0	0.5768

Parallel Tuned Amplifier

Class E operation can be obtained using circuit topologies that include only a parallel-tuned *RLC* circuit. If C_d is a short circuit at the operating frequency, the four circuits shown in Figure 4-57 are equivalent. Note that the same topologies are used in Class A, B, or C amplifiers. However, in a Class E amplifier, the transistor operates as a switch and the design goal is to met the Class E switching conditions given by Equation 4.48.

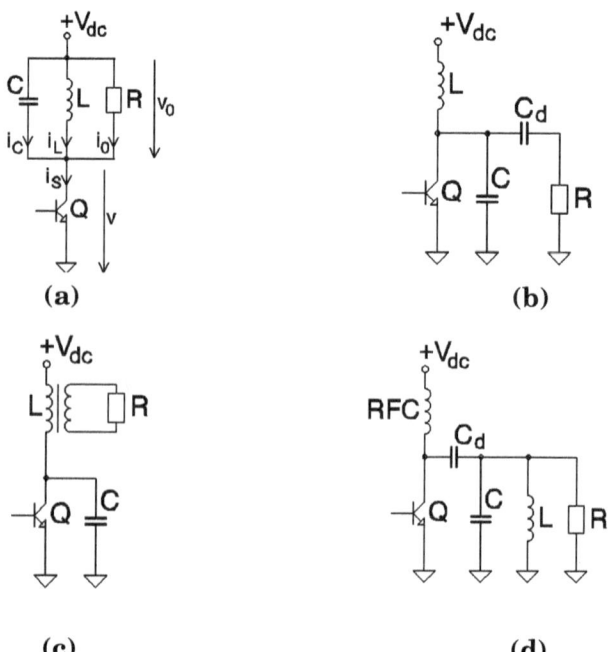

Figure 4-57 *Class E amplifiers with only one capacitor and one inductor in the load network.*

A simple analysis of this amplifier [29, 30] is based on the assumptions given in Section 4.1 (ideal switch, ideal components, the total capacitance in parallel at the switch is independent of the collector voltage). Typical waveforms for the circuit in Figure 4-57 (a) (operating optimally) are shown in Figure 4-58. During the ON state of the switch, the collector voltage is zero, and the output voltage is V_{dc}, which means, $i_C = 0$, $i_0 = V_{dc}/R$, and i_L increases linearly. The collector current $i_S = i_0 + i_L + i_C$ also increases linearly. During the OFF state, the collector current is zero and i_0, i_C, and i_L are damped sine wave pieces.

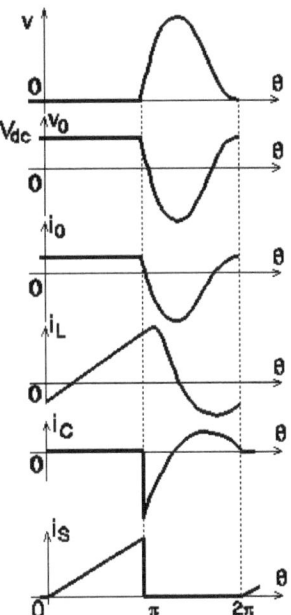

Figure 4-58 *Waveforms in the circuit of Figure 4-57(a) (D = 0.5, optimum operation).*

If V_{dc}, R, and D are given, Equation 4.48 yields the two unknowns L and C. The design and performance values of the circuit are shown in Figure 4-59 [30]. Figure 4-60 depicts the collector voltage and current waveforms for several values of the switch-OFF duty ratio.

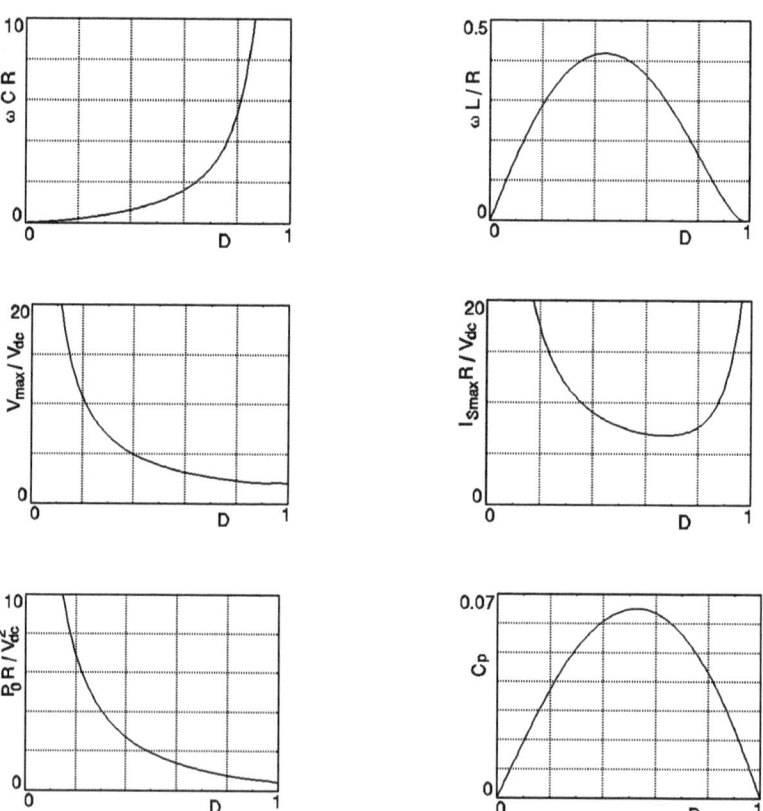

Figure 4-59 *Design and performance values for a Class E amplifier with only one capacitor and one inductor in the load network versus switch-OFF duty ratio.*

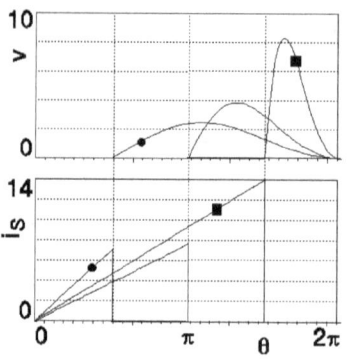

Figure 4-60 *Collector voltage and current waveforms for D = 0.25 (—■—), D = 0.5 (—), and D = 0.75 (—•—);V_{dc} = 1 V, R = 1 Ω.*

The curves in Figure 4-59 show that

- The resonant frequency of the tuned circuit must be always higher than the switching frequency to obtain optimum operation. The difference between the two frequencies decreases as D increases.

- The inductance L decreases toward zero when the switch duty ratio approaches its extreme values. The collector current increases quickly in these situations.

- The power output capability is lower than in the series-tuned, shunt capacitor circuit. At $D = 0.5$, $C_P \approx 0.0649$.

- If $D \to 1$, then $L \to 0$, $C \to \infty$, and $Q \to \infty$, and the resonant frequency of the tuned circuit is equal to the switching frequency. The collector voltage becomes a sinusoidal waveform with an amplitude equal to V_{dc} and the output power is $V_{dc}^2/(2R)$. This behavior is similar to that of a Class C amplifier when the conduction angle approaches zero.

Finally, note that this simple Class E circuit may be used in applications in which the harmonic content and the phase-modulation noise of the output are not important. Some examples of applications include DC/DC power converters, generation of sparks, arcs, or plasmas, providing RF energy for heating, or for communications jamming. The harmonic content of the output can be calculated easily and is high due to the flat bottom of the voltage waveform. The phase noise at the output is considerably higher than in other Class E configurations because the operation of the switch is passed directly to the output. This is explained by the fact that that there is no LC resonant action during the ON state and the timing of each cycle of operation is reestablished by the switch opening in each RF period [13].

Circuit Without Shunt Capacitor

The circuit in Figure 4-61 uses the DC feed inductance instead of the shunt capacitance for energy storage [31, 32]. The Class E switching conditions for this amplifier are given by Equation 4.89, and can be satisfied only if the capacitance in parallel at the switch is negligible. If the circuit operates optimally, the switching losses at turn-OFF are eliminated. The advantage of this topology is that it can use a thyristor as a switch at low frequencies and very high power [32].

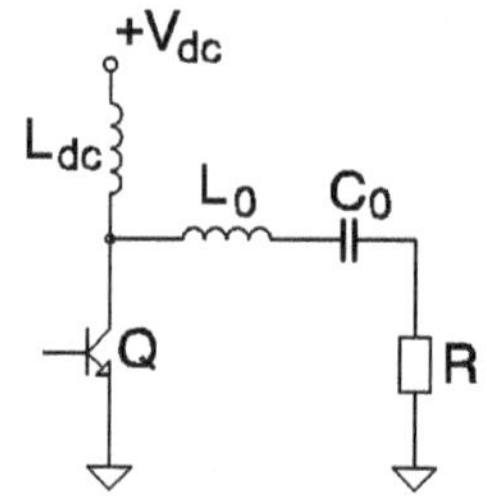

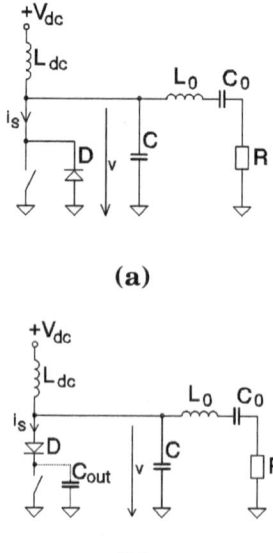

Figure 4-61 Class E amplifier without a shunt capacitor.

Circuit With a Series or Anti-Parallel Diode at Switch

The basic circuits of Class E amplifiers with anti-parallel or series diodes at the switch are shown in Figure 4-62. Illustrative waveforms are shown in Figure 4-63. The effect of the diode can be understood by looking at the voltage and current waveforms presented in Figure 4-63 (a) for an amplifier without a diode. These waveforms require a switch able to pass negative currents and support negative voltages. Note that such a switch cannot be implemented with only one active device; it requires a combination of transistors and diodes.

(a)

(b)

Figure 4-62 Class E amplifier (a) with anti-parallel diode, (b) with series diode.

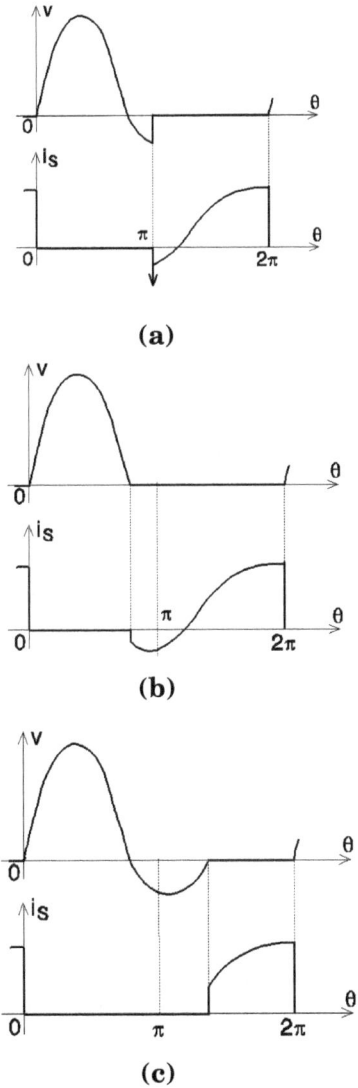

Figure 4-63 *Waveforms in Class E amplifier (a) without diode, (b) with anti-parallel diode, and (c) with series diode.*

An anti-parallel diode at the switch [Figure 4-62(a)] modifies the waveforms as suggested in Figure 4-63 (b). The diode does not allow the collector voltage to drop to a negative level; it is clamped at about –0.7 volt, and the current pulse discharging the shunt capacitor at turn-ON is eliminated. Therefore, the OFF state is shortened and the circuit is forced to switch at zero voltage across the switch. If the collector voltage in the circuit without a diode does not have negative values, the introduction of the anti-parallel diode does not affect the waveforms. A series diode at the switch [Figure 4-62(b)] modifies the waveforms, as shown in Figure 4-63(c). The diode does not allow the collector current to drop to a negative level and the ON state is shortened. As in the previous case, the current pulse discharging the shunt capacitor at turn-ON is eliminated.

Reference [18] presents a detailed analysis of Class E circuits, including a series or anti-parallel diode at the switch. The circuits remain highly efficient when the switch duty ratio (or the switching frequency) varies between certain limits. This feature may be exploited for DC/DC converters. Note that for RF applications,

- MOSFETs include an anti-parallel parasitic diode between source and drain. This diode prevents the drain voltage from becoming negative. MESFETs and BJTs do not include any parasitic diode.

- BJTs can withstand negative collector voltages (in the OFF state). However, if the collector potential drops below the base potential, the base-to-collector diode opens, and the transistor enters the inverted mode and may dissipate a large amount of power or even be destroyed. A series diode is required in the circuit if the bipolar transistor must withstand a large negative voltage. A series diode is also required in circuits with MOSFETs that must withstand negative drain voltages.

- BJTs do not pass negative collector currents. If this is required, an anti-parallel diode must be used in the circuit. An anti-parallel diode may be useful in practical circuits with BJTs to protect the transistor if mistuning occurs.

- The circuits with series diodes have a major disadvantage [see Figure 4-62(b)]. The output capacitance of the transistor (C_{out}) is charged at the maximum value of the collector voltage during the OFF state. The series diode does not allow discharge of this capacitance until the switch turns ON. The stored energy is dissipated in the transistor, which may cause appreciable power loss at high frequencies. Keep in mind that the maximum collector voltage in an optimally operated Class E circuit is about 3.5 V_{dc}. Thus, the power loss is approximately $(1/2)C_{out}f(3.5\ V_{dc})^2$.

Circuit With Multiple Resonance

A possible topology for a circuit with multiple resonance is illustrated in Figure 4-64. The parallel-tuned circuit, L_2C_2, resonates on the second harmonic of the switching frequency, reducing its level across the load. A similar circuit with the parallel resonator tuned to the third harmonic is known as a Class F_3 amplifier and is discussed in Chapter 5.

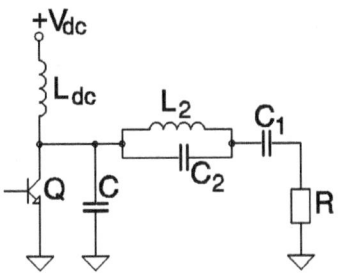

Figure 4-64 *Class E amplifier with a second harmonic resonator.*

Circuits for Impedance Matching

Several simple circuits for impedance matching at the output of the Class E amplifier are presented in Figure 4-65 [1, 2]. Standard solutions such as Pi or T matching circuits, other filters, or wideband transformers can also be used, depending on the application.

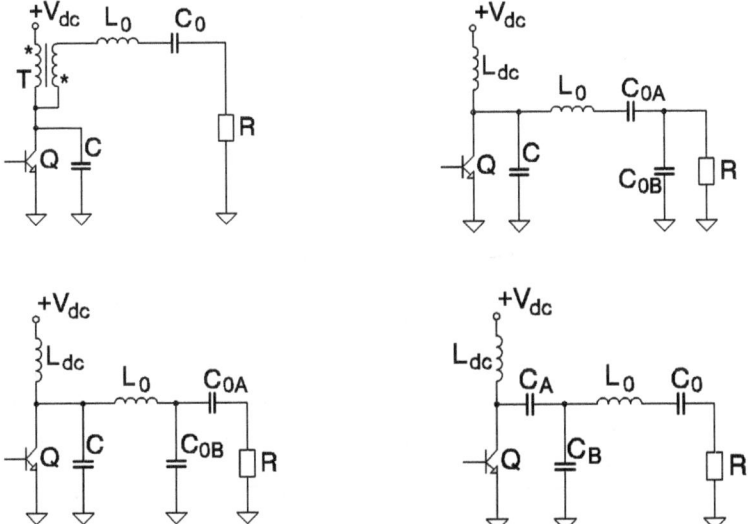

Figure 4-65 *Impedance matching circuits for Class E amplifiers.*

4.6   Class E Frequency Multiplier

The basic Class E configuration presented in Figure 4-4 can be used for frequency multiplication if the series-tuned circuit resonates on the desired harmonic of the switching frequency. If the other simplifying assumptions given in Section 4.1 remain valid, the analytical description of the circuit [33] is similar to that presented for the Class E amplifier. Several results are given below.

Figure 4-66 shows the variation of the design values and circuit performance versus the switch-OFF duty ratio D, for a frequency doubler operating with $\eta = 1$ and $\xi = 0$ (that is, the Class E switching conditions given by Equation 4.68 are satisfied). Figure 4-67 shows the collector voltage and current waveforms, for several values of D. Note that the susceptance of the shunt capacitor is defined at the switching frequency; the load angle ψ (and the net reactance X) is defined at the desired harmonic frequency. The optimum operation of the frequency doubler occurs for $D = 0.25$. In addition to the advantages of Class E operation, this provides maximum power output capability[6] and allows a single FET to function as the switch. (If the circuit is properly tuned, the switch must withstand positive voltages and pass both positive and negative currents.) The design and performance values of the frequency doubler (for optimum operation) are summarized in Table 4-5.

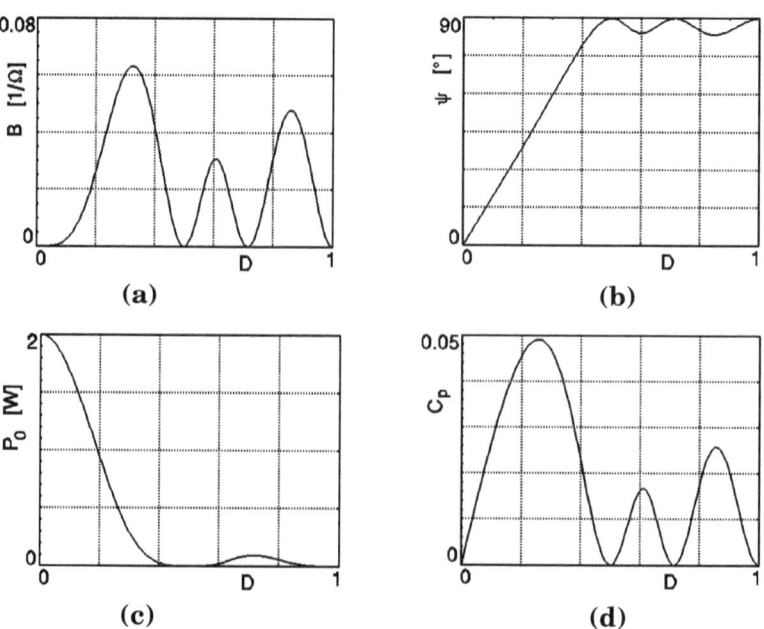

Figure 4-66 (a), (b) Design values of B and ψ, (c) output power P_0, and (d) power output capability C_P, versus D (frequency doubler, $\xi = 0$, $V_{dc} = 1$ V, $R = 1$ Ω, $\eta = 1$).

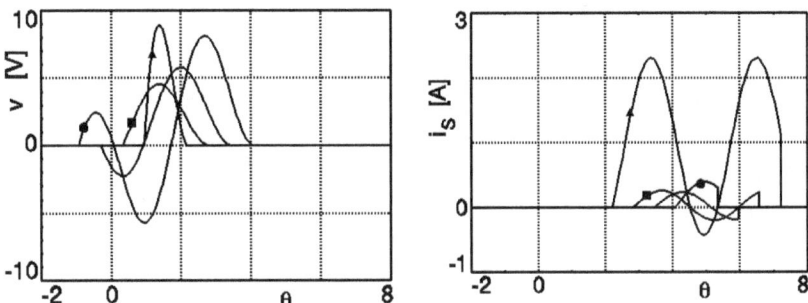

Figure 4-67 *Waveforms of collector voltage and current with D as a parameter (frequency doubler, $\xi = 0$, $V_{dc} = 1$ V, $R = 1$ Ω, $\eta = 1$). —▲— D = 0.2; —■— D = 0.4; ——— D = 0.6; —•— D = 0.8.*

The results can be extended easily for any multiplication order, N [33]. As shown in Figure 4-68, maximum power output capability occurs near $D = 0.5/N$, and this operating point makes it possible to use a FET as the switching device. The design values and the resulting operating parameters are presented in Table 4-5.

Table 4-5 *Design values and operating parameters for Class E frequency multipliers ($V_{dc} = 1$ V, $R = 1$ Ω).*

Parameter	Doubler D=0.25	Doubler D=0.255	Tripler D=0.5/3	Multiplier (N) D=0.5/N
B [1/Ω]	0.0459	0.04795	0.0204	$0.1836/N^2$
ψ [°]	49.05	50.15	49.05	49.05
C_P	0.04905	0.0491	0.0327	$0.0981/N$
P_0 [W]	0.5768	0.5418	0.5768	0.5768
V_{max} [V]	7.124	6.975	10.686	$3.562N$
I_{smax} [A]	1.651	1.583	1.651	1.651

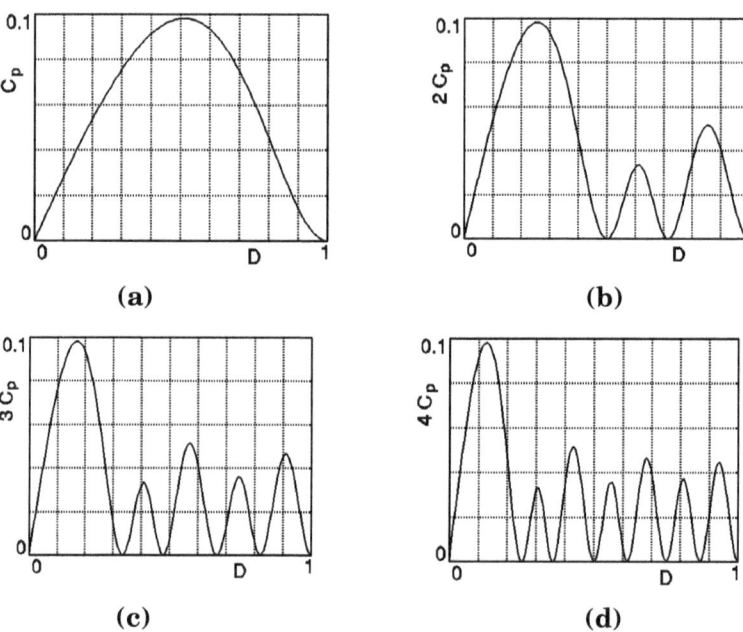

Figure 4-68 *Power output capability versus D (a) amplifier (N = 1), (b) frequency doubler (N = 2), (c) frequency tripler (N = 3), and (d) frequency quadrupler (N = 4).*

Although the maximum power output capability is obtained in circuits using an RF choke, Class E frequency multipliers can operate in circuits with a finite DC feed inductance [27]. Note also that Class E switching conditions can be obtained in practical circuits only for values of Q exceeding a certain value [20]. Other interesting properties of Class E frequency multipliers are presented in References [7, 8, 20, 21, 22, and 23].

4.7   CAD of the Class E Amplifier

Class E circuits may be simulated using a general-purpose circuit simulator (such as SPICE), which allows for the evaluation of time domain waveforms. Although there are several advantages (for example, the circuits may be modeled with any degree of complexity), they are counterbalanced by the fact that the simulations are very slow (the steady-state response usually must be simulated), and it is almost impossible to optimize the design. Efficient simulation requires specialized programs like HEPA-PLUS developed by Design Automation, Inc [28].

HEPA-PLUS can be used to simulate almost all known topologies of untuned and tuned single-ended and push-pull PAs. The program generates directly (and rapidly) the steady-state periodic response of the circuit, and the cycle-by-cycle transient response, from any specified starting conditions. The HEPA-PLUS outputs are

- time-domain graph plots and frequency-domain spectra of all important circuit voltage and current waveforms,

- tabulation of DC input power, output power, efficiency, inefficiency, power dissipation in each circuit element (including the parasitic power losses in each inductor and capacitor), and transistor peak voltage and current stresses, in normal and inverse polarities. Note that many of these results are not available (directly) in SPICE simulators and some of them are very difficult to estimate after a simulation is performed.

The HEPA-PLUS program allows for the sweeping of any circuit parameter between specified limits and the plotting of any two output variables versus the swept parameter. Transfer functions (that is, any computed variable versus any circuit parameter) and data for required control functions (that is, the required value of any circuit parameter versus any other circuit parameter, to maintain a specified output voltage, current, or power) can be computed.

One of the program's powerful capabilities is the ability to adjust the operating parameters and component values automatically to achieve a defined "optimum" performance. The program also evaluates the performance achievable with a candidate transistor in an optimally operated Class E amplifier and creates a preliminary amplifier design that can be used for further optimization.

The generic topology of the single-ended amplifier used in HEPA-PLUS (complete with a switch and a load network) is shown in Figure 4-69. The switch model includes

- a resistor, R_{ON}, modeling the ON resistance ($R_{DS,ON}$ for MOSFET or R_{CEsat} for BJT),

- a voltage source, V_0, modeling V_{CEsat} of BJT,

- a series-RC circuit, $C_{out}R_{cout}$, modeling the output capacitance of the active device and its parasitic resistance (note that both linear and nonlinear models can be used for the output capacitance),

- an optional anti-parallel diode (the parasitic substrate diode of a silicon MOSFET, or an external diode intentionally connected from emitter to collector of a BJT),

- a series inductance, L_Q, modeling the inductances in series with the transistor (the inductance of the emitter or source bonding wires).

The load circuit depicted in Figure 4-69 can be used to represent a non-resonant or a resonant network. Reference [28] presents nine examples of different single-ended topologies included within the generic topology of Figure 4-69. Adjustments can be made by setting the values of certain circuit components to nearly zero or infinity and/or by splitting a component into two parts connected in series or in parallel. Note that all inductors and capacitors have specified loss resistance or unloaded quality factor.

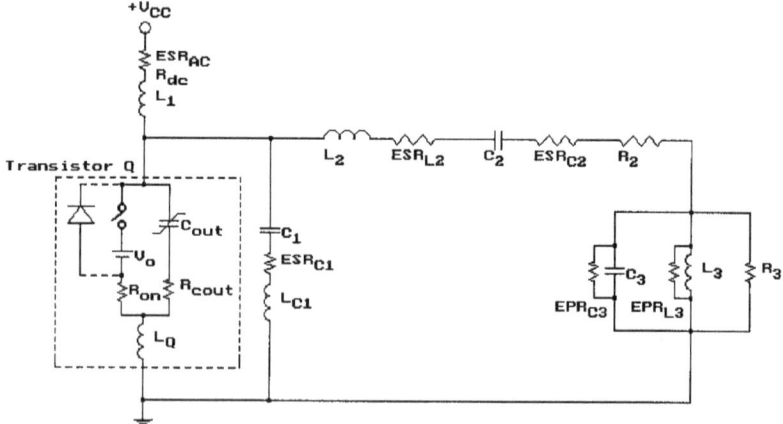

Figure 4-69 *Generic topology of a single-ended power amplifier. (From the HEPA-PLUS User Manual, 1990, courtesy of Design Automation, Inc., Lexington, Massachusetts.)*

Figure 4-70 shows three generic push-pull versions of the circuit in Figure 4-69. For simplicity, the detailed transistor model given in Figure 4-69 is replaced with a switch symbol.

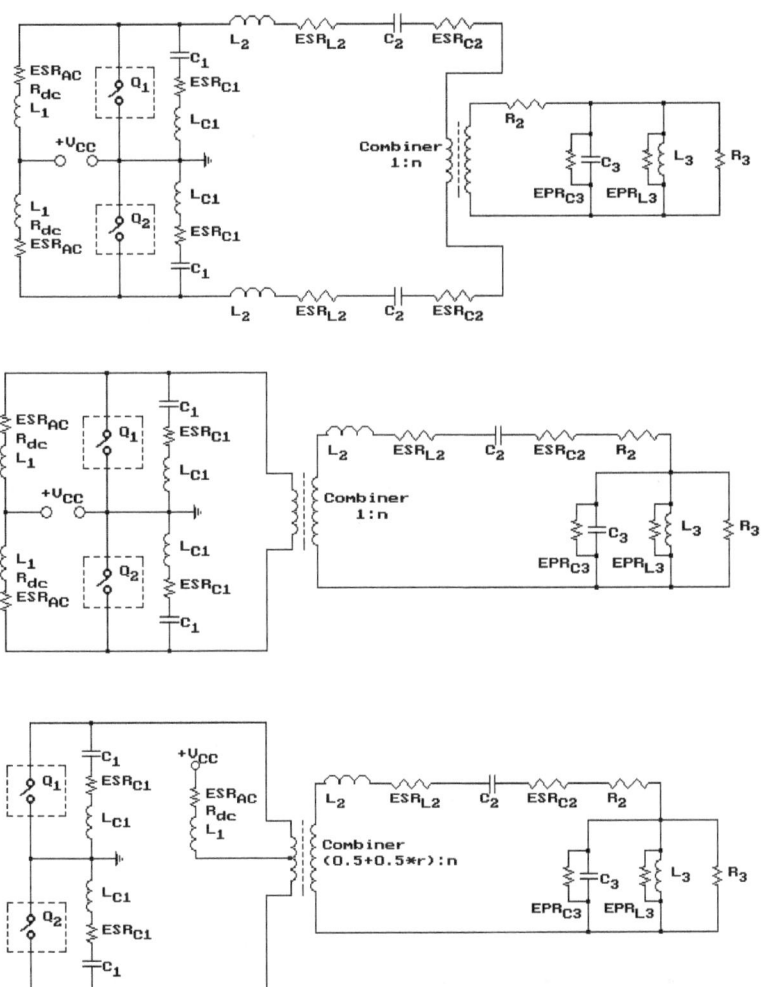

Figure 4-70 *Generic topologies of push-pull power amplifiers. (From the HEPA-PLUS User Manual, 1990, courtesy of Design Automation, Inc., Lexington, Massachusetts.)*

A simple circuit example is presented in Figures 4-71 to 4-73. This single-ended Class E amplifier uses an IRF540 MOSFET operating at 15 MHz [28]. The switch model includes an anti-parallel diode and a linear output capacitance; a 2-watt gate input power from the driver is assumed.

```
┌── High-Efficiency Power Amplifier CAD (HEPA-PLUS/WB)  Sept. 03, 1998 23:23 ──┐
│   15-MHz 110-watt amplifier using IRF540, detuned slightly for example        │
│                   ENTER CIRCUIT PARAMETERS and/or TITLE                        │
│                                                                                │
│   Common-Source TRANSISTOR with diode        LOAD without filter              │
│   Frequency (f).......[Hz]:  1.5E+07   Load location, R3 or R2?  R3            │
│   Duty ratio (D)..........:  0.5       Load resistance..[ohms]:  1.8           │
│   "On" resistance...[ohms]:  0.06      R3/Rload...............:  1             │
│   Vo[V]:0          Lq [H]:  0          L3............[henries]:  7.5E-06       │
│   Transition times: turn-on,turn-off [s]  Qu:150     at freq:  1.5E+07         │
│      on:5E-09         off:  5E-09      C3.............[farads]:  1.5E-11        │
│   Cout...........[farads]:  5.1E-10       Qu:1000    at freq:  1.5E+07         │
│   Cout series resis.[ohms]:  0.3                                               │
│                                                                                │
│                    LOAD NETWORK,  Single-ended                                 │
│   DC supply (Vcc)..[volts]:  22        C2.............[farads]:  6E-09         │
│   L1............[henries]:  2E-06          Qu:1000    at freq:  1.5E+07         │
│      Qu: 150       at freq:  1.5E+07                                           │
│      Rdc............[ohms]:  0.01      Network loaded Q  or L2?  L2            │
│   C1.............[farads]:  7.5E-10    L2............[henries]:  4.7E-08        │
│      Qu: 1000      at freq:  1.5E+07      Qu:200     at freq:  1.5E+07         │
│      Lc1.........[henries]:  0         Network loaded Q.......:  2.3569         │
│   SELECTION: <↑,↓,←,→> ENTRY: ALPHANUM.   COMPUTE: <PgDn>    MAIN MENU: <ESC>  │
└────────────────────────────────────────────────────────────────────────────────┘
```

Figure 4-71 *Entering circuit parameters in HEPA-PLUS, courtesy of Design Automation, Inc., Lexington, Massachusetts.)*

```
┌── High-Efficiency Power Amplifier CAD (HEPA-PLUS/WB)  Sept. 03, 1998 23:23 ──┐
│   15-MHz 110-watt amplifier using IRF540, detuned slightly for example        │
│                   EFFICIENCY, POWERS, and STRESSES                             │
│                            Single-ended                                        │
│      Collector/drain efficiency..........[Pout/Pin]        91.690%            │
│      Collector/drain inefficiency...[(Pin-Pout)/Pin]        8.310%            │
│      Overall efficiency............[Pout/(Pin+Pid)]        90.225%            │
│      DC power input (Pin)...................[watts]        123.14             │
│      Input-drive power (Pid)................[watts]             2             │
│      Power output (Pout)....................[watts]        112.9             │
│      Power loss in L1.......................[watts]        0.33761           │
│      Power loss in L2.......................[watts]        1.3893            │
│      Power loss in C2.......................[watts]        0.11093           │
│      Resistive power loss of transistor & C1[watts]        5.5348            │
│      Turn-off power loss of transistor......[watts]        2.8345            │
│      Turn-on power loss of transistor.......[watts]        0.022922          │
│      Power loss in L3.......................[watts]        0.0019167         │
│      Power loss in C3.......................[watts]        0.0002873         │
│                                                                               │
│      Output voltage, current (at Rload) [V,A]:   14.256      7.9199          │
│   Transistor peak voltages....[volts]:  normal   80.896   inverse   None     │
│   Transistor peak currents..[amperes]:  normal   14.888   inverse   None     │
│                                          "DISPLAY RESULTS" MENU: <PgUp>        │
└────────────────────────────────────────────────────────────────────────────────┘
```

DISPLAYING COMPUTED POWERS...

Figure 4-72 *Powers, efficiency, and switch stresses for the Class E circuit in Figure 4-7, courtesy of Design Automation, Inc., Lexington, Massachusetts.)*

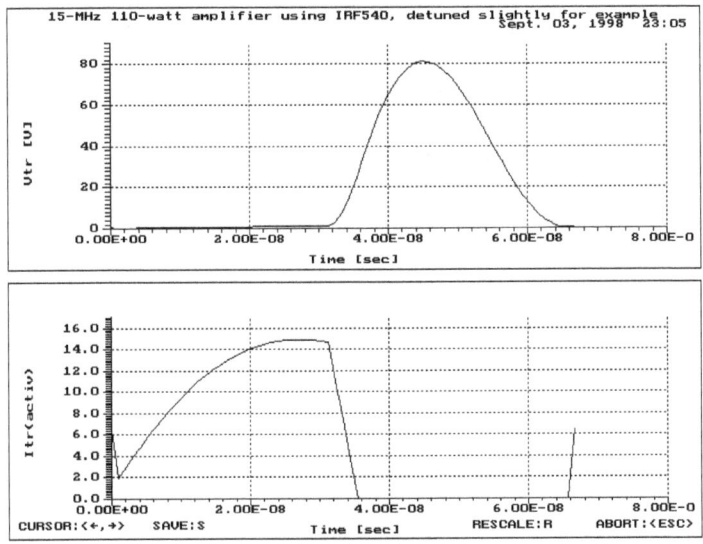

Figure 4-73 *Waveforms of collector voltage and active collector current for the Class E circuit in Figure 4-71. (Courtesy of Design Automation, Inc., Lexington, Massachusetts.)*

4.8   Class E Versus Class B, C, and D Amplifier

Based on the results given above, the Class E amplifier offers a number of advantages when compared with a Class C (or B) amplifier.

1. It has a higher reliability due to lower power dissipation on the active device.

2. Equipment using Class E circuits may be considerably smaller and lighter than equations with Class B or C amplifiers.

3. Higher reproduction quality can be obtained in mass production due to its relatively low sensitivity to component tolerances (including transistor characteristics). Consequently, the switching transistor used in a Class E circuit can be replaced without problems, even with another transistor that has different characteristics.

4. The collector efficiency, as well as the overall efficiency, is higher for the Class E amplifier.

5. Theoretical optimizations for certain goals, such as bandwidth, output power and overall efficiency are possible for Class E PAs.

6. Circuits using Class E amplifiers cost less to design and manufacture.

7. The theoretical operation and the design equations of Class E circuits are well verified in practical designs. Moreover, the design procedures usually need only a few switching transistor parameters, which are often provided in the data sheets or can be measured easily. This situation is completely different for Class C amplifiers. Class C circuits with acceptable performance can be designed easily using the large signal parameters. However, the following points should be taken into consideration:

   • There is no theoretical procedure with which to optimize the circuit for a certain design goal or to evaluate the design trade offs.

   • Some empirical work is needed to optimize the circuit (thus increasing the design cost).

   • The circuit is quite sensitive to variations of the transistor characteristics (and the sensitivity cannot be theoretically estimated).

   • Transistor replacement (even with one that has the same part number, but is from another manufacturer) usually requires changes in the circuit design.

   • Individual tuning is very important for each amplifier.

   • There is no guarantee that the performance obtained in a certain prototype can be duplicated in the mass production process.

Tables 4-6 and 4-7 present experimental data to support this comparison of Class E and Class C circuits [3, 10]. Note that the collector efficiency of Class E circuits in both tables includes the losses on the harmonic-reduction filter, while the collector efficiency of the Class C circuits does not.

Table 4-6 *Experimental results at 27 MHz (MRF472 Motorola transistor) [3,10].*

	Class E		Class C
V_{dc} [V]	13.5	27	12.5
P_0 [W]	5	20	4
η [%]	85	86	65
P_{dis}/P_0	0.18	0.16	0.54
a_n [dBc]	<-67	<-67	-
G_p [dB]	9	11	13
Filter?	Yes	Yes	No

Table 4-7 *Experimental results at 54 MHz (A25-28 CTC transistor) [3, 10].*

	Class E		Class C
P_0 [W]	27	28	25
η [%]	87	92	65
P_{dis}/P_0	0.15	0.09	0.54
G_p [dB]	-	12	13.1
Filter?	Yes	Yes	No

The results given in Chapters 3 and 4 show that a Class D amplifier has several advantages when compared with a Class E amplifier.

• Its power output capability is about 1.623 times higher.

• The power losses on the ON resistance of the transistors are about 1.365 times lower for the CVS and TCVS Class D circuits than for Class E circuits.

• The Class D amplifier has larger bandwidth without retuning or switching of the output filters.

• The push-pull topology used in Class D circuits significantly reduces the even harmonics at the output.

The Class E amplifier provides several important advantages.

• It uses only one switching device, and its reference terminal (usually, emitter or source) is always connected to the ground.

- Output and/or input transformers (which are usually bulky, expensive and introduce additional losses) are not required.

- It is much easier to drive a Class E amplifier (single-ended) than a Class D amplifier (where the timing is critical).

- The power losses due to the output capacitance of the active devices are eliminated, allowing the circuit to operate at much higher frequencies.

- The transistor switching time may be an important part of the RF period (even 10 to 15 percent), without a significant reduction of the collector efficiency.

- The odd harmonics of the voltage across the switch are lower, and the even harmonics may be reduced significantly using a push-pull Class E configuration.

Accordingly, the practical use of Class D amplifiers is advantageous only at low frequencies (megahertz), when the switching times and the power losses due to the output capacitances are negligible. The switching frequency may be somewhat increased (tens of megahertz) if the circuit is Class DE operated. Class E circuits can be used with good performance at much higher frequencies (hundreds of megahertz or even gigahertz) [28, 34].

4.9 A Condition Required to Obtain $\eta = 1$ and $P_0 \neq 0$

A switching-mode circuit (power amplifier or frequency multiplier) may achieve very high collector efficiency even at very high frequencies when the switching times are an important part of the RF period. The waveforms in Figure 4-3 cannot be obtained practically; that is, the switching losses can be significantly reduced in some designs, but cannot be fully eliminated. A fundamental result related to this problem is discussed in References [35] and [36]. According to these references, it is impossible to achieve both unity efficiency and nonzero output power. A presentation of the proof of this statement is beyond the scope of this book; however, the working assumptions and conditions obtained are given below.

The switching-mode circuit includes an ideal switch (that is, zero switching times, zero ON resistance, and infinite OFF resistance) and a load network (containing only time-invariant, passive, and linear components). This circuit can achieve both unity efficiency and non-zero output power only if the waveforms of v and i_S do not overlap, but jpin each other and at joining instants:

1. i_S has a jump and the first derivative of v has a jump, and/or vice versa, or

2. both i_S and v have jumps.

The idealized continuous waveforms depicted in Figure 4-3 (and the associated conditions) cannot be obtained in practical circuits, and the switching losses cannot be eliminated entirely. Practical circuits can only eliminate either the turn-ON losses or the turn-OFF losses.

4.10 Notes

1. Note that any switching-mode PA inherently fulfills this condition and thus is potentially highly efficient.

2. Some Class F circuits [the so-called Class F2 and F3, (see Chapter 5)] may also be described by this definition. However, the main characteristic of Class F amplifiers is that they employ multiple resonators in the load network.

3. It is assumed that an appropriate driving waveform is used to produce the desired switching action of the transistor (alternately, ON and OFF) at the operating frequency, f. Note that the driving waveform may include a frequency or phase modulation desired in the output signal (because the frequency and phase of the output signal track those of the driving signal), but not an amplitude modulation (because there is no relationship between the amplitudes of the driving signal and the output signal).

4. The key waveform is the collector voltage, which is nonzero when the transistor is OFF. The OFF angle is similar to the conduction angle used for the analysis of Class C amplifiers, where the key waveform is that of the collector current, which is nonzero when the transistor is ON.

5. Note that θ_{OFF} only includes the fall time of the collector current. The storage time of BJTs only determines an increase in the ON time and may be compensated for by reducing the switch-ON duty ratio of the driving signal applied to the base.

6. Exact calculations show that the maximum power output capability is obtained for $D = 0.255$ [27]. However, the difference is not significant. The value of $D = 0.25$ is much more convenient for practical applications.

4.11 References

1. Sokal, N. O. and A. D. Sokal. "Class E — A New Class of High-Efficiency Tuned Single-ended Switching Power Amplifiers." *IEEE J. Solid-State Circuits*, vol. SC-10 (June 1975): 168–176.
2. Sokal, N. O. and A. D. Sokal. "High-Efficiency Tuned Switching Power Amplifier." U.S. Patent 3,919,656, November 11, 1975.
3. Sokal, N. O. and A. D. Sokal. "Class E Switching-Mode RF Power Amplifiers-low Power Dissipation, Low Sensitivity to Component Tolerances (including Transistors), and Well-Defined Operation." *RF Design*, vol. 3, no. 7 (July/August 1980): 33–38, 41.
4. Raab, F. H. "Idealized Operation of the Class E-Tuned Power Amplifier." *IEEE Trans. Circuits Syst.*, vol. CAS-24 (December 1977): 725–735.
5. Krauss, H. L., C. V. Bostian, and F. H. Raab. *Solid State Radio Engineering.* New York: John Wiley & Sons, 1980.
6. Sokal, N. O and F. H. Raab. "Harmonic Output of Class E RF Power Amplifiers and Load Coupling Network Design." *IEEE J. Solid-State Circuits*, vol. SC-12 (February 1977): 86–88.
7. Albulet, M. "An Explicit Design Criterion for the RF choke Reactance on the Class E Power Amplifiers and Frequency Multipliers." *The Trans. of the South African Institute of Electrical Engineers*, vol. 85, no. 2 (June 1994): 37–42.
8. M. Albulet and S. Radu. "An Optimum Tradeoff in the RF Choke Design from Class E Amplifiers/Multipliers," *Conf. Proc. RF Expo West '95*, (San Diego, CA) (January/February 1995): 195-206.
9. Raab, F. H. and N. O. Sokal. "Transistor Power Losses in the Class E Tuned Power Amplifier." *IEEE J. Solid-State Circuits*, vol. SC-13 (December 1978): 912–914.
10. Sokal. N. O. "Class E Can Boost the Efficiency." *Electronic Design*, vol. 25, no. 20 (September 1977): 96–102.
11. Kazimierczuk, M. K. "Effects of the Collector Current Fall Time on the Class E Tuned Power Amplifier." *IEEE J. Solid-State Circuits*, vol. SC-18 (April 1983): 181–193.
12. Blanchard, J. A and J. S. Yuan. "Effect of Collector Current Exponential Decay on Power Efficiency for Class E Tuned Power Amplifier." *IEEE Trans. Circuits Syst.*I, vol. 41 (January 1994): 69–72.
13. Raab, F. H. "Effects of Circuit Variations on the Class E Tuned Power Amplifier." *IEEE J. Solid-State Circuits*, vol. SC-13 (April 1978): 239–247.
14. Liou, M. L. "Exact Analysis of Linear Circuits Containing Periodically Operated Switches with Applications." *IEEE Trans. Circuit Theory*, vol. CT-19 (March 1972): 146–154.
15. Kazimierczuk, M. K. and K. Puczko. "Exact Analysis of Class E Tuned Power Amplifier at any Q

and Switch Duty Cycle." *IEEE Trans. Circuits Syst.*, vol. CAS-34 (February 1987): 149–159.

16. Avratoglou, C. P. and N. C. Voulgaris. "A New Method For the Analysis and Design of the Class E Power Amplifier Taking into Account the QL Factor." *IEEE Trans. Circuits Syst.*, vol. CAS-34 (June 1987): 687–691.

17. Avratoglou, C. P., N. C. Voulgaris, and F. I. Ioannidou. "Analysis and Design of a Generalized Class E Tuned Power Amplifier." *IEEE Trans. Circuits Syst.*, vol. 36 (August 1989): 1068–1079.

18. Kazimierczuk, M. K. and K. Puczko. "Class E Tuned Power Amplifier with Antiparallel Diode or Series Diode at Switch, With any Loaded Q and Switch Duty Cycle." *IEEE Trans. Circuits Syst.*, vol. 36 (September 1989): 1201–1209.

19. Smith, G. H. and R. E. Zulinski. "An Exact Analysis of Class E Amplifiers with Finite DC-Feed Inductance at Any Output Q." *IEEE Trans. Circuits Syst.*, vol. 37 (April 1990): 530–534.

20. Albulet, M. "Analysis and Design of the Class E Frequency Multipliers with RF Choke." *IEEE Trans. Circuits and Syst.*, Part I, vol. 42, no. 2 (February 1995): 95–104.

21. Albulet, M. and S. Radu. "Analysis and Design of the Class E Frequency Multipliers Taking into Account the Q Factor." *Int. J. Electronics and Communications* (AEÜ), vol. 49, no. 2 (March 1995): 103–106.

22. Albulet, M. and S. Radu. "Exact Analysis of Class E Frequency Multiplier with Finite DC-Feed Inductance at Any Output Q," *Int. J. Electronics and Communications* (AEÜ), vol. 50, no. 3 (May 1996): 215–221.

23. Albulet, M. and R. E. Zulinski. "Effect of Switch Duty Ratio on the Performance of Class E Amplifiers and Frequency Multipliers." *IEEE Trans. Circuits and Syst.*, Part I, vol. 45, no. 4 (April 1998): 325–335.

24. "Class E RF Power Amplifier Demonstrator — E10-3 User Manual," *Design Automation, Inc.*, Lexington, Massachusetts, 1981.

25. Kazimierczuk, M. K. "Collector Amplitude Modulation of the Class E Tuned Power Amplifier," *IEEE Trans. Circuits Syst.*, vol. CAS-31 (June 1984): 543–549.

26. Albulet, M. and S. Radu. "Second Order Effects in Collector Amplitude Modulation of Class E Power Amplifiers." *Int. J. Electronics and Communications* (AEÜ), vol. 49, no. 1 (January 1995): 44–49.

27. Zulinski, R. E. and J. W. Steadman. "Class E Power Amplifiers and Frequency Multipliers with Finite DC-Feed Inductance." *IEEE Trans. Circuits Syst.*, vol. CAS-34 (September 1987): 1074–1087.

28. *HEPA-PLUS User Manual*, Design Automation, Inc., Lexington, MA, 1990–2001.

29. Sokal, N. O. "Class E High-Efficiency Switching Mode Tuned Power Amplifier With Only One Inductor and One Capacitor in Load Network-Approximate Analysis." *IEEE J. Solid-State Circuits*, vol. SC-16 (August 1981): 380–384.

30. Kazimierczuk, M. K. "Exact Analysis of Class E Tuned Power Amplifier with Only One Inductor and One Capacitor in Load Network." *IEEE J. Solid State Circuits,* vol. SC-18 (April 1983): 214–221.

31. Kazimierczuk, M. K. "Class E Tuned Power Amplifier with Shunt Inductor," *IEEE J. Solid-State Circuits*, vol. SC-16 (February 1981): 2–7.

32. Avratoglou, C. P. and N. C. Voulgaris. "A Class E Tuned Amplifier Configuration with Finite DC-Feed Inductance and No Capacitance in Parallel with Switch." *IEEE Trans. Circuits Syst.*, vol. CAS-35 (April 1988): 416–422.

33. Zulinski, R. E. and J. W. Steadman. "Idealized Operation of Class E Frequency Multipliers." *IEEE Trans. Circuits Syst.*, vol. SC-21 (December 1986): 1209–1218.

34. Sowlati, T., et al. "Low Voltage, High Efficiency GaAs Class E Power Amplifiers for Wireless Transmitters." *IEEE J. of Solid-State Circuits*, vol. 30 (October 1995): 1074–1080.

35. Molnar, B. "Basic Limitations on Waveforms Achievable in Single-Ended Switching-Mode Tuned (Class E) Power Amplifiers." *IEEE J. Solid-State Circuits,* vol. SC-19 (February 1984): 144–146.

36. Kazimierczuk, M. K. "Generalization of Conditions for 100-percent Efficiency and Nonzero Output Power in Power Amplifiers and Frequency Multipliers." *IEEE Trans. Circuits Syst.*, vol. CAS-33 (August 1986): 805–807.

5

Class F Amplifiers

Several different power amplifier topologies that provide high efficiency are known as Class F amplifiers. Although these topologies may appear distinct from one another, they all use a multiple-resonator output filter. This filter controls the harmonic content of the collector voltage and/or current, by shaping their waveforms to reduce the power dissipated on the active device. This, in turn, increases the collector efficiency.

The literature uses many different names to describe this concept: high-efficiency Class C, Class C using harmonic injection, optimum efficiency Class B, optimally loaded Class B, Class CD, single-ended Class D, multiple-resonator Class C, biharmonic or polyharmonic Class C, and even Class E. Recent literature tends to designate all circuits using a multiple-resonator output filter as Class F amplifiers. Further classification of Class F circuits is provided in Reference [1] and will be used in this chapter. Class F circuits are divided into three categories: Class F1, F2, and F3.

5.1 Class F1 Power Amplifiers

The output filter of a Class F1 amplifier resonates on both the operating frequency and one of its harmonics, usually the second or third. The corresponding circuits are also known as second-harmonic peaking Class F PA, and third-harmonic peaking Class F PA, respectively.

The basic circuit of a Class F1 amplifier employing the third harmonic is shown in Figure 5-1 [1–7]. Figure 5-2 presents the characteristic waveforms of the circuit. This circuit is similar to that of a single-ended Class B or Class C amplifier, except for the addition of a third-harmonic parallel resonator, L_3C_3. A simplified analysis of the circuit [1, 2] is based on the assumption of a 180-degree conduction angle (as in Class B amplifiers), and on the usual assumptions regarding the ideal behavior of components.

- The passive components are ideal (RFC is an ideal RF choke and C_d is a DC blocking capacitor).

- The active device behaves as an ideal controlled-current source. Ignore the saturation voltage and/or the saturation resistance and assume a perfectly linear transfer characteristic of the transistor (i.e., a sinusoidal drive signal determines a sinusoidal collector current).

- The parallel-tuned circuit, LC, which resonates on the fundamental frequency, is ideal, i.e., its impedance is infinite at the fundamental frequency and zero otherwise). The parallel-tuned circuit, L_3C_3, which resonates on the third harmonic, is also ideal (its impedance is infinite at the third-harmonic frequency and zero otherwise). Consequently, the load impedance of the transistor is R at fundamental frequency f, infinite at $3f$, and zero otherwise. Note that the half-sine wave collector current characteristic to the single-ended Class B operation consists only of DC, fundamental frequency, and even harmonics. Therefore, the infinite impedance at $3f$ will not create any impediment.

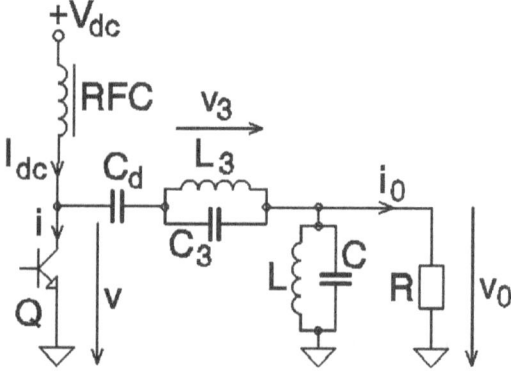

Figure 5-1 *Third-harmonic peaking Class F1 amplifier.*

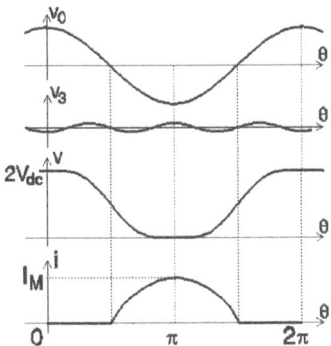

Figure 5-2 *Waveforms in a third-harmonic peaking Class F1 amplifier.*

With this circuit, it is possible to obtain a higher efficiency than in Class B or C amplifiers because the presence of the third harmonic in the collector voltage may cause its waveform to flatten (see Figure 5-2). Thus, the fundamental component of the collector voltage may be greater than its peak value. The collector voltage is flattened if the third harmonic is 180 degrees out of phase with the fundamental component, i.e.,

$$v_0(\theta) = V_0 \cos\theta \qquad v_3(\theta) = -V_3 \cos 3\theta \qquad\qquad 5.1$$

and

$$v(\theta) = V_{dc} + V_0 \cos\theta - V_3 \cos 3\theta = V_{dc} + V_0(\cos\theta - x\cos 3\theta) \qquad 5.2$$

where $x = V_3/V_0$ (note that $x > 0$).

The optimal value of x can be found by setting

$$\frac{dv(\theta)}{d\theta} = 0 \qquad\qquad 5.3$$

The result is

$$\sin\theta = 0 \qquad \sin\theta = \pm\sqrt{\frac{9x-1}{12x}} \qquad\qquad 5.4$$

Therefore

- If $0 < x < 1/9$, the maximum value of $v(\theta)$ occurs in 0 and the minimum value of $v(\theta)$ occurs in π. The peak-to-peak swing of $v(\theta)$ is $2V_0(1-x)$ and decreases with increasing x.

- If $x = 1/9$, the third-order flat maximum value of $v(\theta)$ occurs in 0, and the minimum value of $v(\theta)$ occurs in π. The peak-to-peak swing of $v(\theta)$ is $(16/9)V_0$.

- If $x > 1/9$, a minimum value is obtained in 0 and two adjacent maximum values. A maximum value and two adjacent minimum values are obtained in π. The peak-to-peak swing of $v(\theta)$ increases with increasing x and is given by

$$\frac{2}{3}V_0\sqrt{\frac{(3x+1)^3}{3x}} \qquad 5.5$$

Analysis of these results shows that the peak-to-peak swing of the collector voltage has a minimum value for $x = 1/9$. This corresponds to the maximally flat collector waveform depicted in Figure 5-2. Because the maximum swing allowed for the collector voltage is $2V_{dc}$, the maximum amplitude of the collector voltage is

$$V_0 = \frac{9}{8}V_{dc} \qquad 5.7$$

Thus, the peak output power is given by

$$P_0 = \frac{V_0^2}{2R} = \frac{81}{128}\frac{V_{dc}^2}{R} \approx 0.6328\frac{V_{dc}^2}{R} \qquad 5.8$$

which is about 27 percent greater than for Class B operation.

As for the Class B amplifier, the amplitude of the collector current pulse I_M, and DC input current I_{dc} are calculated as

$$I_M = \frac{V_0}{R} = \frac{9}{8}\frac{V_{dc}}{R} \qquad I_{dc} = \frac{2}{\pi}I_M = \frac{9}{4\pi}\frac{V_{dc}}{R} \qquad 5.9$$

Thus, the collector efficiency is given by

$$\eta = \frac{P_0}{P_{dc}} = \frac{P_0}{V_{dc}I_{dc}} = \frac{9\pi}{32} \approx 88.36\% \qquad 5.10$$

Finally, the power output capability of the circuit can be calculated as

$$C_P = \frac{P_0}{2V_{dc}I_M} = \frac{9}{64} \approx 0.1406 \qquad 5.11$$

which is 12.5 percent greater than for Class B operation.

The circuit employing second-harmonic peaking (see Figure 5-3) is similar to the one shown in Figure 5-2, except that the parallel resonant circuit is now tuned on the second harmonic [1–3, 7, 8]. Figure 5-4 shows the circuit waveforms.

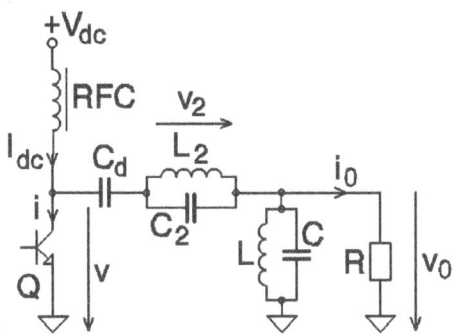

Figure 5-3 *Second-harmonic peaking Class F1 amplifier.*

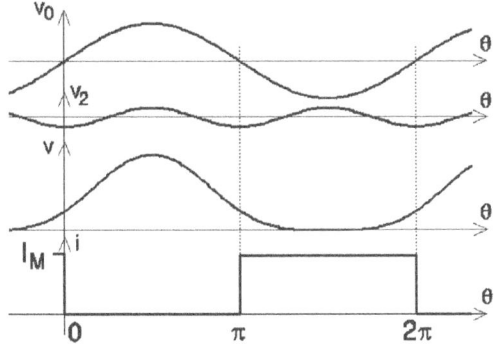

Figure 5-4 *Waveforms in a second-harmonic peaking Class F1 amplifier.*

For this circuit, the transistor also acts as a current source, but the collector current is a square wave (see Figure 5-4) containing only DC, fundamental-frequency, and odd-harmonic components. The load impedance of the transistor is R at fundamental frequency f, infinite at $2f$, and zero otherwise. Collector voltage waveform flattening requires a phase relationship between the fundamental-frequency and second-harmonic components, as shown in Figure 5-4.

$$v(\theta) = V_{dc} + V_0 \sin\theta - V_2 \cos 2\theta = V_{dc} + V_0(\sin\theta - x\cos 2\theta) \qquad 5.12$$

The optimum operating point is

$$x = \frac{1}{4} \qquad V_0 = \frac{4}{3}V_{dc} \qquad I_M = \frac{2\pi}{3}\frac{V_{dc}}{R} \qquad I_{dc} = \frac{I_M}{2} = \frac{\pi}{3}\frac{V_{dc}}{R}$$

$$P_0 = \frac{8}{9}\frac{V_{dc}^2}{R} \qquad \eta = \frac{8}{3\pi} \approx 84.88\% \qquad C_P = \frac{1}{2\pi} \approx 0.1592$$

5.13

The collector efficiency is slightly lower than in the previous case; the power output capability is somewhat higher.

Some observations and practical considerations

1. A greater number of resonators may be included in this design, especially at high frequencies (UHF and above), where it is possible to control several harmonics simultaneously. Theoretical values of collector efficiency and power output capability for various combinations of controlled harmonics are given in Reference [7]. References [9], [10], and [11] provide several practical design examples of Class F1 power amplifiers.

2. Maintenance of the right amount of harmonic content in the collector voltage waveform is critical for Class F1 amplifier performance. In practice, this is achieved by careful tuning and introduction of a certain amount of the desired harmonic into the driving signal. The active device also may be driven into saturation to improve performance.

3. The effects of the saturation voltage and/or saturation resistance of the transistor on the performance of the circuit can be estimated using the same procedure used for Class B and C amplifiers.

5.2 Class F2 Power Amplifiers

If an infinite number of series-connected parallel resonators tuned on the odd harmonics of the operating frequency are included in the third-harmonic peaking circuit of Figure 5-1, theoretical unity efficiency is possible. This hypothetical circuit, which bears an obvious similarity to the third-harmonic peaking Class F1 amplifier, is shown in Figure 5-5 [12].

The practical implementation of this circuit is presented in Figure 5-6 [2, 7, 11, 13] where a quarter-wavelength transmission line replaces the parallel resonators tuned on the odd harmonics of the operating frequency. Figure 5-7 shows the circuit waveforms.

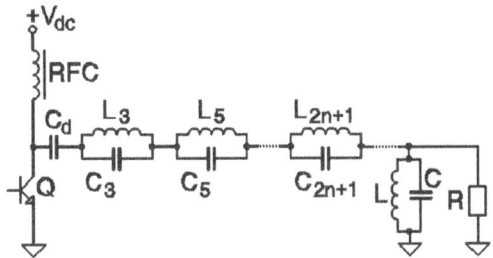

Figure 5-5 *Class F circuit using an infinite number of parallel resonators.*

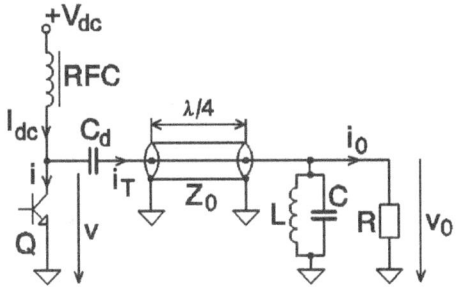

Figure 5-6 *Class F2 amplifier with a parallel-tuned output.*

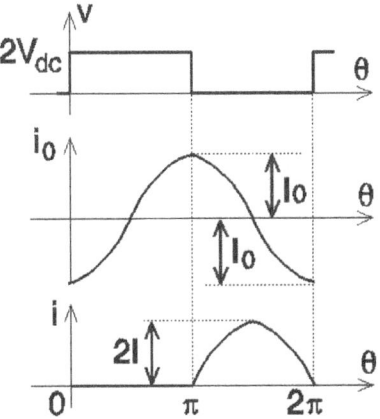

Figure 5-7 *Waveforms in a Class F2 amplifier with a parallel-tuned output.*

A simplified analysis of the Class F2 amplifier is based on the following assumptions [2, 7, 13]:

- The passive components are ideal (RFC is an ideal RF choke and C_d is a DC blocking capacitor).

- A quarter-wavelength lossless transmission line is assumed. The characteristic impedance of the transmission line is Z_0.

- The active device acts as an ideal switch (zero ON resistance, infinite OFF resistance, zero output capacitance, the switching action is instantaneous and lossless), with a switch duty ratio of 50 percent.

- The parallel-tuned circuit LC (which resonates on the fundamental frequency) is ideal: its impedance is infinite at the fundamental frequency and zero otherwise.

At fundamental frequency f, the load impedance of the transmission line is R. Therefore, the input impedance of the transmission line (the load seen by the transistor) is given by

$$R_{in} = \frac{Z_0^2}{R} \qquad 5.14$$

For the harmonic components, the load impedance of the transmission line is zero (short circuit) due to the action of the parallel-tuned circuit. For any odd harmonic, the electrical length of the transmission line is an odd integer multiple of $\lambda/4$, thus the input impedance of the line is infinite (open circuit). For any even harmonic, the electrical length of the transmission line is an integer multiple of $\lambda/2$; thus, the input impedance of the line is zero (short circuit). These considerations show that (see Figure 5-7):

- Collector voltage v consists only of DC, fundamental frequency, and odd harmonics. As a result, because the DC component of v is V_{dc} and $v = 0$ during the ON state ($\pi < \theta < 2\pi$), the collector voltage must be $v = 2V_{dc}$ during the OFF state ($0 < \theta < \pi$).

- Collector current i consists only of DC, fundamental frequency, and even harmonics. As a result, because $i = 0$ during the OFF state ($0 < \theta < \pi$), the collector current must be a half sinusoid during the ON state ($\pi < \theta < 2\pi$).

The collector voltage given by

$$v(\theta) = \begin{cases} 2V_{dc} & 0 \le \theta \le \pi \\ 0 & \pi \le \theta \le 2\pi \end{cases} \qquad 5.15$$

can be expanded into the Fourier series

$$v(\theta) = V_{dc}\left[1 + \frac{4}{\pi}\sum_{n=0}^{\infty}\frac{\sin(2n+1)\theta}{2n+1}\right] \qquad 5.16$$

The fundamental frequency component of $v(\theta)$ determines a sinusoidal output current in quadrature with the fundamental frequency component of $v(\theta)$ due to the $\lambda/4$ transmission line.

$$i_0(\theta) = I_0\sin\left(\theta - \frac{\pi}{2}\right) = \frac{4}{\pi}\frac{V_{dc}}{Z_0}\sin\left(\theta - \frac{\pi}{2}\right) \qquad 5.17$$

Consequently, the output power is given by

$$P_0 = \frac{I^2}{2}R = \frac{8}{\pi^2}\frac{V_{dc}^2}{R_{in}} = \frac{8}{\pi^2}\frac{V_{dc}^2}{Z_0^2/R} \approx 0.8106\frac{V_{dc}^2}{Z_0^2/R} \qquad 5.18$$

The fundamental frequency component of $i_T(\theta)$ (and $i(\theta)$, too is given by

$$I = \frac{4}{\pi}\frac{V_{dc}}{R_{in}} = \frac{4}{\pi}\frac{V_{dc}}{Z_0^2/R} \qquad 5.19$$

Thus, the peak value of the collector current (a half sinusoid waveform) is $2I$. The DC input current and the DC input power are

$$I_{dc} = \overline{i(\theta)} = \frac{2}{\pi}I = \frac{8}{\pi^2}\frac{V_{dc}}{Z_0^2/R}$$

$$P_{dc} = V_{dc}I_{dc} = \frac{8}{\pi^2}\frac{V_{dc}^2}{Z_0^2/R} = P_0 \qquad 5.20$$

The theoretical collector efficiency is 100 percent. Finally, the power output capability is the same value as in second-harmonic peaking Class F1 or Class D circuits.

$$C_P = \frac{P_0}{2V_{dc}2I} = \frac{1}{2\pi} \approx 0.1592 \qquad 5.21$$

Some observations and practical considerations

1. Class F2 circuits are sometimes called "single-ended Class D" amplifiers because of the similarities between the two.

2. The practical performance of the Class F2 amplifier is significantly affected by the switch ON resistance, R_{ON}, and the output capacitance of the transistor, C_{out}. The effect of R_{ON} on the collector voltage waveform is shown in Figure 5-8; the circuit performance is affected as follows:

$$v(\theta) = \begin{cases} 2V_{dc} - 2IR_{ON} \sin\theta, & 0 \le \theta \le \pi \\ -2IR_{ON} \sin\theta, & \pi \le \theta \le 2\pi \end{cases} \qquad I = \frac{4}{\pi} \frac{V_{dc}}{\left(Z_0^2 / R\right) + 2R_{ON}}$$

$$P_0 = \frac{8}{\pi^2} \frac{V_{dc}^2}{Z_0^2 / R} \left[\frac{Z_0^2 / R}{\left(Z_0^2 / R\right) + 2R_{ON}} \right]^2 \qquad P_{dc} = \frac{8}{\pi^2} \frac{V_{dc}^2}{\left(Z_0^2 / R\right) + 2R_{ON}} \qquad 5.22$$

$$\eta = \frac{Z_0^2 / R}{\left(Z_0^2 / R\right) + 2R_{ON}} \qquad C_P = \frac{1}{2\pi} \frac{Z_0^2 / R}{\left(Z_0^2 / R\right) + 2R_{ON}}$$

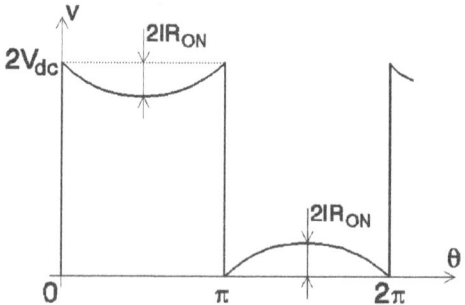

Figure 5-8 *Collector voltage waveform if $R_{ON} > 0$.*

The power losses due to the discharge of the output capacitance of the transistor increase with the operating frequency and limit the usefulness of this circuit at high frequencies.

$$P_d = 2C_{out} f V_{dc}^2 \qquad 5.23$$

3. This circuit presents a very serious impediment at low frequencies: The physical length of the quarter-wavelength transmission line exceeds acceptable practical values. For example, at $f = 25$ MHz, $\lambda/4 = 3$ m (free space). If a typical coaxial cable (such as RG-58) is used, the required physical length is about 2 meters.

4. A pi-matching circuit (or any other circuit equivalent to a parallel resonator) may be used instead of the output parallel-tuned circuit. Such a circuit can provide impedance matching (if required) and filter out unwanted harmonics.

EXAMPLE 5.1

The Class F2 circuit in Figure 5-9 operates at f = 9 MHz and V_{dc} = 30 V. A pi-matching circuit is used to transform the load resistor R_L = 50 Ω into R = 25 Ω, and a 50-ohm transmission line further converts this resistance into R_{in} = 100 Ω. Considering the impedance values in the circuit, it is more convenient to place the RF choke and DC blocking capacitor, as shown in Figure 5-9. If an ideal switch is assumed, the equations above give P_0 = P_{dc}= 7.295 W, I_{dc} = 243.2 mA, and the peak collector current $2I$ = 763.9 mA. If a MOSFET with R_{ON} = 5 Ω is used in this circuit, Equation 5.22 yields P_0 = 6.029 W, P_{dc} = 6.632 W, I_{dc} = 221.1 mA, η = 90.9%, and $2I$ = 694.5 mA. Assuming that the output capacitance of the transistor is C_{out} = 50 pF, the additional power loss that occurs in the circuit is P_d = 0.81 W, decreasing the collector efficiency to below 80 percent.

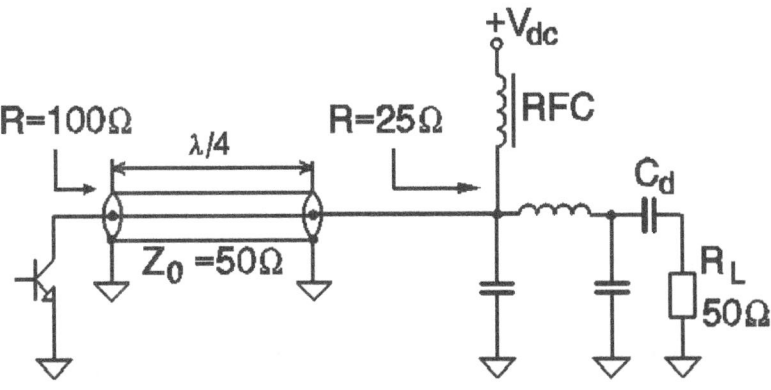

Figure 5-9 *The Class F2 circuit in Example 5.1.*

Another Class F2 configuration is shown in Figure 5-10 [7, 14]. This circuit uses a series-tuned output circuit and is the dual of the Class F2 amplifier presented in Figure 5-6. Figure 5-11 gives the circuit waveforms. Note that this Class F2 circuit may be considered the improvement limit for the second-harmonic peaking Class F1 amplifier presented in Section 5.1.

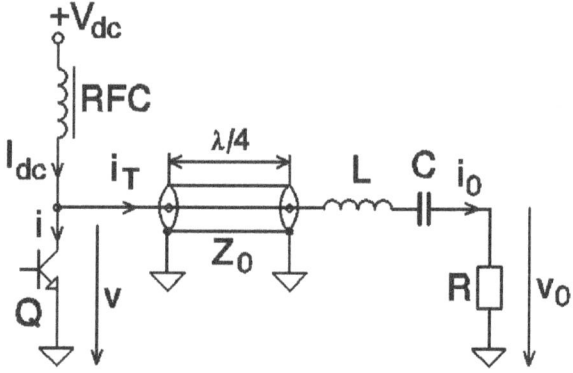

Figure 5-10 *Class F2 amplifier with a series-tuned output circuit.*

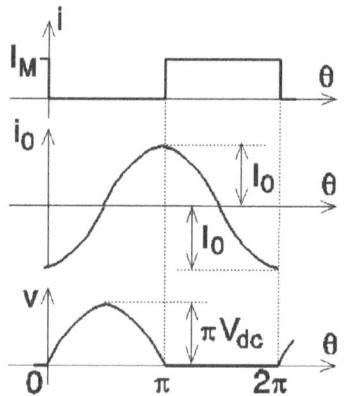

Figure 5-11 *Waveforms in a Class F2 amplifier with a series-tuned output circuit.*

At fundamental frequency f, the load impedance of the transmission line is R and the input impedance of the transmission line is Z_0^2/R. For the harmonic components, the load impedance of the transmission line is infinite (open circuit) due to the action of the series-tuned circuit. Therefore, for any even harmonic, the input impedance of the line is infinite (open circuit), and for any odd harmonic, the input impedance of the line is zero (short circuit). These considerations show that the collector current varies between zero during the OFF state and $I_M = 2I_{dc}$ during the ON state, while the collector voltage is a half sinusoid during the OFF state. The circuit performance is given by

$$I_0 = \frac{\pi}{2} \frac{V_{dc}}{Z_0^2 / R} \qquad P_0 = P_{dc} = \frac{\pi^2}{8} \frac{V_{dc}^2}{Z_0^2 / R} \qquad \eta = 1$$

$$I_M = \frac{\pi^2}{4} \frac{V_{dc}}{Z_0^2 / R} \qquad V_{max} = \pi V_{dc} \qquad C_P = \frac{1}{2\pi}$$

5.24

Some observations and practical considerations

1. Figures 5-10 and 5-11 and Equation 5.24 show some similarities between this Class F2 circuit and the current switching Class D circuit.

2. The practical performance of the Class F2 circuit in Figure 5-10 is significantly affected by switch ON resistance R_{ON} and the series inductance at switch. The power loss due to the series inductance increases with the operating frequency and limits the usefulness of this circuit at high frequencies. The voltage across the switch is zero at turn-ON, thus eliminating the power loss due to the output capacitance of the transistor.

3. A T-matching network (or any other matching network equivalent to a series resonator) may be used instead of the output series-tuned circuit. The T-matching network may provide impedance matching (if required) and filter out the unwanted harmonics.

5.3 Class F3 Power Amplifiers

The basic Class F3 power amplifier circuit is shown in Figure 5-12 [1, 15]. This circuit is a higher-order Class E amplifier with the following characteristics:

* The active device acts as a switch with a 50 percent duty ratio.

* $L_0 C_0$ is a series-tuned circuit; its resonant frequency is close to the switching frequency (as in the case of the Class E circuit).

* $L_3 C_3$ and LC are parallel-tuned circuits that resonate on the third harmonic of the switching frequency.

* The shunting capacitance, C, absorbs the output capacitance of the transistor.

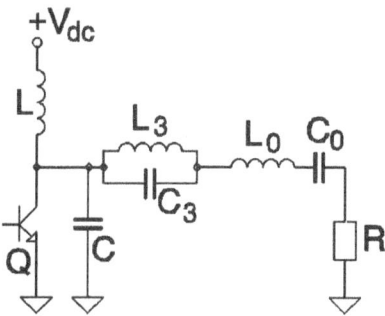

Figure 5-12 *Class F3 amplifier.*

This circuit is designed to satisfy the standard Class E switching conditions given by Equation 4.48. Therefore, the well-known advantages of the Class E operation are also obtained for the Class F3 circuit. The Class F3 circuit has the third harmonic added to the collector voltage waveform, reducing the peak collector voltage and increasing the power output capability. Disadvantages are related to the complications introduced by the increased complexity of the circuit (which require very careful tuning) and additional losses.

5.4   References

1. Sokal, N. O., I. Novak, and J. Donohue. "Classes of RF Power Amplifiers A through S, How They Operate, and When to Use Them." *Proc. RF Expo West* '95, (San Diego, CA) (January/February 1995): 131–138.

2. Krauss, H. L, C. V. Bostian, and F. H. Raab. *Solid State Radio Engineering.* New York: John Wiley & Sons, 1980.

3. Shakhgildyan, V. V. *Radio Transmitters.* Moscow: Mir Publisher, 1981.

4. Tyler, V. J. "A New High-Efficiency High Power Amplifier." *The Marconi Review,* vol. 21, no. 130 (Fall 1958): 96–109.

5. Raab, F. H. "High Efficiency Amplification Techniques." *IEEE Circuits and Systems Newsletter*, vol. 7, no. 10 (December 1975): 3–11.

6. Zivkovic, Z. and A. Markovic. "Third Harmonic Injection Increasing the Efficiency of High-Power RF Amplifiers." *IEEE Trans. Broadcasting,* vol. BC-31, no. 2 (June 1985): 34–39.

7. Raab, F. H. "An Introduction to Class-F Power Amplifiers." *RF Design,* vol. 19, no. 5, (May 1996): 79–84.

8. Zivkovic, Z. and A. Markovic. "Increasing the Efficiency of the High-Power Triode HF Amplifier — Why not with the Second Harmonic?" *IEEE Trans. Broadcasting*, vol. BC-32, no. 1 (March 1986): 5–10.

9. Nishiki, S. and T. Nojima. "Harmonic Reaction Amplifier — A Novel High-Efficiency and High-Power Microwave Amplifier." *IEEE MTT-S Digest,* (1987): 963–966.

10. Toyoda, S. "High Efficiency Amplifiers." *IEEE MTT-S Digest,* San Diego, California (May 1994): 253–256.

11. Sowlati, T., et al. "Low Voltage, High Efficiency GaAs Class E Power Amplifiers for Wireless Transmitters." *IEEE J. of Solid-State Circuits*, vol. 30 (October 1995): 1074–1080.

12. Raab, F. H. "High Efficiency RF Power Amplifiers." *Ham Radio*, vol. 7, no. 10 (October 1974): 8–29.

13. Raab, F. H. "FET Power Amplifier Boosts TransmitterEfficiency." *Electronics*, vol. 49, no. 12 (June 10, 1976): 122–126.

14. Kazimierczuk, M. K. "A New Concept of Class F Tuned Power Amplifier." *Proc. 27th Midwest Symp. on Circuits and Syst.*, Morgantown, West Virginia (June 1984): 425–428.

15. *HEPA-PLUS User Manual,* Design Automation, Inc., Lexington, MA, 1990–1994.

Class S Power Amplifiers and Modulators

The Class S amplifier [1-9] is a switching-mode, high-efficiency (ideally, 100 percent) unit used for amplification of low frequency signals — most often *audio frequency* (AF) signals. High-efficiency amplification of AF signals is especially important for radio transmitters using collector amplitude modulation (AM). As was discussed in Chapters 2, 3 and 4, collector amplitude modulation of Class C or switching-mode PAs is the most efficient way to generate high power *double-sideband* (DSB) AM signals (full carrier). Collector AM may be also used to generate DSB *suppressed-carrier* (SC) AM signals, *vestigial-sideband* (VSB) AM signals, or *single-sideband* (SSB) AM signals, employing the *envelope elimination and restoration* (EER) technique discussed in [10–14]. Note that any band-limited RF signal can be regarded as a simultaneous amplitude and phase modulator of a carrier and be amplified with high efficiency using an EER system.

The main reason for requiring very high efficiency from the modulator (essentially, an AF amplifier) is that the AF power, which must be supplied to the RF amplifier, has the same magnitude order as the RF output power. Consequently, the overall efficiency of the transmitter is significantly affected by the modulator efficiency.

There are two basic solutions for collector amplitude modulation: *transformer-coupled collector modulation* (TCCM) and *series-coupled collector modulation* (SCCM) [1–3, 15–20]. The traditional method of obtaining high-power AM signals (employed by vacuum tube amplifiers) uses a transformer coupled AF PA (usually, Class B operated) to change the supply collector voltage of an RF power amplifier (saturated Class C or Class F1 operated). The block diagram of the TCCM system is presented in Figure 6-1 [1, 2, 15, 17, 20].

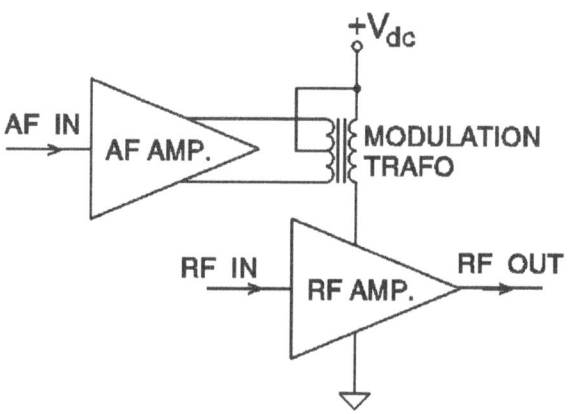

Figure 6-1 *Block diagram of a transformer-coupled collector modulation system.*

In the absence of the modulating signal (carrier mode), the collector supply voltage of the RF PA is V_{dc}. Assuming idealized operating conditions and a single sinusoid modulating signal, the modulation depth, m, of the RF output signal (a DSB-AM signal) may vary between 0 and 1 ($0 \leq m \leq 1$). If a modulating signal is applied at the AF input, the collector supply voltage of the RF PA varies between $(1 - m)V_{dc}$ and $(1 + m)V_{dc}$. The maximum voltage swing occurs for $m = 1$; the collector supply voltage varies between 0 and $2V_{dc}$.

When P_0 represents the RF output power in carrier mode, the DC input power (which must be supplied by the DC power supply, V_{dc}) is P_0/η_{RF}, where η_{RF} is the collector efficiency of the RF PA. If a modulating signal is applied at the AF input, the average RF output power (averaged over a period of the AF signal) is given by

$$P_{0,AV} = \left(1 + \frac{m^2}{2}\right) P_0 \qquad 6.1$$

Consequently, the power that must be provided by the modulator is

$$P_{0,MOD} = \frac{P_{0,AV} - P_0}{\eta_{RF}} = \frac{\dfrac{m^2}{2} P_0}{\eta_{RF}} \qquad 6.2$$

which exceeds $P_0/2$ for $m = 1$. Using η_{AF} for the collector efficiency of the modulator (the AF amplifier) and ignoring the DC power consumption in the other stages of the system (small signal amplifiers, drivers and such), the overall efficiency of the TCCM system is

$$\eta = \frac{P_{0,AV}}{\dfrac{P_0}{\eta_{RF}} + \dfrac{P_{0,MOD}}{\eta_{AF}}} = \eta_{RF} \frac{1 + \dfrac{m^2}{2}}{1 + \dfrac{m^2}{2\eta_{AF}}}$$ 6.3

which for $m = 1$ becomes

$$\eta = \eta_{RF} \frac{3}{2 + \dfrac{1}{\eta_{AF}}}$$ 6.4

Several comparative results for different RF PAs and modulators are provided in Table 6-1.

Table 6-1 *Ideal overall efficiency of collector AM systems (TCCM or SCCM) for 100 percent modulation index.*

RF PA	AF PA	overall efficiency
Class B, $\eta_{RF} = 78.5\%$	Class A, $\eta_{AF} = 50\%$	58.9%
Class B, $\eta_{RF} = 78.5\%$	Class B, $\eta_{AF} = 78.5\%$	71.9%
Class D, E, $\eta_{RF} = 100\%$	Class A, $\eta_{AF} = 50\%$	75%
Class D, E, $\eta_{RF} = 100\%$	Class B, $\eta_{AF} = 78.5\%$	91.6%
Class D, E, $\eta_{RF} = 100\%$	Class S, $\eta_{AF} = 100\%$	100%

Another way of achieving collector amplitude modulation [1–3,15–20] is to use a series-voltage regulator, to produce the collector-supply voltage of the RF PA directly, as shown in Figure 6-2.

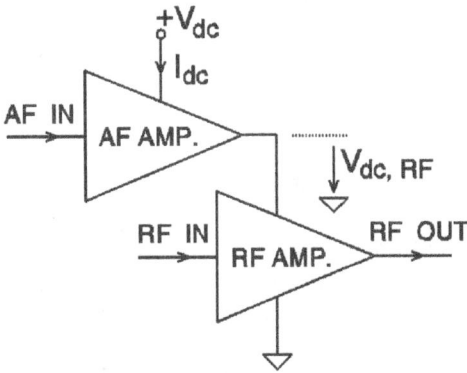

Figure 6-2 *Block diagram of a series-coupled collector modulation system.*

SCCM is advantageous because it does not require a modulation transformer or a modulation choke [17] — both of which are heavy, bulky, expensive, and introduce additional power loss and distortion. Without a modulating signal (carrier mode), the collector supply voltage of the RF PA is $V_{dc}/2$, whereas for $m = 1$, the collector supply voltage varies between 0 and V_{dc}. Assuming that the efficiency of the series-voltage regulator (Class S modulator) is η_{AF}, and ignoring the power consumption in the other stages of the system, the overall efficiency is

$$\eta = \frac{P_{0,AV}}{P_{dc}} = \frac{P_{0,AV}}{P_{0,AV}\ \ \dfrac{1}{\eta_{RF}\ \ \eta_{AF}}} = \eta_{RF}\eta_{AF} \qquad 6.5$$

If the modulator is a Class A amplifier (an emitter follower), its collector efficiency is given by

$$\eta_{AF} = \frac{V_{dc,RF}I_{dc}}{V_{dc}I_{dc}} = \frac{V_{dc,RF}}{V_{dc}} \qquad 6.6$$

Consequently, the overall efficiency of a SCCM system using a Class A modulator is only $\eta_{RF}/2$ in carrier mode. Note that a system employing TCCM achieves an overall efficiency of η_{RF} in carrier mode. If a modulating signal is applied at the AF input, the overall efficiency of the SCCM system using a Class A modulator is given by

$$\eta = \left(1 + \frac{m^2}{2}\right)\frac{\eta_{RF}}{2} \qquad \eta = \frac{3}{4}\eta_{RF} \quad if\ m = 1 \qquad 6.7$$

The numerical values of the overall efficiency for SCCM systems (modulation index of 100 percent) are identical to those obtained for TCCM systems and are given in Table 6-1. These results emphasize the importance of achieving high efficiency in the AF power amplifier to improve the overall efficiency of the AM transmitter.

6.1   Class S Power Amplifier

Class S amplification [1, 2, 3, 4, 6, 8, 9] requires the conversion of the AF signal in a *pulse-width modulated* (PWM) signal. The PWM signal is then amplified using a switching-mode PA (essentially, a two-pole switch). Finally, a high-power AF signal is recovered with a low-pass filter. A possible block diagram of this system is presented in Figure 6-3(a). Figure 6-3(b) illustrates the PWM signal generation. Several techniques can be used to generate PWM signals [2]. The technique presented here uses a comparator to compare the AF input signal with a triangular wave. This triangular wave (with the desired amplitude and linearity) may be supplied from a function-generator integrated circuit or may be obtained integrating a square wave with a simple RC integrator [2].

In the absence of the AF signal at the AF input, the duty ratio of the PWM signal must be 50 percent. This may be achieved by adding an appropriate bias voltage at the AF input of the comparator. To avoid signal distortion, the maximum swing of the AF input signal must not exceed the peak-to-peak amplitude of the triangular wave.

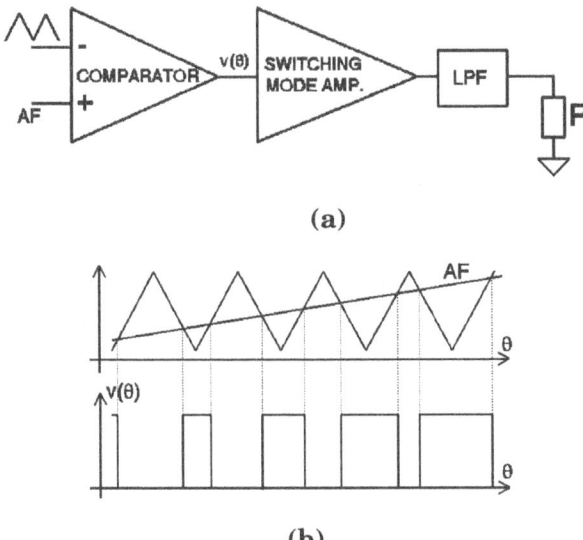

(a)

(b)

Figure 6-3 *(a) Block diagram of high-efficiency AF amplifier and (b) waveforms illustrating PWM generation.*

The circuit configuration of the Class S amplifier is presented in Figure 6-4. (This is a half-bridge topology. As in the case of Class D amplifiers, a full-bridge topology is possible for Class S amplifiers [8].) The circuit topology is somewhat similar to the Class D topology, mainly because the Class S PA uses a two-pole switch (as does the Class D circuit). For this reason, the Class S amplifier is sometimes called a Class D or AF Class D circuit. However, there are several significant differences between the Class D and Class S amplifiers.

- If BJTs are used as switches, anti-parallel diodes are required in the Class S amplifier. If Q_1 and Q_2 are MOSFETs, external anti-parallel diodes are not required.

- The Class S circuit uses a low-pass filter to recover the AF signal into the load. For this reason, the Class S circuits are sometimes called broadband Class D, and the standard Class D circuits (as presented in Chapter 3) are called narrowband Class D circuits [1]. For simplicity, the low-pass filter is represented in Figure 6-4 as a LC filter (C_d is DC blocking capacitor).

- A driving circuit is required to drive the two transistors alternately ON and OFF. However, the two transistors of the Class S amplifier cannot be driven with an input transformer, as in the case of a Class D amplifier, because an input transformer forces a zero average value of the base-to-emitter voltage. For Class D circuits using an input transformer, the switch duty ratio is 50 percent, and the base-to-emitter voltages during the ON and OFF states are equal, but with opposite signs [see Figure 6-5(a)]. The base-to-emitter voltage, V, in Figure 6-5(a) is about 0.7 volt for BJTs and typically 5...10 volts for MOSFETs. For Class S circuits, the switch duty ratio may vary between 0 and 100 percent. Consequently, if an input transformer is used, the base-to-emitter voltage may rise to very high values [see Figure 6-5 (b), where the hatched areas during the ON and OFF states must be equal] destroying the transistors.[1] Class S amplifiers require appropriate driving circuits (discrete logic or integrated circuit) to synthesize the driving signals for transistors [2, 8, 9].

- Q_1 is a pnp BJT or a p-channel MOSFET. A npn BJT or n-channel MOSFET cannot be used for Q_1 because its emitter (or source) would be at a variable potential (0 or V_{dc}), requiring it to apply its drive using an input transformer. However, the switching frequency is usually low in Class S amplifiers (typically 5 to 10 times the maximum frequency component of the AF signal), and pnp BJTs or p-channel MOSFETs are readily available.

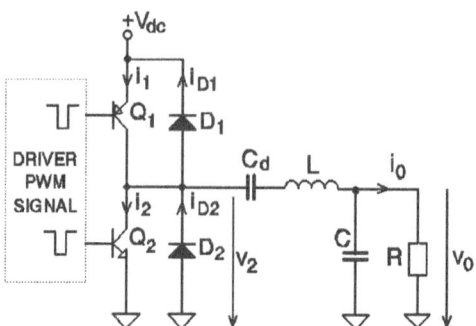

Figure 6-4 *Class S amplifier.*

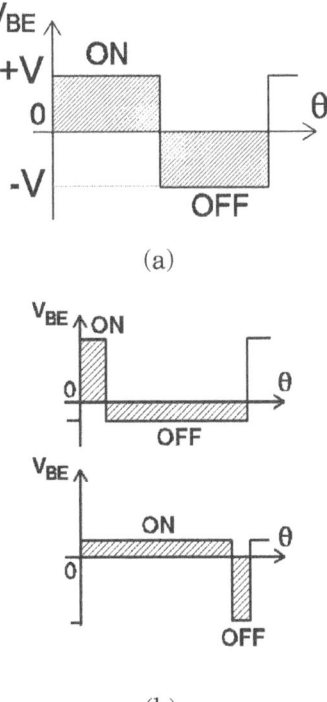

(a)

(b)

Figure 6-5 *Waveforms of base-to-emitter voltage for different values of the switch duty ratio.*

The operation of the Class S amplifier is illustrated using the waveforms of Figure 6-6. Assuming ideal components and an ideal low-pass filter, the output current and voltage are sinusoidal. At any given time, the output current i_0 closes either through Q_1 and D_2 (when $i_0(\theta) > 0$) or Q_2 and D_1 (when $i_0(\theta) < 0$). The voltage $v_2(\theta)$ varies between 0 and V_{dc}; thus, the maximum amplitude of the output voltage across the load is $V_{dc}/2$ and the peak output power is

$$P_0 = \frac{V_{dc}^2}{8R} \qquad 6.8$$

The amplitude of the output current $i_0(\theta) = I\sin\theta$ given by

$$I = \frac{V_{dc}}{2R} \qquad 6.9$$

which is equal to the maximum collector current of the two transistors (and also the maximum direct current through the two diodes). The power output capability is

$$C_P = \frac{P_0}{2V_{dc}I} = \frac{1}{8} = 0.125 \qquad\qquad 6.10$$

If the components are ideal, there are no power losses in the Class S amplifier, and the theoretical collector efficiency is 100 percent. Finally, note that the DC blocking capacitor, C_d, must have a negligible impedance at the lowest AF frequency. It may be eliminated using two DC power supplies ($\pm V_{dc}/2$) and connecting the load to the ground.

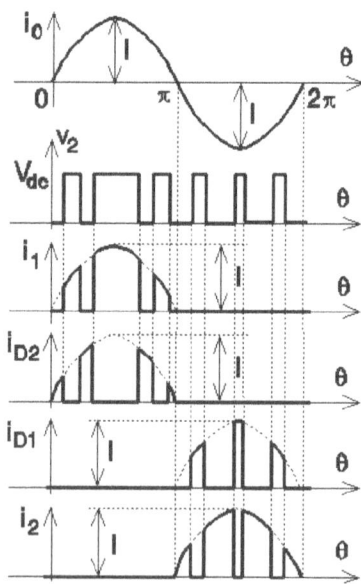

Figure 6-6 *Waveforms in Class S amplifiers.*

6.2 Class S Modulator

The Class S modulator [2–7] is used in SCCM AM systems. Although the following discussion refers to a DSB-AM signal (full carrier), the circuit may be used to amplify any type of RF signal in an EER system.[2]

To generate DSB-AM signals, the modulator must provide an AF voltage varying between 0 and V_{dc} (for a modulation depth $m = 1$) and a DC voltage of $V_{dc}/2$ for the carrier mode. The difference between a Class S amplifier and a Class S modulator is that the amplifier must be able to provide both positive and negative output currents, whereas the modulator must provide only a positive output current. Consequently, the basic circuit of the Class S modulator (see Figure 6-7) includes just one transistor and one diode. R is the equivalent load resistor of the modulator, i.e., the equivalent resistance shown by the RF amplifier to its power supply. This equivalent resistance is almost constant with the DC supply voltage (for any saturated Class C, Class D, or Class E RF amplifier).

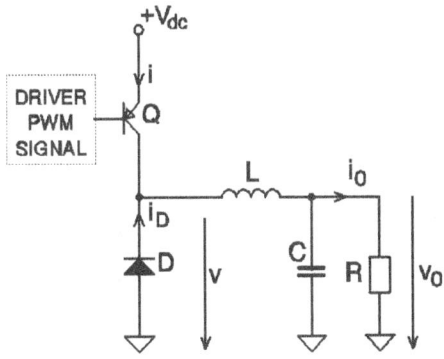

Figure 6-7 *Class S modulator.*

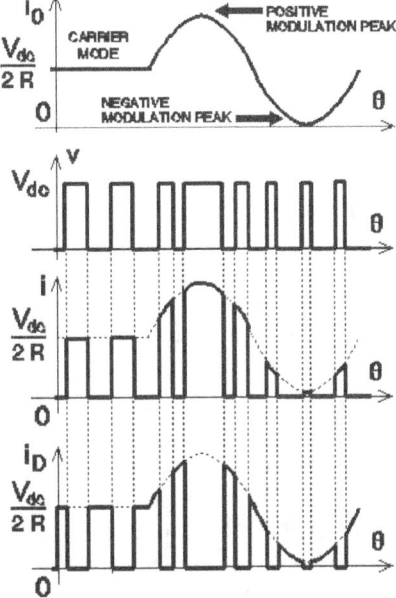

Figure 6-8 *Waveforms in a Class S modulator.*

The operation of a Class S modulator is illustrated by the waveforms in Figure 6-8. Assuming ideal components and an ideal low-pass filter, the output current and voltage are sinusoidal waveforms. At any given time, output current i_0 closes either through Q or D. For a 100 percent modulation depth, voltage $v(\theta)$ varies between 0 and V_{dc}. Therefore, the peak output power is given by

$$P_{0,\max} = \frac{V_{dc}^2}{R} = 4P_{0,MOD} \qquad 6.11$$

where $P_{0,\ MOD}$ is the output power of the modulator in carrier mode (when $v_0(\theta) = V_{dc}/2$). Observing that the maximum amplitude of the output current $i_0(\theta)$ given by

$$I_{\max} = \frac{V_{dc}}{R} \qquad 6.12$$

is equal to the peak collector current of the transistor (and also the peak direct current through the diode), the power output capability is

$$C_P = \frac{P_{0,\max}}{V_{dc}I_{\max}} = 1 \qquad 6.13$$

If the components are ideal, there are no power losses in the Class S modulator and the theoretical collector efficiency is 100 percent.

6.3   Practical Considerations

Shunt Capacitances or Series Inductances at Switches

As with the Class D amplifier, parasitic capacitances in parallel at the switches must be charged and discharged at the instant switching occurs. Unfortunately, this causes power losses. Equation 3.37, which shows the power losses due to the parasitic capacitances in a complementary voltage switching Class D circuit, can be applied to Class S amplifiers and modulators. For the Class S amplifier, the parasitic capacitances must include the diode capacitances (if anti-parallel diodes are used). For the Class S modulator, the two capacitances from Equation 3.37 are not equal: one of them includes the transistor output capacitance, the other includes the diode capacitance.

The waveforms in Figures 6-6 and 6-8 show that currents through the active devices jump at the moment switching occurs, causing additional power losses. This is an important difference with respect to Class D voltage-switching circuits where the switching occurs at zero current. With L_s

representing the series inductance of the active device and with ΔI as the magnitude of the current jump, the energy dissipated at the switching instant is given by

$$E_d = \frac{1}{2} L_S (\Delta I)^2 \qquad 6.14$$

The total power loss caused by this mechanism depends on the actual waveforms in the Class S circuit. Consider, for example, the Class S modulator in Figure 6-9. Assume that its AF output current $i_0(\theta)$ is a periodical waveform. Also, for simplicity's sake, assume that the switching frequency is N times higher than the AF (modulating) frequency (see Figure 6-9).

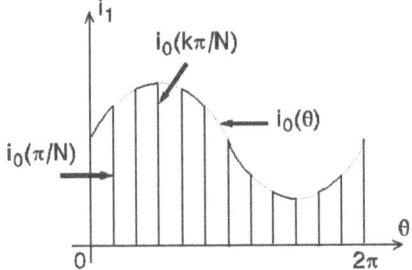

Figure 6-9 *Active devices current waveform in Class S modulator.*

The energy loss caused by the series inductance of one active device during one AF period can be calculated as

$$E_d = \frac{1}{2} L_S \sum_{k=0}^{2N-1} i_0^2 \left(k \frac{\pi}{N} \right) \qquad 6.15$$

If the switching frequency is much higher than the AF frequency, Equation 6.15 may be approximated as

$$E_d = \frac{L_S}{2} \frac{N}{\pi} \sum_{k=0}^{2N-1} \frac{\pi}{N} i_0^2 \left(k \frac{\pi}{N} \right) \approx \frac{N}{2\pi} L_S \int_0^{2\pi} i_0^2(\theta) d\theta \qquad 6.16$$

and the power loss (for only one device) is

$$P_{d1} = \frac{L_S}{2\pi} f \int_0^{2\pi} i_0^2(\theta) d\theta \qquad 6.17$$

where f is the switching frequency.

EXAMPLE 6.1

DSB-AM signal (full carrier), carrier mode. The output current of the modulator is given by

$$i_0(\theta) = \frac{V_{dc}}{2R} \qquad 6.18$$

If the total series inductance of the transistor is equal to the total series inductance of the diode and also equal to L_s, the total power dissipation caused by this mechanism is

$$P_d = 2P_{d1} = \frac{1}{2} L_s f \frac{V_{dc}^2}{R^2} \qquad 6.19$$

EXAMPLE 6.2

DSB-AM signal (full carrier), with $m = 1$. The output current of the modulator and the total power dissipation are given by

$$i_0(\theta) = \frac{V_{dc}}{2R}(1 + \sin\theta)$$
$$P_d = \frac{L_S}{\pi} f \frac{V_{dc}^2}{4R^2} \int_0^{2\pi} (1 + \sin\theta)^2 \, d\theta = \frac{3}{4} L_s f \frac{V_{dc}^2}{R^2} \qquad 6.20$$

EXAMPLE 6.3

DSB-SC AM one-tone signal or SSB two-tone signal. The output current of the modulator and the total power dissipation are given by

$$i_0(\theta) = \frac{V_{dc}}{R} |\sin\theta|$$
$$P_d = \frac{L_S}{\pi} f \frac{V_{dc}^2}{R^2} \int_0^{2\pi} |\sin\theta|^2 \, d\theta = L_s f \frac{V_{dc}^2}{R^2} \qquad 6.21$$

Saturation Voltage of Transistors

The effect of the saturation voltage of BJTs (and of the voltage drop across the diodes) on Class S amplifiers or modulators is similar to that presented for the complementary voltage switching Class D circuits in Chapter 3. A more precise evaluation should take into account that the voltage drop across a direct-biased diode is somewhat larger than the saturation voltage of a switching transistor.

ON Resistance of Transistors

Assuming that the transistor(s) and the diode(s) in a Class S circuit have the same ON resistance, R_{ON}, a resistance equal to R_{ON} appears to be permanently connected in series with the load resistance. Consequently, the effect of R_{ON} on Class S amplifiers or modulators is quite similar to that presented for the complementary voltage-switching Class D circuit in Chapter 3.

Transition Times

The effect of the finite switching time of the active devices can be estimated assuming a linear variation of the current and voltage across the switches during transitions (see Figure 6-10). Both current and voltage have jumps during transition. This is an important difference with respect to Class D circuits where switching always occurs either at zero current or zero voltage.

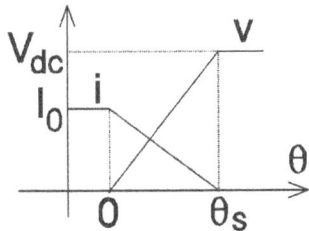

Figure 6-10 *Active devices current and voltage waveforms during ON to OFF transition.*

θ_s is the angular switching time ($\theta_s = 2\pi f t_s$, where f is the switching frequency and t_s is the switching time) and I_0 is the value of the switched current. The energy dissipated by the active device during one transition can be calculated as

$$E_d = \frac{1}{2\pi f} \int_0^{\theta_s} v(\theta) i(\theta) d\theta = \frac{1}{2\pi f} \int_0^{\theta_s} V_{dc} \frac{\theta}{\theta_s} I_0 \left(1 - \frac{\theta}{\theta_s}\right) = \frac{V_{dc} I_0 \theta_s}{12\pi f} \qquad 6.22$$

Assuming that the AF output current is described by the periodic function $i_0(\theta)$, the power dissipated by the active device is given by[3]

$$P_{d1} = \frac{f}{N} \frac{V_{dc}\theta_s}{12\pi f} \frac{N}{\pi} \sum_{k=0}^{2N-1} \frac{\pi}{N} \left| i_0\left(k \frac{\pi}{N}\right)\right| \approx \frac{V_{dc}\theta_s}{12\pi^2} \int_0^{2\pi} |i_0(\theta)| d\theta \qquad 6.23$$

EXAMPLE 6.4
Class S amplifier with maximum AF output current swing, $I = V_{dc}/(2R)$. The total power dissipation due to the finite switching times is given by

$$P_d = 2P_{d1} = \frac{V_{dc}^2 \theta_s}{3\pi^2 R} \qquad 6.24$$

EXAMPLE 6.5
Class S modulator, carrier mode or DSB-MA signal, with $m = 1$. The total power dissipation due to the finite switching times is given by

$$P_d = \frac{V_{dc}^2 \theta_s}{6\pi R} \qquad 6.25$$

EXAMPLE 6.6
Class S modulator, DSB-SC MA one-tone signal or SSB two-tone signal. The total power dissipation due to the finite switching times is given by

$$P_d = \frac{2V_{dc}^2 \theta_s}{3\pi^2 R} \qquad 6.26$$

Distortion of the Output Signal in Class S Circuits

The AF output signal of a Class S circuit is inherently distorted during the amplification process. Obvious causes are the nonlinearity associated with different circuits and signals and their nonideal character. However, a specific distortion mechanism cannot be avoided using the PWM technique, even if the components of the circuit and the signals are ideal [2]. In the analysis below, assume perfectly timed pulses produced by the pulse-width modulator, ideal switches in the Class S circuit, and an ideal output low pass filter. If a sinusoidal AF voltage is applied at the input of the pulse-width modulator, the voltage applied to the low-pass filter is a rectangular waveform (see Figure 6-11), with the pulse width given by

$$y = \frac{\pi}{2} + \pi \frac{V_0}{V_{dc}} \sin\theta_m \qquad 6.27$$

where $V_0 \le V_{dc}/2$ is the amplitude of the sinusoidal voltage obtained at the output of the low-pass filter (across the load resistor), $\theta_m = 2\pi f_m t$, and f_m is the frequency of the AF signal. Figure 6-3 (b) suggests that when a PWM signal is generated comparing the AF signal with a triangular wave, the middle of the generated pulse remains in an almost fixed position (the approximation is better as the switching frequency is larger compared to the AF frequency). Consequently, the middle of the pulse is chosen as a reference in Figure 6-11.

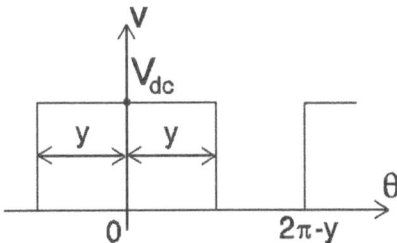

Figure 6-11 *Waveform of voltage across D ($\theta = 2\pi ft$, f is the switching frequency).*

If the modulating signal is a constant voltage, y is constant with time. In this case, the voltage applied at the input of the low pass filter is a periodic function which may be expanded into the Fourier series

$$v(\theta) = V_{dc} \frac{y}{\pi} + \frac{2V_{dc}}{\pi} \sum_{k=1}^{\infty} \frac{\sin ky}{k} \cos k\theta \qquad 6.28$$

If y varies in time as indicated by Equation 6.27, this Fourier series becomes

$$v(\theta) = \frac{V_{dc}}{2} + V_0 \sin\theta_m + \frac{2V_{dc}}{\pi} \sum_{k=1}^{\infty} \frac{1}{k} \sin\left(k\frac{\pi}{2} + k\pi\frac{V_0}{V_{dc}}\sin\theta_m\right)\cos k\theta =$$

$$= \frac{V_{dc}}{2} + V_0 \sin\theta_m +$$

$$+ \frac{2V_{dc}}{\pi} \sum_{k=1}^{\infty} \frac{1}{k} \sin k\frac{\pi}{2} \cos\left(k\pi\frac{V_0}{V_{dc}}\sin\theta_m\right)\cos k\theta +$$

$$+ \frac{2V_{dc}}{\pi} \sum_{k=1}^{\infty} \frac{1}{k} \cos k\frac{\pi}{2} \sin\left(k\pi\frac{V_0}{V_{dc}}\sin\theta_m\right)\cos k\theta$$

6.29

Because

$$\cos\left(x\sin\theta_m\right) = J_0(x) + 2\sum_{k=1}^{\infty} J_{2k}(x)\cos 2k\theta_m$$

6.30

$$\sin\left(x\sin\theta_m\right) = 2\sum_{k=1}^{\infty} J_{2k-1}(x)\cos(2k-1)\theta_m$$

where $J_k(x)$ are the Bessel function of the first kind,

$$v(\theta) = \frac{V_{dc}}{2} + V_0 \sin\theta_m +$$

$$+ \frac{2V_{dc}}{\pi} \sum_{k=1}^{\infty} \frac{1}{k} \sin k\frac{\pi}{2} J_0\left(k\pi\frac{V_0}{V_{dc}}\right)\cos k\theta +$$

$$+ \frac{4V_{dc}}{\pi} \sum_{k=1}^{\infty} \frac{1}{k} \sin k\frac{\pi}{2} \sum_{i=1}^{\infty} J_{2i}\left(k\pi\frac{V_0}{V_{dc}}\right)\cos 2i\theta_m \cos k\theta +$$

6.31

$$+ \frac{4V_{dc}}{\pi} \sum_{k=1}^{\infty} \frac{1}{k} \cos k\frac{\pi}{2} \sum_{i=1}^{\infty} J_{2i-1}\left(k\pi\frac{V_0}{V_{dc}}\right)\sin(2i-1)\theta_m \cos k\theta$$

According to Equation 6.31, the spectrum of the signal applied at the low-pass filter input includes

- the DC component, $V_{dc}/2$,
- the desired component with the amplitude V_0,
- the switching frequency and its odd harmonics,

$$V_k = \frac{2V_{dc}}{k\pi} J_0\left(k\pi\frac{V_0}{V_{dc}}\right), \quad k = 1, 3, 5, \ldots$$

6.32

- spurious products $kf \pm if_m$, where k and i are non-zero integers, with different parities,

$$V_{k,i} = \frac{2V_{dc}}{k\pi} J_i\left(k\pi \frac{V_0}{V_{dc}}\right) \quad \begin{matrix} k = 1, 3, \ldots \, and \quad i = 2, 4, \ldots \\ or \\ k = 2, 4, \ldots \, and \quad i = 1, 3, \ldots \end{matrix} \qquad 6.33$$

Some of these spurious products may fall into the passband of the output filter, causing distortion in the load. The magnitude of the most important products (normalized to the desired signal, V_0, and expressed in dB) is represented in Figure 6-12 [2] as a function of the normalized output AF voltage, V_0/V_{dc}. These results show that

1. The distortion increases as the output voltage increases. A decrease in V_0 with respect to V_{dc} is clearly impractical because it reduces the modulation depth of the AM signal. In practical applications, V_0 is limited only because of the difficulties involved in handling very narrow rectangular pulses.

2. The distortion decreases as the switching frequency increases. For example, if $f > 5\,f_m$, the spurious components falling in the passband of the output filter are at least 40 dB below the desired signal. Choosing the switching frequency is a tradeoff among several practical considerations:

   - Reducing the switching frequency simplifies the requirements for the pulse-width modulator, the driver, and the active devices used in the Class S circuit. The power losses decrease with the switching frequency, thus increasing the efficiency of the circuit.

   - Increasing the switching frequency reduces the distortion and simplifies the requirement imposed to the output filter.

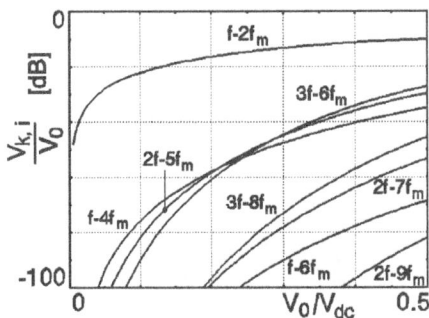

Figure 6-12 *Distortion inherent in PWM.*

6.4   Notes

1.   BJTs used for switching applications usually withstand inverse emitter-to-base voltages of 5...6 volts. MOSFETs usually withstand direct and inverse gate-to-source voltages of 20...30 volts.

2.   A Class S amplifier may be also used in a TCCM system. However, the use of a Class S modulator in a SCCM system is much more convenient: the Class S modulator is simpler than the Class S amplifier and the SCCM system eliminates the need for a modulation transformer.

3.   Because $i_0(\theta)$ may have negative values, $|i_0(\theta)|$ must be used in Equation 6.23.

6.5 References

1. Terman, F. E. *Radio Engineering.* New York: McGraw-Hill, 1947.

2. Clarke, K. K. and D. T. Hess, *Communication Circuits: Analysis and Design.* Boston: Addison-Wesley, 1971.

3. B. Norris, *MOS and Special-Purpose Bipolar Integrated Circuits & RF Power Transistor Circuit Design.* New York: McGraw-Hill, 1976.

4. Krauss, H. L., C. V. Bostian, and F. H. Raab, *Solid State Radio Engineering.* New York: John Wiley & Sons, 1980.

5. Shakhgildyan, V. V. *Radio Transmitter.* Moscow: Mir Publisher, 1981.

6. Bowick, C. *RF Circuit Design.* Indianapolis: Howard W. Sams & Co, 1982.

7. DeMaw. D. *Practical RF Design Manual.* Englewood Cliffs, NJ: Prentice-Hall, 1982.

8. Smith, J. *Modern Communication Circuits.* New York: McGraw-Hill, 1986.

9. Shakhgildyan, V. V. *Radio Transmitter Design.* Moscow: Mir Publisher, 1987.

10. Kahn, L. R. "Single-Sideband Transmission by Envelope Elimination and Restoration." Proc. IRE, vol. 40, no.7 (July 1952): 803–806.

11. N. O. Sokal and A. D. Sokal, "Amplifying and Processing Apparatus for Modulated Carrier Signals," *U.S. Patent 3,900,823* (August 19, 1975).

12. F. H. Raab, Envelope-Elimination and Restoration System Concepts,"

Proc. RF Expo East '87, Boston, MA (November 1987): 167–177.

13. Raab, F. H. and D. J. Rupp, "High-Efficiency Single-Sideband HF/VHF Transmitter Based upon Envelope Elimination and Restoration," *Proc. Sixth Int. Conf. HF Radio Syst. Tech. (HF'94)*, York, UK (July, 1994): 21–25.

14. Raab, F. H. "Envelope-Elimination-and-Restoration System Requirements." *Proc. of RF Technology Expo*, Anaheim, CA (Februray 1988): 499–512.

15. Raab, F. H. "High Efficiency RF Power Amplifiers," *Ham Radio*, vol. 7, no. 10, (October 1974): 8–29.

16. Raab, F. H. "High Efficiency Amplification Techniques." *IEEE Circuits and Syst. Newsletter*, vol. 7, no. 10, (December 1975): 3–11.

17. Raab, F. H. and D. J. Rupp, "Class-S High-Efficiency Amplitude Modulator." *RF Design*, vol. 17, no. 5 (May 1994): 70–74.

18. Sokal, N. O., I. Novak, and J. Donohue. "Classes of RF Power Amplifiers A Through S, How They Operate, and When to Use Them." *Proc. RF Expo West '95*, San Diego, CA, (January / February 1 1995): 131–138.

19. Danz, G. E. "Class-D Audio II Evaluation Board (HIP4080AEVAL2)." *Harris Semiconductor*, Melbourne, FL, AN9525.2, (March 1996).

20. Pauly, D. E. "High Fidelity Switching Audio Amplifiers Using TMOS Power MOSFETs." *Motorola Semiconductor Products, Inc., Phoenix, AZ, Application Note* AN1042.

7

RF Power Transistors

R F power amplifiers use a wide range of active devices, such as vacuum tubes and transistors (usually BJTs or MOSFETs). Vacuum tubes, still widely used in high-power RF amplifiers, have the following important advantages [1, 2]:

- Tubes are available in a wide range of frequencies (up to several GHz) and power ratings (up to several megawatts).

- Their parameters are almost constant over a wide frequency range.

- They exhibit high power gain (up to 15...20 dB). This is especially true for tetrodes in a common-cathode circuit.

- Tubes exhibit high efficiency when operated in Class C or F1.

- They have low harmonic content when operated in Class A or Class AB push-pull circuits.

- They exhibit low nonlinear distortion when used in plate amplitude modulation circuits.

Modern ceramic tubes also offer high reliability, long service life, and low power consumption for the filament (heater) circuit. Vacuum tube cathodes lose their emission capability gradually and exhibit a slow degradation in performance that allows for timely replacement. New techniques used in vacuum-tube radio transmitters, such as pulse step modulation or dynamic carrier control, allow these transmitters to achieve remarkable performance.

Bipolar and MOS transistors are used increasingly in high-power RF amplifiers for powers up to more than tens of kilowatts. They display important advantages over the vacuum tubes, including higher reliability and longer life, small size and weight, lower cost, easier cooling, and low input and output impedances (reducing parasitics).

Today's technologies only allow for powers of 300 to 600 watts per unit, but the use of bridge combiners to join the output powers of a large number of modules allows construction of solid-state PAs or radio transmitters with higher output power. Although vacuum tubes are still the best choice for PAs operating in tough conditions (penetrating radiation, high temperature, or a wide temperature range), recent applications look toward transistors and especially MOS transistors.

7.1   Bipolar Junction Transistors

The output power that can be achieved by high-frequency bipolar transistors is limited by factors, such as the intrinsic limitation of bipolar technologies expressed by the equation $V_{CB,max}f_T$ = constant, or the ability to evacuate the heat dissipated in the die.

The best way to implement high output power is to use a large number of chips connected in parallel. Extreme must be taken to avoid unequal current and power sharing, which could create local hot spots and lead to thermal runaway. Several techniques are used to handle this problem, but the primary method involves the use of "ballast" emitter resistors, i.e., small resistors placed in series with each emitter cell to ensure that the current is divided evenly among the individual cells. "Interdigitated," "overlay," and "network" structures are commonly used (see Figures 7-1 to 7-3) [3, 4, 5].

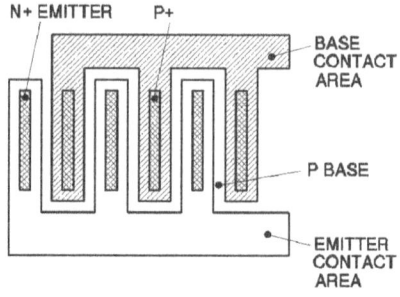

Figure 7-1 Interdigitated structure [6].

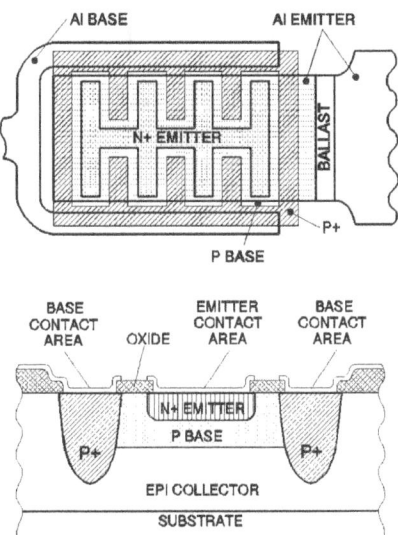

Figure 7-2 *Overlay structure [6].*

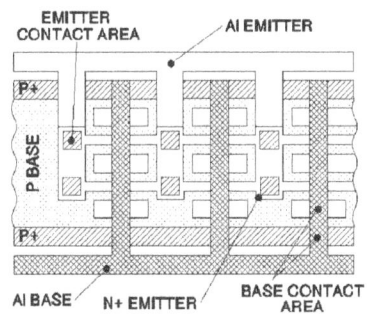

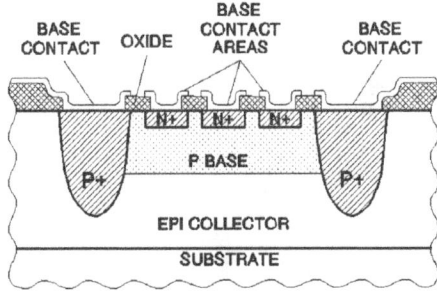

Figure 7-3 *Network structure [6].*

An internal impedance network is sometimes included in the package (see Figure 7-4) and can be employed with both monolithic and hybrid integrated circuit techniques. These devices, called *controlled-Q transistors* [7] are easier to match into circuit networks and provide higher gain and more consistent RF parameters than their "conventional" counterparts.

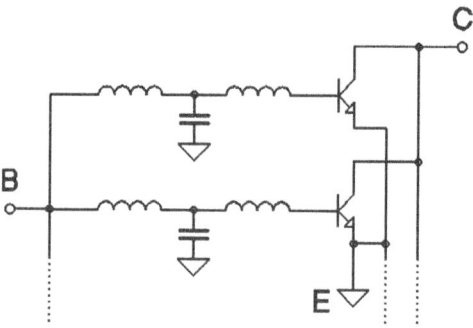

Figure 7-4 *Controlled Q BJT [7].*

RF power BJTs typically use the *stripline opposed emitter* (SOE) package [8]. They are available in two types: stud-mounted (see Figure 7-5) and flange-mounted (see Figure 7-6). The main advantages of the SOE package are the low inductance strip line leads (easily interfaced with the microstrip lines used in the matching networks) and the low base-to-collector parasitic capacitance. The package body is a beryllium oxide (BeO) disc, a material with high thermal conductivity. More details and practical recommendations for mounting SOE transistors on a heatsink are given in Reference [8}.

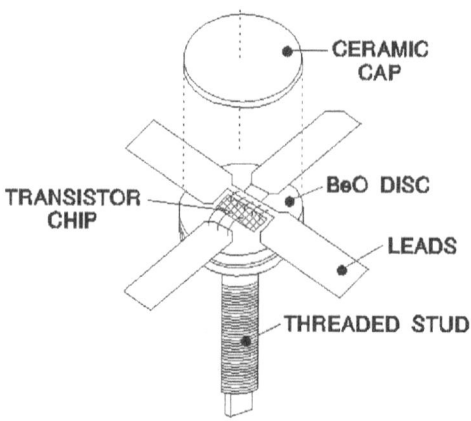

Figure 7-5 *Stud-mounted SOE package [8].*

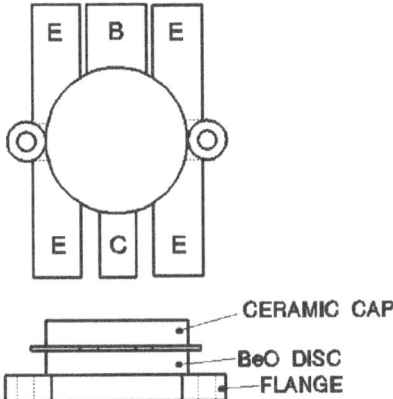

Figure 7-6 *Flange-mounted SOE package.*

7.2 MOS Transistors

There are two types of field-effect transistors:

* *junction field-effect transistor* (JFET) and
* *metal-oxide semiconductor field-effect transistor* (MOSFET).

Although their operating principles are similar (in both cases, the drain current is controlled by an electric field), considerable differences exist in technologies, device characteristics and performance, and circuit design. In the mid-1960s, the first field-effect transistors (depletion-mode JFETs) became available for low-power (5 to 10 watts), low frequency (30 to 50 MHz) applications [9]. The practical use of RF power amplifiers was limited by poor RF performance. A considerable advancement occurred in the 1970s with vertical MOS technology, which allowed for a denser die layout that yielded excellent RF and switching characteristics. Further developments led to new processes with improved performance, lower costs, and ease of manufacture. The double-diffused MOS (DMOS) process is used by most manufacturers today for both RF and power-switching FETs.

Basic MOSFET

Figure 7-7 presents a cross section of an enhancement-mode, n-channel, small-signal MOSFET [10].

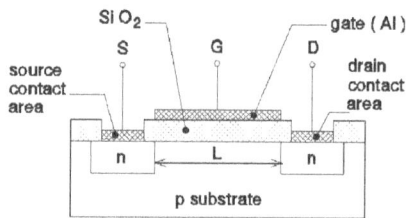

Figure 7-7 *Enhancement-mode n-channel MOSFET.*

The design of this structure uses planar technology and starts from a p-type substrate. Two separate low-resistivity, n-type regions are diffused into the substrate; these are the source and drain of the transistor. The entire surface is then covered with a silicon oxide (SiO2) insulating layer.[1] Next, holes are cut into the oxide layer to allow metallic contact with the source and drain. Another metallic layer (aluminum) covering the channel region forms the transistor gate. This metal area may be regarded as the top plate of a capacitor formed with the oxide layer (the dielectric) and the semiconductor channel (the bottom plate). Note that the drawing in Figure 7-7 does not show the real proportions in a small signal MOSFET. A typical transistor uses a substrate area of about 150 μm^2, the depth of the source and drain regions is 5...10 μm, and the length of the channel (*L* in Figure 7-7) is 10...20 μm. The oxide layer is about 0.1 μm.

Without gate voltage, the source and drain are practically isolated because of the two p-n junctions connected back to back. A positive gate potential (positive charges at the gate) induces negative charges on the substrate side. This negative charge increases as the gate voltage increases, until the area beneath the oxide becomes an n-type semiconductor region. This n-channel allows the current to flow between drain and source, and the drain current is controlled by the gate voltage (in other words, channel conductance G_C is controlled by the gate voltage). This interdependent condition is illustrated in Figure 7-8.

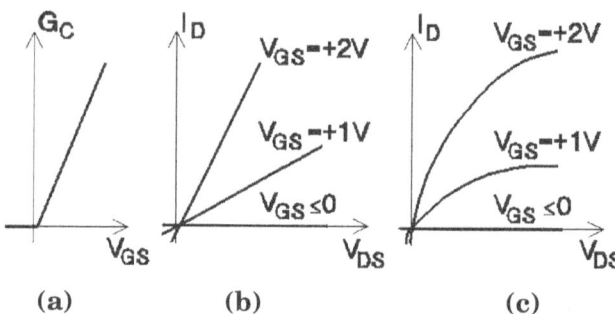

Figure 7-8 *(a) Channel conductance G_C versus V_{GS} (b), (c) Static characteristics of I_D versus V_{DS}, with V_{GS} as a parameter.*

The curves in Figures 7-8(b) and (c) (I_D as a function of V_{DS}, with V_{GS} as a parameter) show the MOSFET drain characteristics. If the drain-to-source voltage is low in comparison to the gate-to-source voltage, the intensity of the electrical field in the oxide dielectric is almost constant along the channel. The channel conductance linearity depends on V_{GS} (see Figures 7-8(a) and (b). If the drain-to-source voltage increases, the intensity of the electric field varies along the channel: it is higher near the source (where the voltage across the capacitor is V_{GS}) and lower toward the drain (where the voltage across the capacitor decreases). The variation of the channel conductance with the gate voltage becomes nonlinear [Figure 7-8(c)], and for a given value of V_{GS}, the drain current tends to saturate with increasing V_{DS}. This saturation occurs when the voltage across the gate capacitor at the drain end of the channel becomes zero and the induced charge is also zero. A further increase of the drain voltage does not affect the drain current.

The MOS transistor presented above is the so-called enhancement-mode n-channel MOSFET. This transistor has negligible drain current flow for negative or zero gate-to-source voltage. The drain current occurs only if V_{GS} exceeds a threshold value, V_T ($V_T > 0$ for n-channel MOSFETs). The depletion-mode n-channel MOSFET is somewhat similar, except that a moderately resistive n-channel is diffused between the source and drain. Consequently, the drain current flows at zero gate voltage. If a negative gate voltage is applied, the channel begins to suffer carrier depletion, and when $V_{GS} = V_T$, the drain current becomes zero. For positive gate voltages, the channel carriers are enhanced, as in the case of an enhancement-mode MOSFET. These characteristics are illustrated in Figure 7-9 [10]. For both enhancement-mode and depletion-mode MOSFETs, the n-channel structure may be changed to a p-channel structure by reversing the material types.

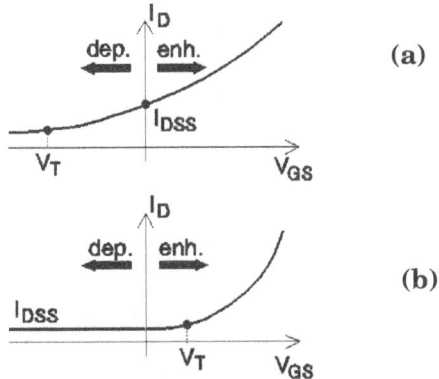

Figure 7-9 *Transfer characteristics of (a) depletion/enhancement type MOSFET, and (b) enhancement type MOSFET.*

The following equations describe the n-channel MOSFET; the p-channel MOSFET is described by similar equations.

a) If $V_{DS} << V_{GS}\text{-}V_T$, then

$$G_C = \frac{I_D}{V_{DS}} = \frac{\varepsilon \mu_e h}{LW}(V_{GS} - V_T)$$ 7.1

where V_T is the threshold voltage, G_C is the channel conductance, ε is the dielectric constant of the oxide layer, μ_e is the electron mobility in the channel region, h is the channel width, L is the channel length, and W is the thickness of the oxide layer.

b) If $V_{DS} \leq V_{GS} - V_T$ (and $V_{GS} - V_T > 0$), then

$$I_D = \frac{\varepsilon \mu_e h}{LW}\left[(V_{GS} - V_T)V_{DS} - \frac{V_{DS}^2}{2}\right]$$ 7.2

c) If $V_{DS} > V_{GS} - V_T$ (and $V_{GS} - V_T > 0$), then

$$I_D = \frac{\varepsilon \mu_e h}{LW}\frac{(V_{GS} - V_T)^2}{2}$$ 7.3

LDMOS Transistors

The conventional MOSFET structure cannot be used for power RF or switching applications requiring high drain voltages, very low capacitances and drain-to-source resistance because intrinsic limitations of the conventional technology do not allow the channel to be shortened beyond a certain limit. Moreover, for the conventional MOSFET structure, the reverse biased body-to-drain junction depletes the body region more than the drain region (because the drain region is more heavily doped than the body region). Consequently, large values of L are required to obtain high breakdown voltages.

The lateral double-diffused MOS (LDMOS) structure shown in Figure 7-10 [11] eliminates many problems associated with the conventional MOS structure. Channel length L is now given by the difference between two subsequent diffusions. Therefore, it can be well controlled and will have good reproducibility. The body region is more heavily doped than the drift region, which allows the structure to withstand large drain voltages without significantly affecting the electrical channel length of the transistor.

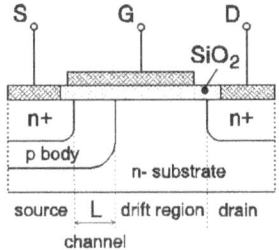

Figure 7-10 *LDMOS structure on n substrate.*

Analysis of the LDMOS operation shows that the square law $I_D(V_{GS})$, a fundamental characteristic of the conventional MOS transistor, remains valid only for gate-to-source voltages slightly exceeding the threshold voltage. Short-channel MOSFETs present an approximately linear relationship between the drain current and the gate voltage (see Figure 7-11); consequently, transconductance g_{fs} is almost constant with the drain current, for large drain currents (see Figure 7-12). The output characteristics of the short-channel MOSFET are illustrated in Figure 7-13(b). For comparison, Figure 7-13(a) shows the corresponding characteristics for a conventional MOSFET.

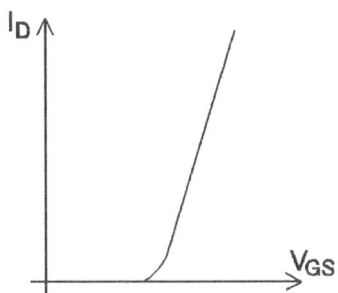

Figure 7-11 *Transfer characteristic of short-channel MOSFET showing linear $I_D(V_{GS})$ relationship.*

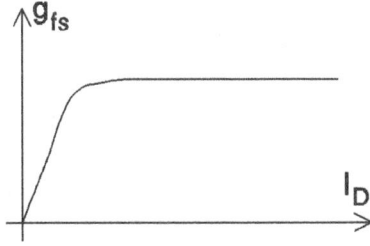

Figure 7-12 *Transconductance versus drain current for a short-channel MOSFET.*

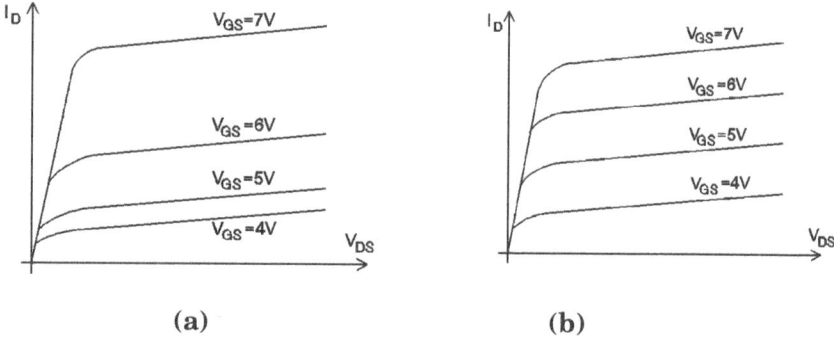

(a)   (b)

Figure 7-13 *Output characteristics of (a) standard MOSFET, and (b) short-channel MOSFET.*

Vertical MOS Transistors

The vertical MOSFET structure is similar to the double-diffused epitaxial bipolar transistor shown in Figure 7-14. Power MOSFETs used in the mid-1970s relied on a vertical channel structure, as shown in Figure 7-15 [3, 9, 12–15]. Individual small FETs (on the same die) were connected in parallel. These "V-groove" FETs had low capacitances and ON-resistance, but were difficult to manufacture, especially when it came to etching the sharp V-groove and obtaining a uniform coverage of the oxide and metal layers. The use of a "U-groove" (that is, a V-groove with a flat bottom) helped to some extent, but the process was still difficult to control.

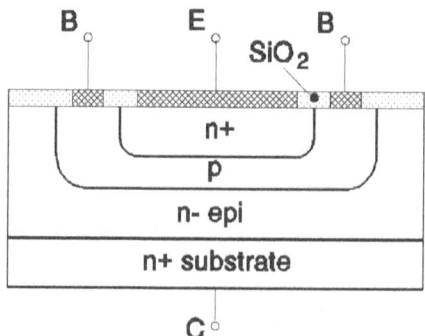

Figure 7-14 *Cross section of a double-diffused epitaxial bipolar technology.*

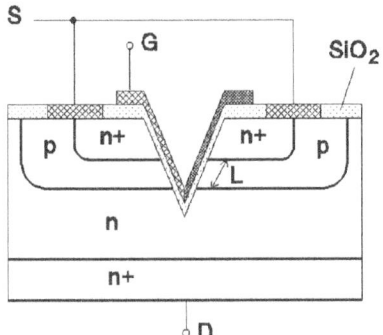

Figure 7-15 *Cross section of a vertical MOS.*

Most manufacturers today use the DMOS (double-diffused MOS) structure shown in Figure 7-16 [9, 11, 16, 17, 18]. The resulting vertical-channel structure is similar to the V-groove design, but the manufacturing process is much easier.

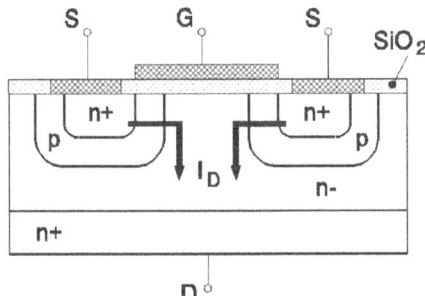

Figure 7-16 *Cross section of a vertical DMOS.*

Characteristics of Vertical MOS Transistors

This section presents the most important characteristics of vertical MOSFETs for RF power or switching applications. For comparison, the similar characteristics of BJTs are shown.

a. Bias

BJTs require bias only when operated in Class A or AB. The bias source must be able to provide relatively large currents (see Chapter 2). Usually, MOS transistors must be biased, even when operated in Class C or in switching mode. However, the bias source must provide a negligible current so it can be a simple resistor divider (see Chapter 2).

b. Drive (for switching-mode operation)

In switching-mode amplifiers, the driving signal must be sufficient to ensure that the transistors are alternately cut off or fully saturated. Depending on the circuit topology, a sinusoidal or rectangular waveform may be required to drive the transistor(s). (See Chapter 3 for a more detailed discussion).

c. Switching time

The BJT is a minority carrier device that is holes in a npn transistor and electrons in a pnp transistor. The MOSFET is a majority carrier device that are electrons in an n-channel transistor or holes in a p-channel transistor. The storage times associated with the minority carriers in BJTs are not present in MOSFETs, and consequently, the MOSFET switching times (both ON→OFF and OFF→ON) are much faster. The switching times (see Chapter 3) are mainly related to the charging/discharging of the MOSFET's capacitance.

d. Voltage drop (in the ON state)

The voltage drop across a BJT in the ON state (V_{CEsat}) is lower than the voltage drop across a MOSFET carrying the same current and having similar geometry and electrical size. Consequently, the BJTs have less total power loss than comparable MOSFETs at low frequencies, where the conduction losses in the ON state are dominant. Note that the ON resistance of MOSFETs increases significantly with the junction temperature increasing (see Figure 7-17), resulting in an increase in the conduction power loss. The switching power loss dominates at high frequencies and thus, the MOSFETs have superior performance over BJTs (see Figure 7.18) [11].

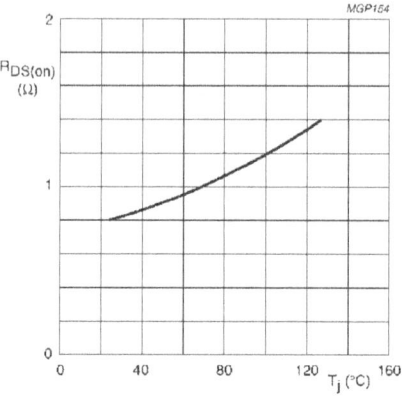

Figure 7-17 *Drain-source ON-state resistance as a function of junction temperature, typical values. (From BLF 244 VHF power MOS transistor — Product Specification, September 1992, courtesy of Philips Semiconductors.)*

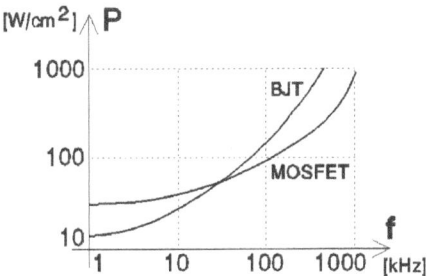

Figure 7-18 *Total power loss versus frequency for similar MOS and bipolar transistors [11].*

e. OFF-resistance

In the OFF state, the leakage current of a MOSFET is comparable to the leakage current of a BJT with a similar geometry and electrical size.

f. Linearity

Whether MOSFETs are more linear than BJTs or not is disputed in literature [19, 20]. MOS transistors seem to be superior, although they usually require higher bias current than bipolars to achieve good linearity. High-order intermodulation distortion (IMD) of MOSFETs (9th and up) is clearly superior. The emitter-ballast resistors used in bipolars (not required in MOSFETs) have the main effect in high-order IMD because of the nonlinear feedback through the capacitances from the collector [20]. A secondary IMD effect is played by the feedback capacitances (base-to-collector or gate-to-drain). BJTs have a larger feedback capacitance that is highly nonlinear with the collector voltage. This nonlinear internal feedback also produces high-order IMD.

g. Capacitances

Gate-to-source capacitance C_{GS} is mainly a MOS capacitance and less dependent on the drive level. In contrast, the base-to-emitter capacitance of BJTs (mostly a p-n junction capacitance) is largely dependent on the transistor state (ON, OFF) and the drive level.

The gate-to-drain (C_{GD}) and drain-to-source capacitances (C_{DS}) are both combinations of MOS and diode capacitances and vary significantly with the voltage across them.

MOSFET data sheets usually present (see Figures 7-19 and 7-20):

- C_{ISS} = input capacitance (with the drain shorted to source),
  $C_{ISS} = C_{GS} + C_{GD}$
- C_{OSS} = output capacitance (with the gate shorted to source),
  $C_{OSS} = C_{DS} + C_{GD}$

- C_{RSS} = feedback capacitance,

$C_{RSS} = C_{GD}$

Note: C_{GD}, C_{GS}, and C_{DS} are sometimes useful for modeling purposes. Their values can be obtained from C_{ISS}, C_{OSS}, and C_{RSS}, which are given in the data sheet as follows:

$C_{GS} = C_{ISS} - C_{RSS}$

$C_{DS} = C_{OSS} - C_{RSS}$

$C_{GD} = C_{RSS}$

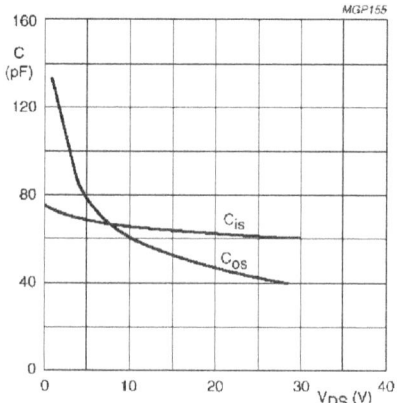

Figure 7-19 *Input and output capacitance as functions of drain-source voltage, typical values. (From BLF 244 VHF power MOS transistor — Product Specification, September 1999, courtesy of Philips Semiconductors.)*

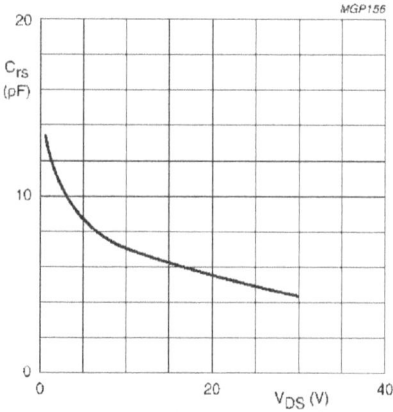

Figure 7-20 *Feedback capacitance as a function of drain-source voltage, typical values. (From BLF 244 VHF power MOS transistor — Product Specification, September 1992, courtesy of Philips Semiconductors.)*

For switching-mode amplifiers (where the transistor is either OFF or driven into saturation), the variation of the input capacitance versus gate and drain voltage is of particular interest. The curves in Figure 7-21 [21] show that the input capacitance reaches a sharp peak during the OFF-to-ON transitions. This is caused by the drain-to-gate capacitance multiplied by the transistor gain and reflected to the gate.

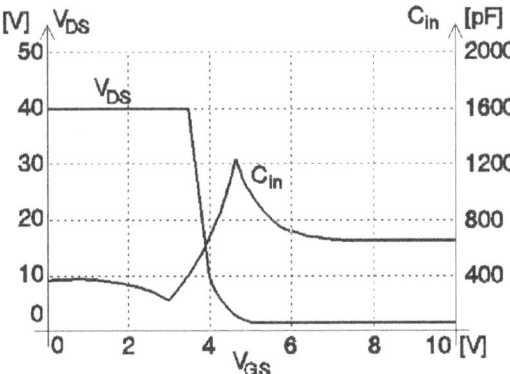

Figure 7-21 *Typical MOSFET input capacitance versus VGS and VDS (Motorola MRF150) [21].*

A more conservative approach when designing a driver circuit for a high-frequency switching-mode power MOSFET takes into consideration the peak value of the input capacitance. A more accurate design technique is based on the voltage versus gate charge characteristics of the transistors [22].

h. Noise factor

The *noise factor* (NF) of MOSFETs is usually two to three times smaller than the noise factor of comparable BJTs. This is mainly because the MOS transistors do not exhibit the shot noise characteristic to BJTs. If the source impedance is matched for optimum noise figure, an NF of 1.5...2 dB can be achieved. However, most PAs are matched to optimize the output power or power gain, and in this case NF ≈ 4...5 dB.

i. Input and output impedances

The large signal input impedance of MOSFETs is capacitive up to very high frequencies, and usually a little larger than that of a comparable BJT. The output impedances of comparable MOSFETs and BJTs are nearly the same, depending upon the DC supply voltage and the output power. (See Chapter 2, Section 2.5 for a more detailed discussion of large signal impedances.)

j. Power gain

Because of their higher transconductance, lower feedback capacitance, and higher input impedance, MOS transistors achieve a power gain exceeding the power gain obtained by similar BJTs by 4 to 5 dB. For example, at f = 150 MHz, P_0 = 45 W, and V_{dc} = 28 V, the power gain of a MRF171 MOSFET is 15 dB, while the power gain of a MRF315 BJT is only 11 dB. An important advantage of MOSFETs is that the power gain can be controlled by changing the bias voltage of the gate; a gain variation of 15 to 20 dB may be obtained (useful to control the output power or for circuit protection).

k. Unity gain frequency

The unity gain frequency of a MOSFET may be defined as the ratio of the transconductance G_{FS} to the input capacitance C_{ISS}. For instance, a BLF244 MOS transistor has G_{FS} = 0.87 S (at I_D = 0.75A) and C_{ISS} = 59 pF (at V_{DS} = 28 V); with these values, f_T = 2.35 GHz. Note that this parameter is not usually given in the data sheets of MOS transistors. For an equivalent geometry and electrical size, the unity gain frequency of a MOSFET can be two to five times higher than for a BJT. Figure 7-22 presents the variation of the unity gain frequency with the drain or collector current, for (a) a 150-watt RF power MOSFET, (b) a switching power MOSFET with approximately equal gate periphery, and (c) a BJT with the same basic die geometry [19].

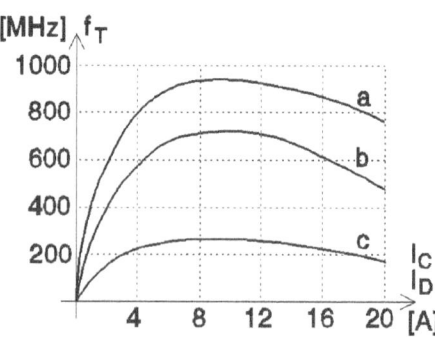

Figure 7-22 *Unity gain frequency versus collector or drain current [9, 19].*

l. Thermal stability

Figure 7-23 presents a typical MOSFET transfer characteristic — the drain current versus the gate-source voltage, with the junction temperature T_j as a parameter (BLF 244 VHF power MOSFET). For V_{GS} < 3 V, I_D is negligible. For V_{GS} > 3 V, the drain current increases with the square of V_{GS}, up to $I_D \approx$ 1 A. For 1 A < I_D < 4 A, I_D increases linearly with V_{GS} (this linear variation is a characteristic of short channel MOSFETs). The transconductance starts to decrease at large drain currents (in this case, over 4 amps).

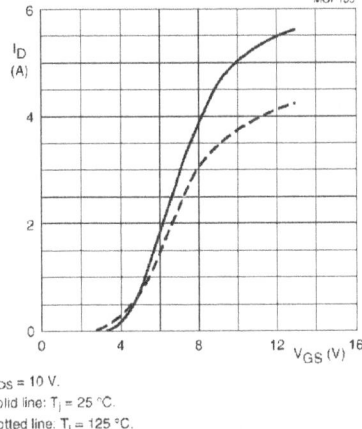

Figure 7-23 *Drain current as a function of gate-source voltage, typical values. (From BLF 244 VHF power MOS transistor — Product Specification, September 1992, courtesy of Philips Semiconductors.)*

When junction temperature T_j increases, the I_D (V_{GS}) curve rotates around $I_D \approx 0.7...0.8$ A. Consequently, this particular transistor can be used in Class A amplifiers biased at $I_D \approx 0.7...0.8$ A without temperature compensation. If the device is used in Class AB amplifiers (whose quiescent current is considerably lower), temperature compensation is required. The curve in Figure 7-24 can be used to design the compensation circuit.

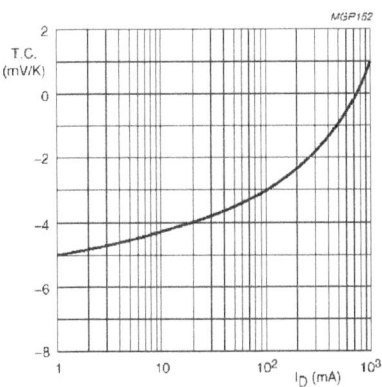

Figure 7-24 *Temperature coefficient of gate-source voltage as a function of drain current, typical values. (From BLF 244 VHF power MOS transistor-Product Specification, September 1992, courtesy of Philips Semiconductors.)*

For large drain currents, the temperature coefficient of I_D becomes negative (see Figure 7-25), ensuring the thermal stability of MOSFETs and protecting them from thermal runaway.

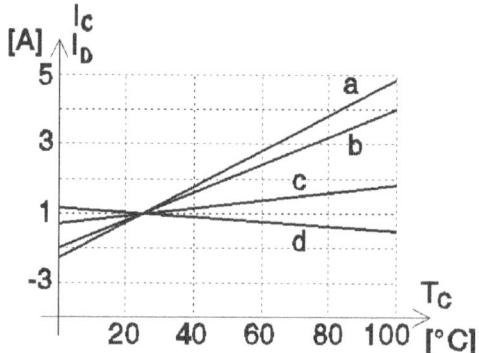

Figure 7-25 *I_C and I_D versus case temperature at constant I_B or V_{GS}, respectively; (a) BJT, $I_C = 0.1$ A, (b) MOSFET, $I_D = 0.1$ A, (c) BJT, $I_C = 10$ A, and (d) MOSFET, $I_D = 10$ A [19].*

m. Safe operating area

Figure 7-26 shows the safe operating areas of power MOS and bipolar transistors [11, 16]. As illustrated in Figure 7-26(a), BJTs can withstand a significantly higher voltage at low currents. However, MOSFETs can handle larger currents at moderately high voltages. The safe operating areas for packaged MOSFETs and BJTs are given in Figure 7-26(b), where

*   a is the chip/bond wire limitation on current,
*   b is the package dissipation limitation,
*   c is the chip limitation on voltage,
*   d is the increase in MOSFET safe operating area performance over BJT due to the thermally induced secondary breakdown.

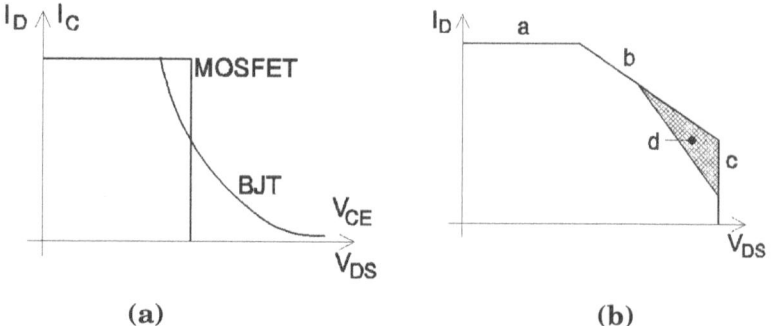

(a)   (b)

Figure 7-26 *The safe operating area limits of (a) power MOSFET and BJT chips, and (b) assembled power MOSFET and BJT [11,16].*

The characteristics discussed in the last two sections show that, overall, the MOS transistors are superior to BJTs in terms of ruggedness.

P-Channel MOSFETs

Because the resistivity of p-type silicon is much higher than that of n-type silicon, the characteristics of p-channel MOSFETs are considerably different from those of their n-channel counterparts. Moreover, this difference in resistivity does not allow construction of a p-channel MOSFET that is electrically complementary in all respects to an n-channel device. As discussed in the previous chapters, truly electrically complementary MOSFETs are extremely useful in Class B or D power amplifiers. An immediate consequence of the higher resistivity is that the p-channel transistors require a larger active area to achieve the same ON resistance and current rating as a similar n-channel device. This is a major obstacle to the construction of high-power, high frequency p-channel MOSFETs (for RF or switching applications). Many parameters, such as voltage ratings, ON-resistance, gate threshold voltage V_T, and transconductance G_{FS} may be almost identical or reasonably well matched (see Table 7-1).

Table 7-1 *Parameters of two typical complementary p- and n-channel MOSFETs (International Rectifier's IRF9130 and IRF120 [23].*

	n-channel IRF120	p-channel IRF9130
Chip area	8.04 mm^2	13.25 mm^2
V_{DSS} (max.)	100 V	–100 V
R_{ON}	0.3 Ω	0.3 Ω
I_{Dmax} (continuous)	6 A	–8 A
I_{Dmax} (pulsed)	15 A	–30 A
V_T	2...4 V	–2...–4 V
G_{FS}	2.5 S	3.5 S
C_{ISS}	450 pF	500 pF
C_{OSS}	200 pF	300 pF
C_{RSS}	50 pF	100 pF
R_{thjc}	3.12 °C/W	1.67 °C/W

Other parameters that are closely related to the die area, such as capacitances, safe operating area, current ratings, and thermal resistance R_{thjc} are considerably different [23]. Temperature variations of the threshold voltage, ON-resistance, and transconductance are slightly different and may raise additional problems in practical applications.

7.3 Notes

1. An additional silicon nitrate layer (not shown in Figure 7-7) shields the oxide layer from contamination. The sodium ions (found in all environments) may produce oxide contamination, resulting in long-term instability and changes in device characteristics [10].

7.4   References

1. Terman, F. E. *Radio Engineering.* New York: McGraw-Hill, 1947.
2. Shakhgildyan, V. V. Radio Transmitter. Moscow: Mir Publisher, 1981.
3. Krauss, H. L., C. V. Bostian, and F. H. Raab. *Solid State Radio Engineering.* New York: John Wiley & Sons, 1980.
4. Dye, N. E. and D. Schnell. *"RF Power Transistors Catapult into High-Power Systems."* Article Reprint AR179, Motorola Semiconductor Products, Inc., Phoenix, Arizona.
5. Gauen, K. "Understanding and Predicting Power MOSFET Switching Behavior." *Application Note* AN1090, Motorola Semiconductor Products, Inc., Phoenix, Arizona.
6. *Motorola RF Device Data,* Vol. II, 1991, Motorola Semiconductor Products, Inc., Phoenix, Arizona.
7. "Controlled-Q RF Technology — What it Means, How It's Done." *Engineering Bulletin EB-19,* Motorola Semiconductor Products, Inc., Phoenix, Arizona.
8. Danley, L. "Mounting Stripline-Opposed-Emitter(SOE) Transistors." *Application Note* AN555, Motorola Semiconductor Products, Inc., Phoenix, Arizona.
9. Granberg, H. O. "RF Power FETs — Their Characteristics and Applications. Part I and II." Article Reprint AR346/D, Motorola Semiconductor Products, Inc., Phoenix, Arizona.
10. "Field Effect Transistors in Theory and Practice." *Application Note* AN211A, Motorola Semiconductor Products, Inc., Phoenix, Arizona.
11. Blanchard, R. "Bipolar and MOS Transistors: Emerging Partners for the 1980s." *Technical Article* TA82-3, Siliconix, Inc., Santa Clara, California.
12. Norris, B. *MOS and Special-Purpose Bipolar Integrated Circuits & RF Power Transistor Circuit Design,* New York: McGraw-Hill, 1976.
13. Bowick, C. *RF Circuit Design.* Indianapolis: Howard W. Sams & Co, 1982.
14. Smith, J. *Modern Communication Circuits,* New York: McGraw-Hill, 1986.
15. Oxner, E. "Meet the Power-MOS-FET Model." *RF Design,* vol. 2, (January/February 1979): 16–22.
16. Gauen, K. "Designing with TMOS power MOSFETs." *Application Note* AN913, Motorola Semiconductor Products, Inc., Phoenix, Arizona.
17. Hilbers, A. H. "RF Power MOS-Transistors for the HF and VHF Range." *Application Report* NCO 8601, Philips Semiconductors, Eindhoven, The Netherlands.
18. Oxner, E. "MOSPOWER Semiconductor." *Technical Article* TA82-2, Siliconix, Inc., Santa Clara, California.
19. Granberg, H. O. "RF Power MOSFETs." *Article Reprint* AR165S, Motorola Semiconductor Products, Inc., Phoenix, Arizona.
20. Granberg, H. O. "Power MOSFETs Versus Bipolar Transistors." *Application Note* AN860, Motorola Semiconductor Products, Inc., Phoenix, Arizona.

21. Granberg, H. O. "Applying Power MOSFETs in Class D/E RF Power Amplifier Design." *Article Reprint* AR141, Motorola Semiconductor Products, Inc., Phoenix, Arizona.

22. Evans, A. and D. Hoffmann. "Dynamic Input Characteristics of a MOSPOWER FET Switch." *Application Note* AN79-3, Siliconix, Inc., Santa Clara, California.

23. Clemente, S. "An Introduction to International Rectifier p-Channel HEXFETs," *Application Note* 940B, International Rectifier, Inc., El Segundo, California.

Bibliography

Chudobiak, M. J. "The Use of Parasitic Nonlinear Capacitors in Class E Amplifiers." *IEEE Trans. Circuits Syst.* I vol. 41 (December 1994): 941–944.

Clemente, S., B. R. Pelly, and R. Ruttonsha. "Current Ratings, Safe Operating Area, and High Frequency Switching Performance of Power HEXFETs." *Application Note* 949B, International Rectifier, Inc., El Segundo, California.

Czarkowski, D. and M. K. Kazimierczuk. "Simulation and Experimental Results for Class-D Series Resonant Inverter." *IEEE Int. Telecom. Energy Conf.* (1992): 153–159.

Data Sheet BLF278 VHF Push-Pull Power MOS Transistor. Philips Semiconductors, Eindhoven, The Netherlands, 1996.

"Driving MOSPOWER FETs." *Application Note* AN79-4, Siliconix, Inc., Santa Clara, California.

Gauen, K. "Paralleling Power MOSFETs in Switching Applications." *Application Note* AN918, Motorola Semiconductor Products, Inc., Phoenix, Arizona.

Grant, D. "HEXFET III: A New Generation of Power MOSFETs." *Application Note* 966A, International Rectifier, Inc., El Segundo, California.

Hughes, D. W. "UHF Power Transistors For Use in Military Electronics Systems." *RF Design*, vol. 14, no. 6 (June 1991): 68–74.

HEPA-PLUS User Manual, Design Automation, Inc., Lexington, MA, 1990–2001.

Kazimierczuk, M. K. "Class E Tuned Power Amplifier with Nonsinusoidal Output Voltage." *IEEE J. Solid-State Circuits*, vol. SC-21, (August 1986): 575–581.

Lohrmann, D. "Boost Class-D RF Amplifier Efficiency." *Electronic Design*, vol. 16, no. 1 (1968): 96–99.

Moline, D. "MIMP Analyzes Impedance Matching Networks." *RF Design*, vol. 16, no. 1 (January 1993): 30, 34, 39–40, 42.

Page, D., F. W. D. Hindson, and W. J. Chudobiak. "On Solid-State Class D Systems." *Proc. IEEE*, vol. 53 (1965): 423–424.

Pitzalis, O., R. E. Horn, and R. J. Baranello. "Broadband 60-W HF Linear Amplifier." *IEEE J. Solid-State Circuits,* vol. SC-6 (June 1971): 93–103.

Shinoda, K. M. Fujii, M. Matsuo, T. Suetsugu, and S. Mori. "Idealized Operation of Class DE Frequency Multipliers." *IEEE Int. Symp. Circ. Syst.*, Atlanta, GA, vol. 1 (May 1996): 581–584.

Sokal, A. D. "A Remark on 'Exact Analysis of Linear Circuits Containing Periodically Operated Switches with Applications.'" *IEEE Trans. Circuits Syst.,* vol. CAS-24 (October 1977): 575–577.

Sokal, N. O., I. Novak and L. Drimusz. "Tradeoffs in Practical Design of Class-E High-Efficiency RF Power Amplifiers." *Proc. RF Expo East '93*, Tampa, FL (October 1993): 100–117.

Teale, G.S.M. "Parasitic Oscillation in VHF Power Amplifiers." *Mullard Technical Communications*, no. 90 (November 1967): 279–284.

Index